MILLIMETER WAVE COMMUNICATION SYSTEMS

IEEE Press
445 Hoes Lane
Piscataway, NJ 08854

A complete list of titles in the IEEE Press Series on Digital and Mobile Communication appears at the end of this book.

MILLIMETER WAVE COMMUNICATION SYSTEMS

Kao-Cheng Huang
Zhaocheng Wang

John B. Anderson, *Series Editor*

A JOHN WILEY & SONS, INC., PUBLICATION

Published by John Wiley & Sons, Inc., Hoboken, New Jersey.
Published simultaneously in Canada.

Library of Congress Cataloging-in-Publication Data:

Huang, Kao-Cheng.
Millimeter wave communication systems / Kao-Cheng Huang, Zhaocheng Wang.
p. cm.
ISBN 978-0-470-40462-1 (hardback)
1. Milimeter waves. 2. Millimeter wave communication systems. 3. Gigabit communications–Equipment and supplies. 4. Radio–Receivers and reception. I. Wang, Zhaocheng, 1968- II. Title.
TK5103.4835.H83 2010
621.384–dc22

2010013929

Printed in the United States of America.

10 9 8 7 6 5 4 3 2

CONTENTS

PREFACE

This book presents millimeter wave communication system design and analysis at the level to produce an understanding of the interaction between a wireless system and its front end so that the overall performance can be predicted. Gigabit wireless communications require a considerable amount of bandwidth, which can be supported by millimeter waves. Millimeter wave technology has come of age, and at the time of writing the standards of IEEE 802.15.3c, *WiGig, Wireless HD*™, and the European Computer Manufacturers Association have recently been finalized. The technology has attracted new commercial wireless applications and new markets such as high-speed download and wireless high definition TV. This book emphasizes both the front end and system for gigabit wireless communications, with an emphasis on wireless communications at the 60 GHz ISM band as well as the *E* band. We review the particular requirements for this application and address the design and feasibility of millimeter wave system, such as transceivers, antennas, equalizers, beam steering and beam forming. The applications of these systems are discussed in light of different forthcoming wireless standards. Examples of designs are presented and the analysis of their performance is detailed. In addition, the book includes a bibliography of current research literature and patents in this area.

Millimeter Wave Communication Systems endeavours to offer a comprehensive treatment of the advanced communication system based on the electronic consumer applications, providing a link to applications of computer-aided design tools and advanced methods and technologies. The major features of this book include discussion of many novel millimeter wave communication system configurations with new design techniques and concepts.

Although it contains some introductory material, this book is intended to provide a collection of millimeter wave front-end design data for communication system designers and front-end designers. The book should also act as a reference for postgraduate students, researchers, and engineers in millimeter wave engineering. It can also be used for millimeter wave teaching. A summary of each chapter follows.

Chapter 1 gives an overview of millimeter wave background and recent developments in personal communication systems. Characteristics of millimeter wave are reviewed and analyzed in the context of IEE802.15.3c-related communication systems. System requirements and challenges are explicitly studied.

Chapter 2 summarizes modern digital communication modulation techniques, such as OOK, PSK, FSK, QAM, and OFDM. The objective is to make a smooth transition between millimeter wave techniques and signal processing.

Chapter 3 presents both classical and modern transceiver architectures. As millimeter wave needs a lot of DC power to build wireless links, this issue is vital for personal communication and there is a strong demand to develop low-power transceivers. The transceiver block is studied in this chapter to discover low DC power solutions.

Chapter 4 focuses on antenna issues of communication systems. As antennas need to satisfy the system requirement, parameters must find trade-offs between beamwidth, coverage, and the number of users. Several types of millimeter wave antennas are discussed and their pros and cons in terms of feasibility and flexibility are compared.

Chapter 5 explores multiple input and multiple output for millimeter wave applications. Multiple antenna issues such as mutual coupling and spatial diversity are tackled. Multiple transceivers issues such as channel knowledge, receiver diversity are also discussed. Signal-to-noise ratio of the whole system is analyzed.

Chapter 6 presents new aspects of advanced diversity for millimeter wave wireless systems, including spatial diversity, temporal diversity, frequency diversity, and the combination of them.

Chapter 7 gives an overview of the advanced concept of beam steering and beam forming for millimeter wave wireless systems. Various frame structures to support beam steering and beam forming with reduced hardware complexity are introduced. The idea of beam steering using regular antenna radiation patterns and the concept of beam steering using irregular antenna radiation patterns are summarized.

Chapter 8 provides an overview of a single-carrier frequency domain equalizer with and without decision feedback equalization. Various frame structures, adaptive channel estimation methods, and a plurality of equalizer structures are demonstrated.

The authors are obliged to many people. First of all, we wish to acknowledge the valuable review comments from Prof. Sheng Chen, IEEE Fellow, University of Southampton, U.K., and Prof. Tarief Elshafiey, head of the electrical and computer engineering department, October University for Modern Sciences and Arts, Cairo, Egypt. Also, the authors acknowledge the copyright permission from IEEE (U.S.) and Su Khiong Yong (U.S.).

The authors also wish to thank Dr. Chao Zhang at Tsinghua University, China, for authority Chapter 2.

The authors are indebted to many researchers for their published works, which were rich sources of reference. Our sincere gratitude extends to Jeanne Audino and other IEEE editors for their support in writing the book. The help provided by Mary Mann and other members of the staff at John Wiley & Sons is most appreciated.

In addition, Kao-Cheng Huang would like to thank Prof. David J. Edwards, University of Oxford (U.K.), Prof. Mook-Seng Leong, National University of Singapore (Singapore), Prof. Predrag Rapajic, University of Greenwich (U.K.), Prof. Rüdiger Vahldieck, ETH (Switzerland), the late Prof. Ban-Leong Ooi, National University of Singapore (Singapore), Dr. David Haigh, IEEE Fellow, Imperial College (U.K.), Prof. Francis Lau, Hong Kong Polytechnic University (China), Prof. Hsueh-Man Shen, New York University (U.S.), Dr. Chris Stevens, University of Oxford, (U.K.), Dr. Jia-Sheng Hong, Heriot-Watt University (U.K.) and Dr. John Thornton, University of York (U.K.) for their many years of support.

LIST OF ABBREVIATIONS

AAS	antenna asymmetric system
ADC	analog-to-digital converter
AGC	automatic gain controller
AP	access points
APSK	amplitude and phase shift keying
ASK	amplitude shift keying
ASS	antenna symmetric system
AWGN	additive white Gaussian noise
AWV	antenna-array weight vector
BER	bit error rate
BPSK	binary phase shift keying
CE	channel estimation
CFR	crest factor reduction
CFR	channel frequency response
CMOS	complementary metal-oxide-semiconductor
CP	cyclic prefix
CPFSK	continuous phase frequency shift keying
CP-SC	cyclic prefix single carrier
CRC	cyclic redundancy check
CSF	channel sounding frame
CSI	channel state information
CTP	candidate transmission path
DAC	digital-to-analog converter
DBPSK	differential binary phase shift keying
DEV1	device 1
DEV2	device 2
DF	data frame
DFE	decision feedback equalization
DFT	discrete Fourier transformation
DPD	digital predistortion
DPSK	differential binary phase shift keying
DQPSK	differential quadrature phase shift keying
DSP	digital signal processing
ECMA	European Computer Manufacture Association
EGC	equal gain combining
EHF	extremely high frequency
ESNR	effective SNR
FB	feedback

FCC	Federal Communications Commission
FD-DFE	frequency domain decision feedback equalization
FDE	frequency domain equalization
FDE-NP	frequency domain linear equalizer using time domain noise prediction filter
FD-LE	frequency domain linear equalizer
FDM	frequency division multiplexing
FEC	forward error correction
FER	frame error rate
FF	feedforward
FFT	fast Fourier transform
FM	frequency modulation
FOQPSK	Feher offset quadrature phase shift keying
FPC	Fabry-Perot cavity
FSK	frequency shift keying
GSM	global system for mobile communications
HDR	high data rate
HDTV	high-definition television
HPBW	half-power beam width
IBDFE	iterative block decision feedback equalization
ICI	inter-carrier interference
IDFT	inverse discrete Fourier transform
IF	intermediate frequency
IFFT	inverse fast Fourier transform or inverse fast Fourier transformation
ISI	inter-symbol interference
ISM	industrial, scientific and medical
LAN	local area network
LDF	long data frame
LDR	low data rate
Len_CDP	length of channel delay profile
LNA	low noise amplifier
LO	local oscillator
LOS	line-of-sight
LSB	least significant bit
LTCC	low temperature cofired ceramics substrates
LTE	long-term evaluation
LUT	look-up table
MAC	media access control
MAM	M-ary amplitude modulation
MC	multiple-carrier
MFB	matched filter bound
MIMO	multiple-input-multiple-output
MLSE	maximum likehihood sequence estimation
MMIC	monolithic microwave integrated circuit
MMSE	minimum mean squared error

MRC	maximum ratio combining
MSB	most significant bit
MSE	mean squared error
MT	mobile terminal
MTBF	mean time between failure
NLOS	non-LOS
OBO	output back-off
OFDM	orthogonal frequency division multiplexing
OOK	on/off keying
OQPSK	offset quadrature phase shift keying
PA	power amplifier
PAPR	peak-to-average power ratio
PER	packet error rate
PHY	physical layer
PLL	phase-locked loop
PN	pseudo noise
PPM	pulse position modulation
PS	portable station
PSDS	phase spectrum demodulation schemes
PSK	phase shift keying
PSMS	phase spectrum modulation schemes
QA	quantized amplitude
QAM	quadrature amplitude modulation
QoS	Quality of service
QPS	quantized phase shift
QPSK	quadrature phase shift keying
RF	radio frequency
RMS	root mean square
RRC	root raised cosine
RSSI	received signal strength indicator
Rx	receiver
SC	single-carrier
SC-FDE	single-carrier frequency domain equalizer
SC-FDMA	single-carrier frequency division multiple access
SDF	short data frame
SDR	software defined radio
SFD	start of frame delimiter
SISO	single-input-single-output
SNR	signal-to-noise ratio
SOQPSK	shaped offset quadrature phase shift keying
SPDT	single-pole double-throw
SPI	six-port interferometer
SVD	singular value decomposition
TDD	time division duplex
Tx	transmitter

UEP	unequal error protection
ULA	uniform linear arrays
UW	unique word
UWB	ultrawideband
UW-SC	unique word single carrier
VCO	voltage controlled oscillator
VSWR	voltage standing wave ratio
Wireless HD	Wireless high definition
WLAN	wireless local area networks
WPAN	wireless personal area network
XOR	exclusive-OR

1

MILLIMETER WAVE CHARACTERISTICS

Corresponding to the progress of multimedia technology and data storage technology, a high data rate of 10 Gbit/s is expected, driven by the increasing memory capacity in wireless/mobile devices. It seems clear that the demand for a high data rate and high-integrity services will continue to grow in the foreseeable future, especially at the V band (40–75 GHz) and W band (75–111 GHz). In this chapter we introduce the basic characteristics of millimeter wave communication and its areas of application.

In very broad terms, millimeter wave technology can be classified as occupying the electromagnetic spectrum that spans between 30 and 300 GHz, which corresponds to wavelengths from 10 to 1 mm. In this book, we will focus on the 60 GHz industrial, scientific, and medical (ISM) band (unless otherwise specified, the terms 60 GHz and millimeter wave will be used interchangeably), which has emerged as one of the most promising candidates for multi-gigabit wireless indoor communication systems.

This chapter is organized as follows: Section 1.1 describes the characteristics of millimeter waves. Section 1.2 describes the channel performance of millimeter waves. Using these characteristics and performance, one can achieve the concept of gigabit wireless communications as shown in Section 1.3. Section 1.4 presents the development of millimeter wave standards in different countries. Section 1.5 describes interoperability, convergence, and co-existence of millimeter wave standards with other wireless local area network (LAN) standards.

Millimeter Wave Communication Systems, by Kao-Cheng Huang and Zhaocheng Wang

1.1 MILLIMETER WAVE CHARACTERISTICS

Before beginning an in-depth discussion of millimeter wave communication systems, it is important to understand the characteristics of millimeter waves. Millimeter waves are usually considered to be the range of wavelengths from 10 to 1 mm. This means they are larger than infrared waves or x-rays, for example, but smaller than radio waves or microwaves. The millimeter-wave region of the electromagnetic spectrum corresponds to the radio band frequency range of 30–300 GHz and is also called the extremely high frequency (EHF) range. The high frequencies of millimeter waves, as well as their propagation characteristics (i.e., the ways they change or interact with the atmosphere as they travel), make them useful for a variety of applications, including the transmission of large amounts of data, cellular communications, and radar. This section presents the benefits of 60 GHz technology and its characteristics. It can be used for high-speed Internet, data, and voice communications, offering the following key benefits:

1. Unlicensed operation—No license from the Federal Communications Commission is required. (Note: details on the 60 GHz unlicensed band can be found in Section 1.4.)
2. Highly secure operation—Resulting from short transmission distances due to oxygen absorption, narrow antenna beamwidth, and no wall penetration.
3. High level of frequency re-use enabled—The communication needs of multiple cells within a small geographic region can be satisfied.
4. Fiber optic data transmission speeds possible—7 GHz (in the U.S.) of continuous bandwidth available compared to less than 0.3 GHz at the other unlicensed bands.
5. Mature technology—This spectrum has a long history of being used for secure communications.
6. Carrier-class communication links enabled—60 GHz links can be engineered to deliver “five nines” (99.999%) availability if desired. For outdoor applications, such as back bones or by-pass bridges, a huge rain margin should be considered.

Two aspects of the characterizations of millimeter wave can be discussed: free space propagation and loss factors.

1.1.1 Free Space Propagation

As with all propagating electromagnetic waves, for millimeter waves in free space the power falls off as the square of the range. When the range is doubled, the power reaching a receiver antenna is reduced by a factor of four. This effect is due to the spherical spreading of the radio waves as they propagate. The frequency and distance dependence of the loss between two isotropic antennas can be expressed in absolute numbers by the following equation (in dB):

$$L_{\text{free space}} = 20\log_{10}\left(4\pi\frac{R}{\lambda}\right)\ (dB) \tag{1.1}$$

where $L_{\text{free space}}$ is the free-space loss, R is the distance between the transmitting and receiving antennas, and λ is the operating wavelength. This equation describes line-of-sight (LOS) wave propagation in free space. It shows that the free space loss increases when the frequency or range increases. Also, that millimeter wave free space loss can be quite high even for short distances. It suggests that the millimeter-wave spectrum is best used for short-distance communication links.

When the distance of the link $R = 10\,m$, path loss can be calculated using $(4\pi R/\lambda)^2$. The path losses of different unlicensed bands are listed below.

Unlicensed Bands	Path Loss
2.4 GHz	60 dB
5 GHz	66 dB
60 GHz	88 dB

The loss difference between 60 GHz and other unlicensed bands already pushes system design to the limit. One way to cover this extra 22 dB (i.e., $88 - 66 = 22$) of loss is with a high gain antenna and architecture.

Another way of expressing the path loss is the Friis Equation (H.T. Friis, 1946). It gives a more complete accounting for all the factors from the transmitter to the receiver (as a ratio, linear units) [1]

$$P_{RX} = P_{TX} G_{RX} G_{TX} \frac{\lambda^2}{(4\pi R)^2 L} \tag{1.2}$$

where G_{TX} = transmitting antenna gain, G_{RX} = receiving antenna gain, λ = wavelength (in the same units as R), R is the line-of-sight distance separating the transmitting and receiving antennas, and L = system loss factor (≥ 1).

Here, the factor $G_{RX}\frac{\lambda^2}{4\pi}$ is the effective area of the receiving antenna, and so shows why there is a wavelength dependency in (1.1) and (1.2).

1.1.2 Millimeter-Wave Propagation Loss Factors

In addition to the free-space loss, which is the main source of transmission loss, there are also absorption loss factors, such as gaseous losses and losses from rain (or other micrometeors) in the transmission medium. The factors that affect millimeter wave propagation are given in Figure 1.1.

Atmospheric losses means that transmission losses occur when millimeter waves traveling through the atmosphere are absorbed by molecules of oxygen, water vapor, and other gaseous atmospheric constituents. These losses are greater at certain frequencies, coinciding with the mechanical resonant frequencies of the gas molecules.

The H_2O and O_2 resonances have been studied extensively for the purpose of predicting millimeter wave propagation characteristics. Figure 1.2 shows an expanded plot of the atmospheric absorption versus frequency at an altitude of 4 km and at sea level, for water content of 1 gm/m^3 and 7.5 gm/m^3, respectively (the former value represents relatively dry air while the latter value represents 75% relative humidity at a temperature of 10°C).

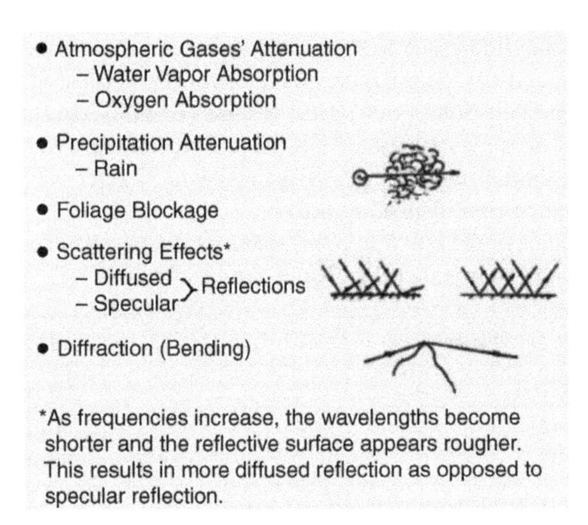

Figure 1.1. Propagation effects influencing millimeter-wave propagation [2] (© 2005 IEEE)

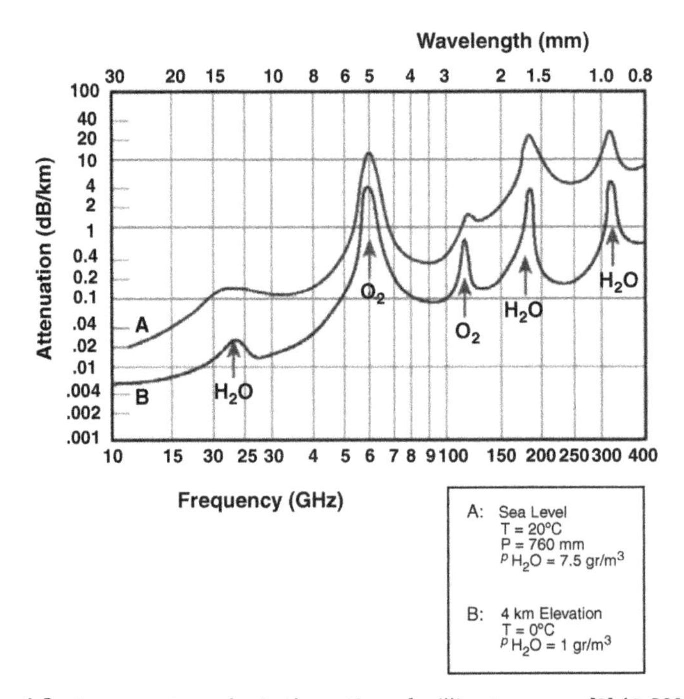

Figure 1.2. Average atmospheric absorption of millimeter waves [2] (© 2005 IEEE)

1.2 CHANNEL PERFORMANCE AT 60 GHz

Knowing the propagation characteristics and channel performance of millimeter waves is the first step to plan for a millimeter wave communication system. While signals at lower frequency bands, such as Global System for Mobile Communications (GSM) signals, can propagate for many kilometers and more easily penetrate buildings, millimeter wave signals can only travel a few kilometers, or less, and suffer from high transmission losses in the air and solid materials. However, these millimeter wave propagation characteristics can be very advantageous in some applications such as wireless personal area networks. Millimeter waves can be used to establish more densely packed communication links, thus providing very efficient spectrum utilization, by means of frequency reuse, and thus they can increase the overall capacity of communication systems. The characteristics of millimeter wave propagation are summarized in this section, including free space propagation and the effects of various physical factors on propagation.

The main challenges in utilizing a 60 GHz channel can be described as follows:

- High loss, as calculated from Friis equation (1.2)
- Human shadowing
- Non-line-of-sight propagation, which induces random fluctuations in signal level, known as multipath fading, as shown in Figure 1.3
- Doppler shift is non-negligible at pedestrian velocities
- Noise

The Federal Communications Commission (FCC) limits the equivalent isotropically radiated power (EIRP) of a 60 GHz communication link to +40 dBm. The

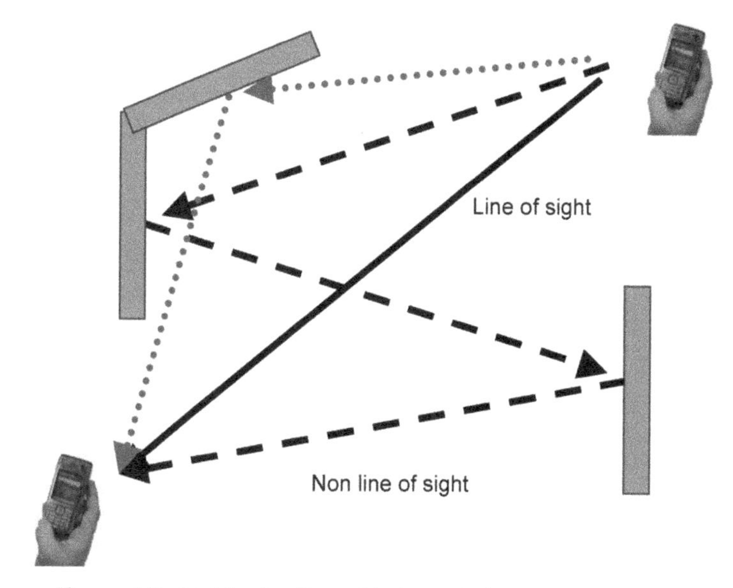

Figure 1.3. Multipath effect of indoor wireless communications

transmitter's power and path loss can be limiting factors for a high-speed wireless link. However, antenna directivity can be used to increase the power gain in the desired direction.

Assuming a simple line-of-sight free-space communication link, where both the received power P_{RX} and noise figure for the receiving system are 10 dBm and 10 dB, respectively, three different lengths are used for the distance R between the transmitter and receiver: *10, 15*, and *20 m*. Also, if we assume single omnidirectional antennas at the transmitter and receiver, the received power P_{RX} at a distance R decreases with an increase in frequency, and a 10-dB human shadowing loss is assumed for a system operating at 60 GHz.

Even when the bandwidth is unlimited, the received power P_{RX} is still limited by the Shannon Additive White Gaussian Noise (AWGN) capacity, as given by

$$C = BW \cdot \log_2\left(1 + \frac{P_{RX}}{BW \cdot N_0}\right) \approx 1.44 \frac{P_{RX}}{N_0} \quad \text{when} \quad BW \rightarrow \infty \tag{1.3}$$

The result is shown in Figure 1.4. As we can see, it is very unlikely that we could use an omnidirectional antenna to achieve Gbps data rate when human shadowing exists.

When the transceiver has $P_T = 10$ dBm, the noise figure $NF_{RX} = 6$ dB, and the environment has a human shadowing loss of 18 dB, we need the antenna gain of α to be in the range of 10–15 dB for 1 Gbps at 60 GHz. The results for other values of α can be found in [3, 4]. This means the total antenna gain has to be approximately 30 dB at least.

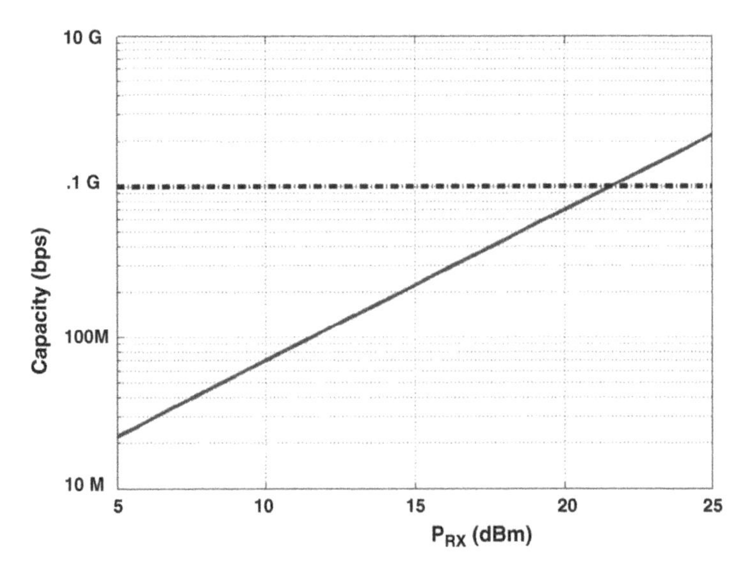

Figure 1.4. Shannon limit with distance $d = 10$ m between a transmitting omnidirectional antenna and a receiving omnidirectional antenna [3]

If we ignore the human shadowing loss, a clear path exists between the transmitter and receiver. When a 60 GHz system with the following parameters is defined,

Tx Power, P_T	10 dBm
Noise figure, NF	6 dB
Implementation loss, IL	6 dB
Thermal noise, N_0	174 dBm/MHz
Bandwidth, BW	1.5 GHz
Distance, R	20 m
Path loss at 1 m, PL_0	57.5 dB

we can calculate the ratio of the signal power to noise power at the *Rx* (in dB), by

$$\text{SNR} = P_T + G_{Tx} + G_{Rx} - PL_0 - PL(R) - IL - (KT + 10\log_{10}(BW) - NF) \tag{1.4}$$

where G_{Tx} and G_{Rx} denote the transmitting and receiving antenna gains, respectively, and P_{Tx} denotes the transmitter power. Inserting Equation (1.4) into the Shannon capacity formula of Equation (1.3), the maximum achievable capacity in AWGN can be computed. Figure 1.5 shows the Shannon capacity limit of an indoor office for LOS and non-LOS (NLOS) cases, using an omni-omni antenna setup [12]. It was found that for the LOS condition, data rate can go up to 5 Gbps. On the other hand, the operating distance for an NLOS condition is limited to below 3 m, although the NLOS capacity decreases more drastically as a function of distance.

To improve the capacity for a given operating distance, one can either increase the bandwidth or signal-to-noise ratio (SNR) or both. It can also be seen from Figure 1.5 that increasing the bandwidth used by more than 4 times only significantly improves the capacity for distances below 5 m. Beyond this distance, the capacity for the 7-GHz

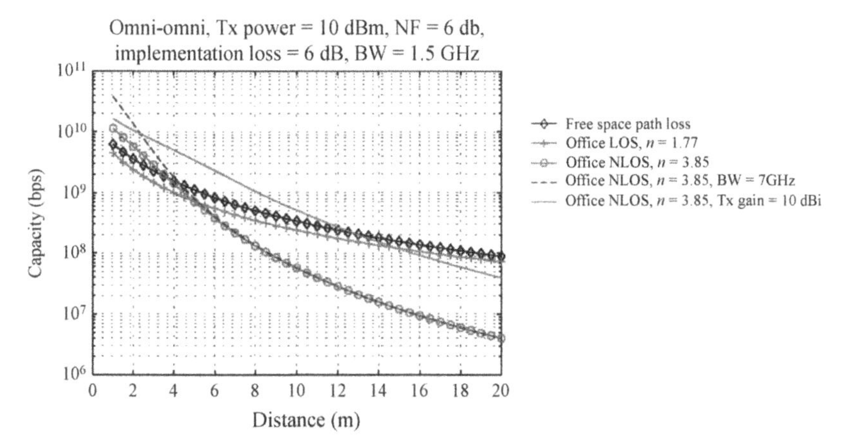

Figure 1.5. Shannon capacity limits for the case of indoor office using omni-omni antenna setup (Reproduced by permission of ©2007 S. K. Yong and C.-C. Chong [12])

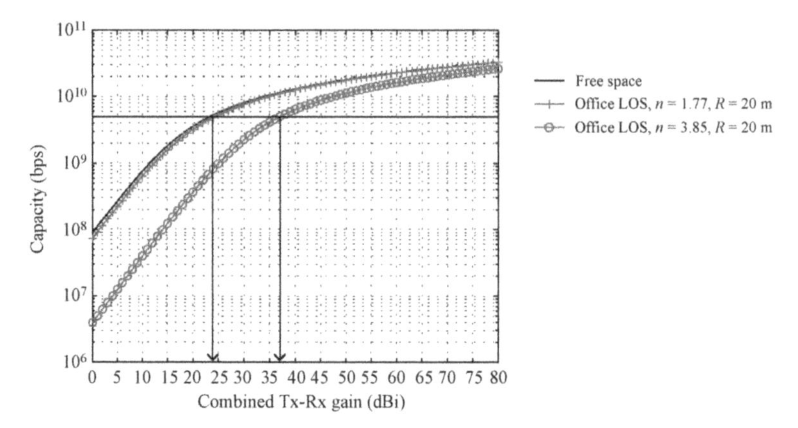

Figure 1.6. The required combined Tx-Rx antenna gain to achieve a target capacity (Reproduced by permission of ©2007 S. K. Yong and C.-C. Chong [12])

bandwidth is only slightly above the case of 1.5-GHz bandwidth, since the SNR at the Rx is reduced considerably at longer distance due to higher path loss. On the other hand, the overall capacity over the considered distance increases notably if a 10-dBi transmit antenna gain is employed as compared to the omnidirectional antenna for both 1.5-GHz and 7-GHz bandwidths. This clearly shows the importance of antenna gain in providing a very high data application at 60 GHz, which is not possible to be provided with omni-omni antenna configuration. But the question remains, how much gain is required?

To answer this question, the capacity as a function of combined Tx and Rx gain for operating distance at 20 m is plotted as depicted in Figure 1.6. In order to achieve 5 Gbps data rate at 20 m, combined gains of 25 dBi and 37 dBi were required for LOS and NLOS, respectively [12], when there was no human shadowing. These seem to be practical values, since they combine the Tx and Rx gains. However, to achieve the same data rates in a noisy channel, a higher gain is needed to overcome the interference.

Thus, directional antennas are required for Gigabit wireless communications. We can have different configurations for the access points (AP) and the mobile terminal (MT) based on different applications, as shown in Figure 1.7.

Considering the 60 GHz measurement setup shown in Figure 1.8, the synthesizer has a maximum output power of 0 dBm at 65 GHz. The coaxial cables have a maximum transmission loss of 6.2 dB/m at 60 GHz. The conversion loss of the subharmonic mixer is assumed to be 40 dB and its noise figure is 40 dB, while voltage standing wave ratio (VSWR) is 2.6:1. The noise floor for the spectrum analyzer is assumed to be −130 dBm.

The dynamic range can be measured as a function of the total antenna gain and the distance between antennas at 60 GHz. The results are shown in Figure 1.9.

Multipath propagation occurs when waves emitted by the transmitter travel along a multiplicity of different paths and interfere with waves traveling in a direct line-of-sight path, or with other multipath waves. Fading is caused by the destructive interference of these waves [5]. This phenomenon occurs because waves traveling along different paths may be completely out of phase when they reach the antenna, thereby canceling each

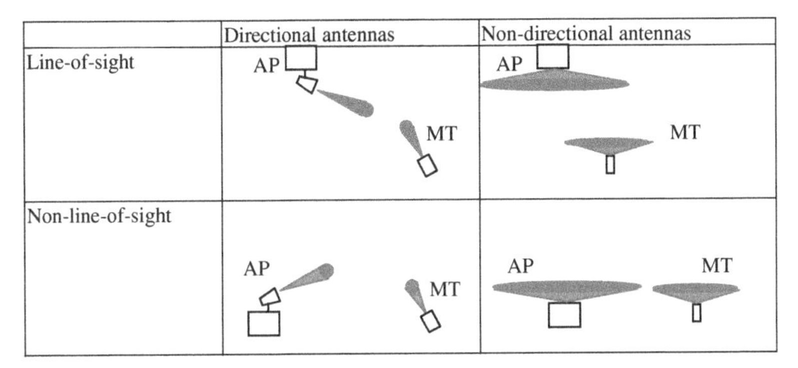

Figure 1.7. Classification of millimeter wave links according to the antenna beamwidth of the access point (AP) and mobile terminal (MT), in respect to the existence of a line-of-sight path. The radiation beamwidth is shown in gray

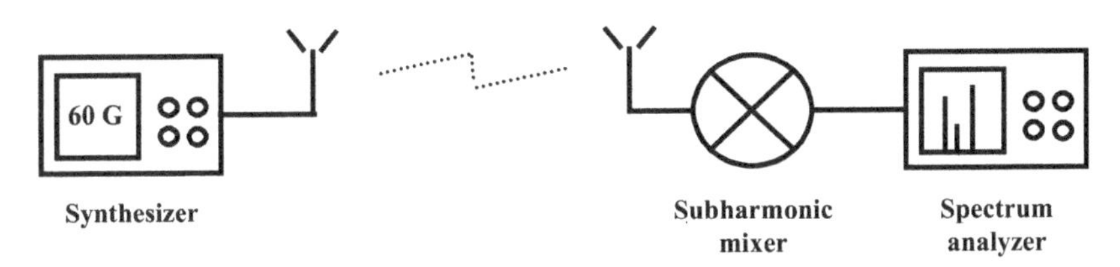

Figure 1.8. Channel measurement setup

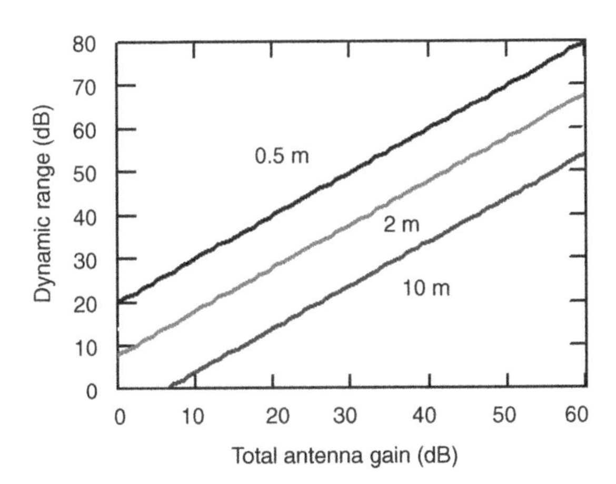

Figure 1.9. Dynamic range as a function of total antenna gain and the distance between antennas at 60 GHz

other out and forming an electric field null. One method of overcoming this problem is to transmit more power (either omnidirectionally or directionally). In an indoor environment, multipath propagation is always present and tends to be dynamic (constantly varying), due to the movement of scatterers. Severe fading due to indoor multipath propagation can result in a signal reduction of more than 30 dB. It is therefore essential to provide an adequate link margin to overcome this loss when designing a wireless system. Failure to do so will adversely affect the reliability of the link. The amount of extra radio frequency (RF) power radiated to overcome this phenomenon is referred to as a fade margin. The exact amount of fade margin required depends on the desired reliability of the link, but a good rule of thumb is 20–30 dB.

In channel measurements, as shown in Figure 1.10, we compare antennas with different beamwidths. For antennas with a narrow beamwidth, a notch appears in the frequency response of channel measurement. For antennas with a broad beamwidth, the notch in the frequency response becomes severe. In the extreme case, if an antenna beam is as sharp as a laser, this notch will not exist in the frequency response.

The notch step is affected by the delay time, while the notch depth is affected by the difference in the path gains (or losses). In addition, the notch position is affected by the difference in the lengths of the propagation paths.

Based on these observations, three solutions can be considered to minimize the notch effect. One would be to employ a narrow beam antenna to reduce reflected paths and achieve a smaller notch depth and wider notch period. However, this would involve solving the tracking resolution and the tracking speed. In addition, the selection of suitable equalizers having good performance should be investigated. Finally, we could use precise source tracking or space diversity to avoid the notch effect. A few implementation issues, such as tracking algorithm and dual antenna implementation, need to be tackled.

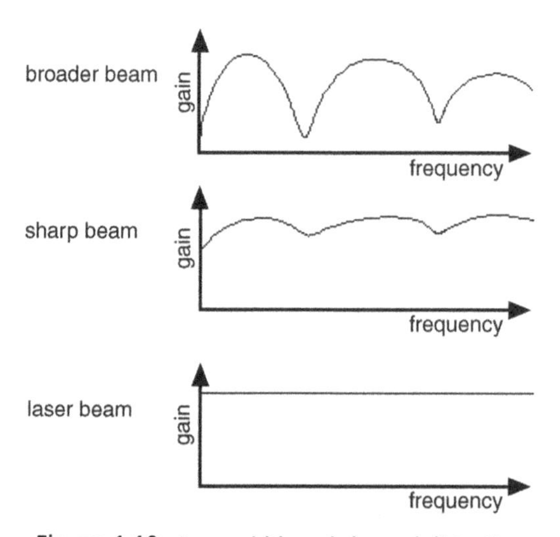

Figure 1.10. Beamwidth and channel distortion

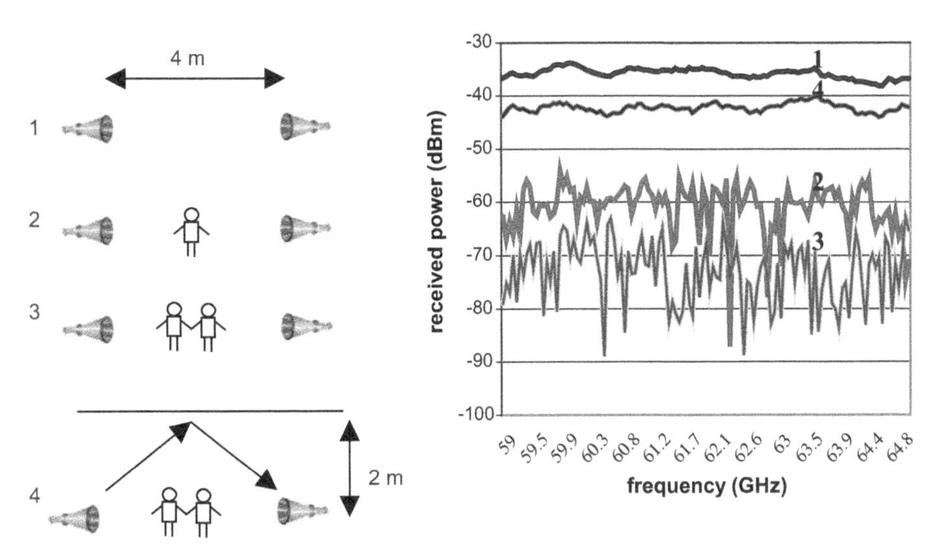

Figure 1.11. Indoor channel measurement at 60 GHz for NLOS

In an office environment, the reflection characteristics of interior structures have been studied [6]. Furthermore, the human shadowing problem has been investigated and the results are shown in Figure 1.11. When 0 dBm power at 60 GHz is transmitted via a 10 dBi gain transmitting antenna to a receiving antenna with a 10 dBi gain at a distance of 4 m, the spectrum shows that the received power is approximately −35 dBm when there is no human shadowing (Case 1). If there is a human body between the two antennas, the signal is reduced to a range of between −55 dBm and −65 dBm (Case 2). If there are two human bodies between the two antennas, the signal is reduced to a range of between −65 dBm and −80 dBm (Case 3). If we change the beam direction of the 10 dBi transmitting antenna so that the signal can bounce off the concrete ceiling, at a height of 2 m, so that it is reflected to the receiver, the received signal is increased to −42 dBm (Case 4). This shows that reflected propagation at 60 GHz can be used for NLOS wireless communications.

The transmitting and receiving antennas have a gain of 10 dBi. No. 1 shows a LOS scenario. No. 2 shows a person standing between the two antennas. No. 3 shows two people standing between the two antennas. No. 4 shows an NLOS wireless link, which improves the received signal power.

1.3 GIGABIT WIRELESS COMMUNICATIONS

After knowing the millimeter wave channel performance, we can start thinking of its application in wireless communication. In wired communication, the adoption of each successive generation of Ethernet technology has been driven by economics, performance demand, and the rate at which the price of the new generation has approached the old one. As the cost of 100 Mbps Ethernet dropped and approached the previous cost of

10 Mbps Ethernet, users rapidly moved to the higher performance standard. In 2007, the industry announced the availability of 10-Gb Ethernet over copper wiring [7]. Also, gigabit Ethernet became affordable (below $200) for server connections, and desktop gigabit connections have come within $10 or less of the cost of 100 Mbps technology. The serial 100-Gb Ethernet PHY (physical layer) based on advanced modulation techniques such as differential quadrature-phase-shifted-keying (DQPSK) have been experimentally demonstrated worldwide [8]. Consequently, gigabit Ethernet has become the standard for servers, and systems are now routinely ordered with gigabit network interface cards. Mirroring events in the wired world, as the prices of wireless gigabit links approach the prices of 100 Mbps links, users are switching to the higher-performance product, both for traditional wireless applications and for applications that only become practical at gigabit speeds.

In terms of a business model, wireless communications have pointed toward the need for wireless USB (Universal Serial Bus) 2.0, gigabit speeds, and longer-range connectivity as applications emerge for home audio/video (A/V) networks, high-quality multimedia, and voice and data services. Previously wireless local area networks (WLANs) only offered peak rates of 54 Mb/s, but 150–300 Mb/s, such as IEEE 802.11n, have become available. However, even 500 Mb/s is inadequate when faced with the demand for higher access speed from rich media content and competition from 10-Gb/s wired LANs. In addition, future home A/V networks will require Gb/s data rates to support multiple high-speed high-definition A/V streams (e.g., carrying uncompressed high-definition video at resolutions of up to 1,920 × 1,080 progressive scan, with latencies ranging from 5 ms to 15 ms) [9].

Based on the technical requirements of applications for high-speed wireless systems, both the industry and the standardization bodies need to take the following issues into account:

1. Pressure to increase the data rate will persist. More specifically, there is a need for advanced domestic applications, such as high-definition wireless multimedia, which demand higher data rates.
2. Data streaming and download/memory back-up times for mobile and personal devices will also place demands on shared resources and user models point to very short dwell times for these downloads.

Some approaches, such as IEEE 802.11n, enhance data rates by evolving the standards of the existing wireless LANs to increase data rates to speeds that are up to 10 times faster than IEEE 802.11a or 802.11g. IEEE 802.16 (WiMAX) can provide broadband wireless access up to 50 km with speeds around 70 Mbps. Others, such as ultra-wideband (UWB), are pursuing much more aggressive strategies, such as sharing spectra with other users. Another approach that will no doubt be taken will be the time-honored strategy of moving to higher, unused, and unregulated millimeter wave frequencies.

Despite the fact that millimeter wave technology has been established for many decades, the millimeter wave systems that are available have mainly been deployed for military applications. With the advances in process technologies and low-cost integration solutions, this technology has started to gain a great deal of momentum from academia, industry, and standardization bodies.

Although the IEEE 802.11n standard will improve the robustness of wireless communications, only a modest increase in wireless bandwidth is provided and the data rate is still lower than 1 Gbps. Importantly, 60 GHz technology offers various advantages over the currently proposed or existing communications systems. One of the deciding factors that makes 60 GHz technology attractive is the establishment of a (relatively) huge unlicensed bandwidth (up to 7 GHz), which is available worldwide. Spectrum allocation is mainly regulated by the International Telecommunication Union (ITU). The details of band allocation around the world can be found in Section 1.4.

While this is comparable to the unlicensed bandwidth allocated for UWB purposes (~2 GHz–10 GHz), the 60 GHz band is continuous and less restricted in terms of power limits (also there are fewer existing users). This is due to the fact that the UWB system is an overlay system, and thus subject to different considerations and very strict regulation. The large band at 60 GHz is, in fact, one of the largest unlicensed spectral resources ever allocated. This huge bandwidth offers great potential in terms of capacity and flexibility and makes 60 GHz technology particularly attractive for gigabit wireless applications. Although 60 GHz regulations allow a much higher transmission power compared to other existing WLAN (e.g., maximum 100 mW for 802.11 a/b/g, as defined in Table 1.1) and wireless personal area network (WPAN) systems, the higher transmission power is necessary to overcome the higher path loss at 60 GHz.

In addition, the typical 480 Mbps bandwidth of the UWB cannot fully support broadcast video, requiring the recompression of packets. This forces manufacturers to utilize expensive encoders and install more memory in their systems, in effect losing video content and adding latency in the process. Therefore, 60 GHz technology could actually provide better resolution, with less latency and cost for television, DVD players, and other high-definition equipment, compared to the UWB.

IEEE standard 802.15.3C was published in September 2009 [10]. It allows very high data rate over 2 Gbit/s applications. Optional data rate in excess of 3 Gbit/s can be provided.

Figure 1.12 shows the development and trends for wireless standards. Advanced wireless technology should always adopt a timeline, or milestones, to increase data rates by 5 to 10 times every three or four years to keep pace with the projected demand.

TABLE 1.1. Transmission Power Comparison for Different Wireless Standards

	Maximum Transmission Power
802.11 a	40 mW
802.11 b/g	100 mW
802.15.3c	500 mW

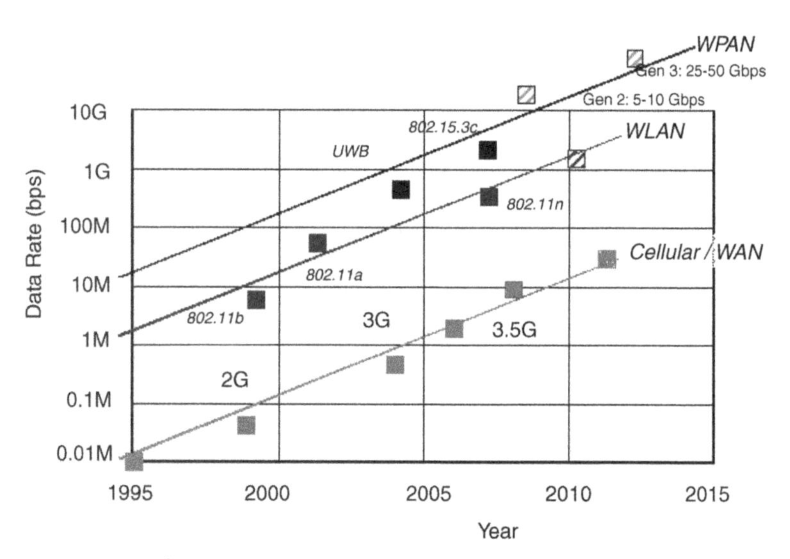

Figure 1.12. Data rate projections over time [11]

Although the high path loss at 60 GHz seems to be a disadvantage, it essentially confines the 60 GHz power and system operation to within a room in an indoor environment. Hence, the effective interference levels for 60 GHz are less severe than for those systems located in the congested 2–2.5 GHz and 5–5.8 GHz regions. In addition, higher frequency reuse can also be achieved over a very short distance in an indoor environment, thus allowing a very high throughput network. The compact size of a 60 GHz receiver also permits multiple antenna solutions at a user terminal, which are otherwise difficult, if not impossible, at lower frequencies. Compared to a 5-GHz system, the form factor of millimeter wave systems is approximately 140 times smaller and can be conveniently integrated into consumer electronic products. Thus, it will require new design methodologies to meet modern communication needs.

Designing a very-high-speed wireless link that offers good quality of service and range capability constitutes a significant research and engineering challenge. Ignoring fading for the moment, we can, in principle, meet the 1 Gbps data rate requirement if the product of the bandwidth (in units of Hz) and spectral efficiency (in b/s/Hz units) equals 10^9. As we shall describe in the following sections, a variety of cost, technology, and regulatory constraints make such a solution very challenging.

Despite the various advantages offered, millimeter wave–based communications suffer from a number of critical problems that must be resolved. Figure 1.13 shows the data rates for various WLAN and WPAN systems. Since there is a need to distinguish between different standards for broader market exploitation, IEEE 802.15.3c is positioned to provide gigabit rates and longer operating ranges. At these rates and ranges, it will be a nontrivial task for millimeter wave systems to provide a sufficient power margin to ensure a reliable communication link. Furthermore, the delay spread of the channel under consideration is another limiting factor for high-speed transmission. Large delay

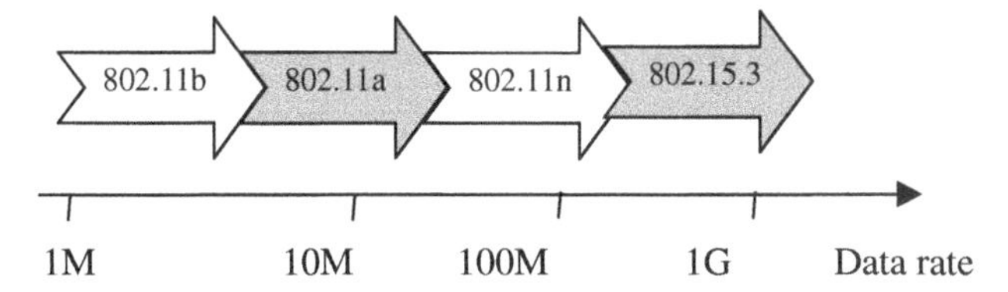

Figure 1.13. Data rates requirements for WLAN and WPAN standards and applications. Millimeter wave technology, that is, IEEE 802.15.3c, aims at very high data rates [12]

spread values can easily increase the complexity of the system beyond the practical limit for equalization [12].

Millimeter wave technology is certainly the first choice to build gigabit wireless communications. To understand its features, we can compare it with other technologies such as wireless optical LANs and UWB.

Wireless optical LANs also compete as a communication technology that is able to offer a significant unregulated spectrum. Diffuse optical networks use wide-angle sources and scatter from surfaces in the room to provide an optical "ether" similar to that obtained using a local radio transmitter [13]. This produces coverage that is robust to blocking, but the multiple paths between the source and receiver cause dispersion of the channel, thus limiting its performance. The optical transmitters require extremely high power, and dynamic equalization is needed for high-bandwidth operation.

Within buildings or in spaces with limited coverage, optical networks have the potential to offer significant advantages over radio approaches. One approach is to use direct LOS paths between the transmitter and receiver [14]. These can provide data rates of hundreds of megabits per second and above, depending on the particular parameters. However, the coverage area provided by a single channel can be quite small. Thus, providing area coverage and the ability to roam presents a major challenge. LOS channels can be blocked, as there is no alternative scattered path between the transmitter and receiver, and this presents a major challenge in network design [13]. Multiple base stations within a room would provide coverage in this case, and optical or fixed connections could be used between the stations. JVC offers a commercial LOS system with 10 Mb/s Ethernet connections [15].

In general, optical channels are subject to eye safety regulations, and these are difficult to meet, particularly for LOS channels [13]. Typically, optical LANs work in the near-infrared region (between 700 and 1000 nm), where optical sources and detectors are low cost, and regulations are particularly strict in this region. At longer wavelengths (1500 nm and above) the regulations are much less stringent, although sources at this wavelength and power output are not widely available [16].

As previously mentioned, the other major problem for optical channels is that of blocking. LOS channels are required for high-speed operation and these are by their nature subject to blocking. Within a building, networks must be designed using appropriate geometries to avoid blocking, and this is usually solved by using multiple access points to allow complete coverage [16, 17].

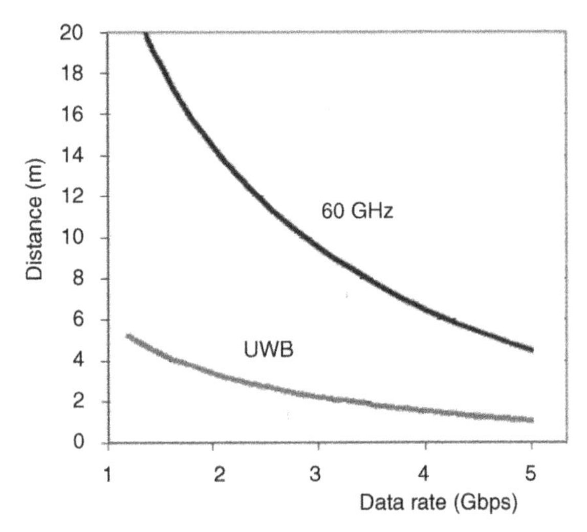

Figure 1.14. Shannon's capacity curve at an occupied bandwidth of 1 GHz for 60 GHz vs. UWB (noise figure is set at 8 dB) [11]

If we assume a 10-mW power input to the antenna, with a 10-dBi gain based on a highly integrated, low-cost design with a steerable beam at 60 GHz, we get the Shannon capacity curve shown in Figure 1.14. The formula used to derive these curves is presented in Equations (1.3) and (1.4) in Section 1.2.

Table 1.2 compares the characteristics of three technologies for gigabit communications: UWB radio, millimeter wave, and wireless optics.

1.4 DEVELOPMENT OF MILLIMETER WAVE STANDARDS

1.4.1 Europe

The European Telecommunications Standards Institute (ETSI) and the European Conference of Postal and Telecommunications Administrations (CEPT) have been working closely to establish a legal framework for the deployment of unlicensed 60 GHz devices [21]. In general, the 59–66 GHz band has been allocated for mobile services without a specific decision on the regulations, as shown in Figure 1.15. In Recommendation T/R 22–03, the CEPT provisionally recommended the use of the 54.25–66 GHz band for terrestrial and fixed mobile systems [22]. However, this provisional allocation was withdrawn in September 2005 [12].

In 2003, the European Radiocommunications Committee (ERC) of the European Conference of Postal and Telecommunications Administrations revised the European Table of Frequency Allocations and Utilizations [23].

The ERC also considered the use of the 57–59 GHz band for fixed services without requiring frequency planning [24]. Later, the Electronic Communications Committee (ECC) of CEPT recommended the use of point-to-point fixed services in the 64–66 GHz band [25]. A few years ago, ETSI proposed 60 GHz regulations to be considered by the

TABLE 1.2. Comparison of Three Technologies for Gigabit Wireless Communications [18–20]

	Millimeter Wave	UWB Radio	Optical Wireless
Advantage	1. High data rates (up to Gbps) 2. Compatible to fiber optic networks at 60 GHz	1. Low power 2. Short range 3. Low data 4. Penetration through obstacles in the transmission path	1. High data rate 2. Unlicensed and unregulated
Challenge	1. Reduce cost 2. Reduce power	1. Matched filter problem 2. Antenna parameter tradeoff	1. Atmospheric loss ranging from 10 dB/km(sunny) to 350 dB/km(foggy) 2. Multi-user application 3. No federal rights or protection for the link.
Peer to peer	Indoor/outdoor	Indoor/outdoor	Indoor/outdoor
Multiple-access	Indoor/outdoor	Indoor	Indoor
Data rate	>1.25 Gbps at 60 GHz ~10 Gbps at 122.5 GHz	500 Mbps within 10 m (FCC)	~1.25 Gbps (peer-to-peer)
Indoor max. range	Room area	76 m (station in commercial building)	7 m(mobile) 10 m (station)
DC power consumption	High	Low	5 VDC, 500 mA (mobile)
Max TX power	500 mW(FCC 15.255)	Maximum output power of 1 Watt spread over spectrum. Maximum power density: −41.3 dBm/MHz (FCC)	Power density should be less than 1 mW/cm^2 (FDA)
Notes	Antenna design is one of the main challenges	1. Infrastructure or peer-to-peer for indoor application 2. Only peer-to-peer for hand-held application (FCC)	Eye safety should be considered.

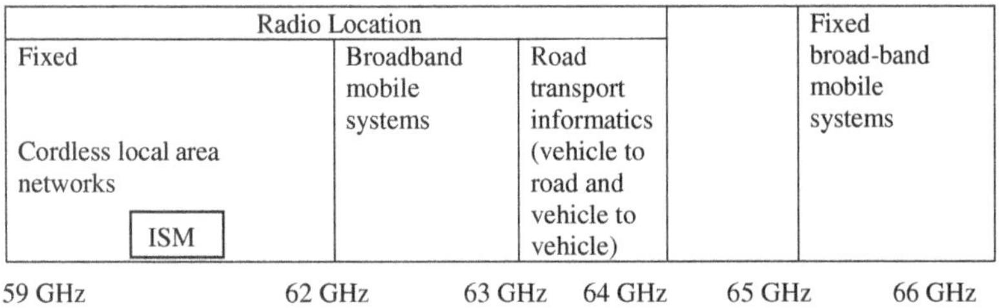

Radio Location
Fixed
Broadband mobile systems
Road transport informatics (vehicle to road and vehicle to vehicle)
Fixed broad-band mobile systems
Cordless local area networks
ISM
59 GHz
62 GHz
63 GHz
64 GHz
65 GHz
66 GHz

Figure 1.15. 60 GHz frequency spectrum in Europe [23]

ECC of the European Conference of Post and Telecommunications Administrations for WPAN applications [26]. Under this proposal, a 9-GHz unlicensed spectrum has been allocated for 60 GHz operation. This band that represents the union of the various previous bands was approved in 2009 [27].

The frequency band being considered is 57–66 GHz. The spectrum allocation is shown in Figure 1.16 and Table 1.3. This is an amalgamation of the bands currently approved for license-exempt use in Japan and the United States, and being proposed for allocation in the Republic of China and the Republic of Korea. The existing etiquette rules, spectrum sharing studies, and other analyses in these countries may be a model for considering the needs of the commercial, military, and scientific uses of these frequencies worldwide. In Germany, the regulatory requirements are that the frequency bands of 57.1–57.8 GHz and 58.6–58.9 GHz be used for time division duplex (TDD) point-to-point connections [28].

57 GHz
58 GHz
59 GHz
60 GHz
61 GHz
62 GHz
63 GHz
64 GHz
65 GHz
66 GHz
Europe (57 – 66 GHz)
Japan (59 – 66 GHz)
Australia (59.4 – 62.9 GHz)
USA (57 – 64 GHz)

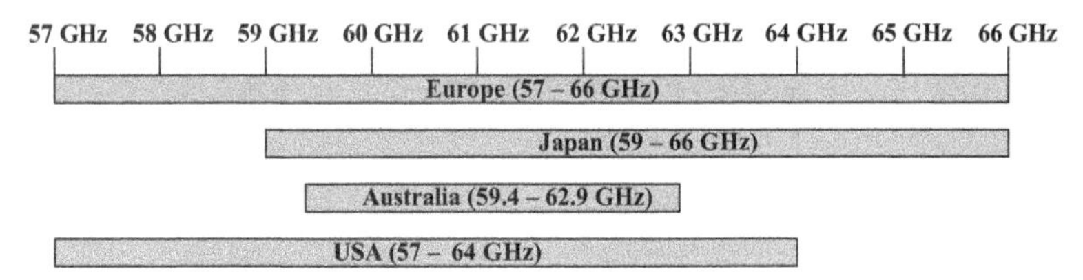

Figure 1.16. Geographically available 60 GHz spectrum and power

TABLE 1.3. International Frequency Allocation at 60 GHz [32]

Region	Unlicensed Bandwidth (GHz)	Max. Tx Power	Max. Antenna Gain	Ref.
Europe	9 GHz (57–66) min 500 MHz	20 mW	37 dBi	[26]
Japan	7 GHz (59–66) max 2.5 GHz	10 mW	47 dBi	[29]
Korea	7 GHz (57–64)	10 mW	To be decided	[30]
Germany	1 GHz (57.1–57.8) (58.6–58.9)	50 mW	Not specified	[28]
USA	7 GHz (57–64)	500 mW	Not specified	[31]

1.4.2 United States

In 2001, the U.S. FCC allocated 7 GHz in the 54–66 GHz band for unlicensed use [31]. In terms of the power limits, the FCC rules allow emissions with an average power density of 9 $\mu W/cm^2$ at 3 m and a maximum power density of 18 $\mu W/cm^2$ at a range of 3 m from the radiating source. These data translate to an average equivalent isotropic radiated power (EIRP) and a maximum EIRP of 40 dBm and 43 dBm, respectively. The FCC also specified a total maximum transmission power of 500 mW for an emission bandwidth greater than 100 MHz. The devices must also comply with the radio frequency (RF) radiation exposure requirements specified in [31]. After taking the RF safety issues into account, the maximum transmission power is limited to 10 dBm. Furthermore, each transmitter must transmit at least one transmitter identification signal within a one-second interval of the signal transmission. It is important to note that the 60 GHz regulations in Canada, which are regulated by Industry Canada Spectrum Management and Telecommunications (IC-SMT) [33], harmonize with the United States.

In 2003, the FCC announced that the frequency bands found at 71–76 GHz, 81–86 GHz, and 92–95 GHz were available for wireless applications [34]. FCC chairman Michael Powell heralded the ruling as opening a "new frontier" in commercial services and products for the U.S. users [35]. The allocation provides the opportunity for a broad range of new products and services, including high-speed, point-to-point WLANs and broadband Internet access at gigabit data rates and beyond.

The significance of the 70, 80, and 90 GHz allocations cannot be overstated. Collectively referred to as the E-band (60–90 GHz), these three allocations are the highest ever licensed by the FCC. Together, the nearly 13 GHz of allocated spectrum is greater than all of the other previously existing commercial wireless spectra combined. The FCC ruling also permits a novel licensing scheme, allowing cheap and fast allocations to prospective users. All this was achieved at unprecedented speed, with barely more than two years from the initial FCC petition to the formal release of the rules.

1.4.3 Japan

In the year 2000, the Ministry of Public Management, Home Affairs, Posts, and Telecommunications (MPHPT) of Japan issued 60 GHz radio regulations for unlicensed utilization of the 59–66 GHz band [29]. However, the 54.25–59 GHz band is allocated for licensed use. The maximum transmission power for this unlicensed use is limited to 10 dBm, with a maximum allowable antenna gain of 47 dBi. Unlike the arrangements in North America, the Japanese regulations specified that the maximum transmission bandwidth must not exceed 2.5 GHz. There was no specification for RF radiation exposure and transmitter identification requirements [29].

1.4.4 Industrial Standardization

Originally, the first international industry standard that covered the 60 GHz band was the IEEE 802.16 (WiMAX) Standard for local and metropolitan area networks [36].

TABLE 1.4. 60 GHz Standards in Japan

Code	Standard Name	Note
ARIB STD-T69 (Jul.2004)	Millimeter-Wave Video Transmission Equipment for Specified Low Power Radio Station	Bandwidth: 1208 MHz
ARIB STD-T69 Revision (Nov. 2005)	Millimeter-Wave Video Transmission Equipment for Specified Low Power Radio Station (Only the part of the revision from Ver.2.0 to Ver.2.1)	Tx power: 10 dBm Rx antenna gain: 0 dBi.
ARIB STD-T74 (May 2001)	Millimeter-Wave Data Transmission Equipment for Specified Low Power Radio Station (Ultra High Speed Wireless LAN System)	Bandwidth: 200 MHz
ARIB STD-T74 Revision (Nov. 2005)	Millimeter-Wave Data Transmission Equipment for Specified Low Power Radio Station (Ultra High Speed Wireless LAN System) (Only the part of the revision from Ver.1.0 to Ver.1.1)	Tx power: 10 dBm Rx antenna gain: 0 dBi.

However, this is a licensed band and is used for LOS outdoor communications for last mile connectivity. In Japan, two standards related to the 60 GHz band were issued by the Association of Radio Industries and Business (ARIB), namely, ARIB-STD T69 and ARIB-STD T74 [37, 38], as shown in Table 1.4. The former is a standard for millimeter wave video transmission equipment for a specified low-power radio station (point-to-point system), while the latter is a standard for a millimeter wave ultra high-speed WLAN for specified low-power radio stations (point-to-multipoint). Both standards cover the 59–66 GHz band defined in Japan.

The interest in 60 GHz radio continued to grow with the formation of the Millimeter Wave Interest Group and the Millimeter Wave Study Group within the IEEE 802.15 Working Group for WPAN. In 2005, the IEEE 802.15.3c Task Group (TG3c) was formed to develop a millimeter wave-based alternative physical layer (PHY) for the existing IEEE 802.15.3 WPAN Standard 802.15.3-2003 [39]. The developed PHY is aimed at supporting a minimum data rate of 2 Gbps over a few meters with optional data rates in excess of 3 Gbps. This is the first standard that addresses multi-gigabit wireless systems and will form the key data rate solution for many applications, especially those related to wireless multimedia distribution. Its channel plan is divided into two groups: full-rate channel plan and half-rate channel plan (see Figure 1.17). The full-rate plan has four full-rate channels in 9-GHz bandwidth. It supports data rates of up to 6.118 Gbps approximately in single-carrier mode or up to 7.3 Gbps in OFDM mode. The half-rate channel plan has four half-rate channels with the same center frequencies as the full-rate channels. It supports data rates of up to 1.530 Gbps using $\pi/2$ BPSK in single-carrier mode.

In 2007, another group, WirelessHD™ (high definition), also released a specification, which uses the unlicensed 60 GHz radio waves to send uncompressed HD video and audio at 5 Gbps at distances of up to 30 feet, or within one room of a house. Its core technology promotes theoretical data rates of up to 20 Gbps, permitting it to scale to

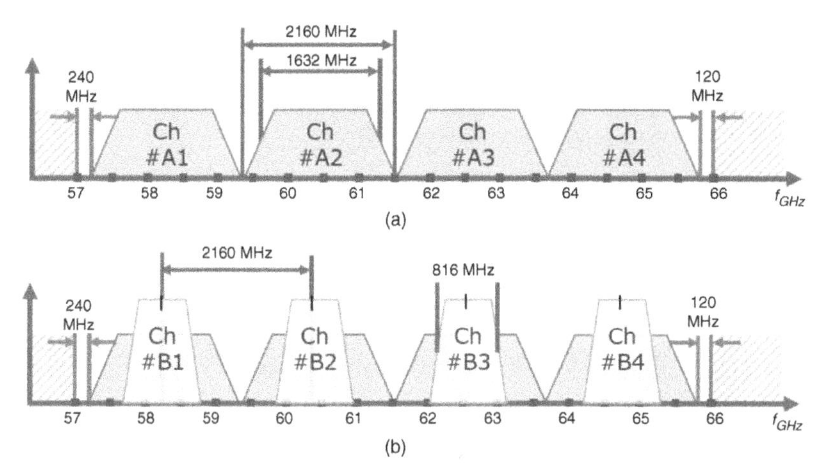

Figure 1.17. (a) Full-rate channel plan, (b) Half-rate channel plan of IEEE 802.15.3c [40] (©2007 IEEE)

higher resolutions, color depths, and ranges. Coexisting with other wireless services, the WirelessHD platform is designed to operate cooperatively with existing, wireline display technologies. The specification maintains high-quality video, ensures the interoperability of consumer electronic devices, provides signal interference protection, and uses existing content protection techniques. The WirelessHD™ group believes that 60 GHz will allow the fast transmission speeds required for high-definition content. The ideal speed for WirelessHD is 5 Gbps and a coverage range of up to 10 m. It is worth noting that IEEE 802.15.3c targets at Fast Data Download whereas wirelessHD targets at wireless entertainment and displays.

In addition, technical committee task group TG20 of the European Computer Manufacturers' Association (ECMA International) is also developing a standard for a 60 GHz PHY and Media Access Control (MAC) short-range unlicensed communications. This standard provides up to 10 Gbps WPAN (including point-to-point) transport for both bulk data transfer and multimedia streaming. TG20 is considering three device types ranging from high-end devices with steerable antennas to low-end devices for cost-effective, short-range, gigabit solutions. This confirms the role of millimeter wave antennas in gigabit communications.

Also in 2007, the National Institute of Information and Communications Technology (NICT) in Japan and 20 companies formed a organization called CoMPA (Consortium of Millimeter-wave Practical Applications). The group, together with IEEE 802.15.3c, had to discuss three key points, namely the system model, the usage model, and the channel model. In 2009, the Wireless Gigabit Alliance also completes multi-gigabit 60 GHz wireless specification to set up wireless multimedia streaming. It supports data transmission rates up to 7 Gbps and supports for beamforming, enabling robust communication at distances beyond 10 m.

Table 1.5 summarizes the potential applications of millimeter wavelength systems as submitted in response to the IEEE Call for Applications (CFA). The proposals are intended to illustrate support for some of the applications listed. The applications have been arranged in the numeric order of the CFA document number (last column) [41].

TABLE 1.5. Possible Applications for Millimeter Wave Communications [41]

No.	Description of Applications	Outdoor	Indoor	IEEE CFA Document Number
1	• Outdoor: Distribution links in apartments, stadium, etc. • Indoor: Ad hoc network	• LOS • P2P and P2MP • Bandwidth: >300 MHz • Range: ≤ 220 m	• LOS • Data rate: ≥ 1 Gbps and ≥ 622 Mbps • Range: ≥ 20 m and ≥ 3 m	04-0352
2	Gigabit Ethernet link, wireless IEEE1394 applications	–	• LOS • Data rate: ≤ 1 Gbps duplex • Range: ≤ 17 m	04-0019
3	Multimedia, information distribution system	–	• LOS • Data rate: ≥ 1 Gbps • Range: ≤ 10 m	04-0098
4	• Outdoor: Fixed wireless access, distribution in stadiums, intervehicle communication, etc. • Indoor: Connecting multimedia devices (wireless home link), ad hoc meeting, heavy content download, distribution system	• LOS • P2P, P2MP • Data Rate: 156 Mbps to 1.5 Gbps • Range: 400 m to 1 km	• LOS • Data Rate: 100 Mbps to 1.6 Gbps • Range: ~ 10 m	04-0118

5	Distribution links in apartments, stadium, etc.	• LOS • P2P • Bandwidth: >300 MHz • Range: ≤220 m	–	04-0153
6	Wireless home video server connected to HDTV, PC, and other video devices	–	• LOS • Data rate: 300 Mbps, 400 Mbps, and 1.5 Gbps uncompressed HDTV data • Range: ≤10 m	04-0348
7	• Replacement for 1394 FireWire • Replacement for USB • Military: future combat systems, secure communication	–	• LOS and NLOS (people) • 100–500 Mbps link, 1 Gbps in 2007 • Short range	04-0665

1.5 COEXISTENCE WITH WIRELESS BACKHAUL

Millimeter wave communication can operate with other wireless/mobile communication standards. For instance, the 802.16 specification applies across a wide range of the RF spectrum, and WiMAX could function on any frequency below 66 GHz [42] (higher frequencies would decrease the range of a base station to a few hundred meters in an urban environment). Although 802.16-2004 and 802.16e have reduced their operating frequencies to the 2–11 GHz range, a millimeter wave link could still work as its wireless backhaul solution. In fact, two of the best solutions are the millimeter-wave V-band and E-Band for supporting WiMAX operations in the enterprise and urban markets. The options for wireless backhaul, simply put, are 60 GHz and 80-GHz high-power radios. The short range of these products is, relative to other wireless backhaul options, the strength of the solution. Short propagation range is due to the oxygen molecules absorb electromagnetic energy at 60 GHz. This shortcoming of 60 GHz is used as an advantage to reuse the same frequency in short distance applications.

Free-space optical backhaul is another means of high-bandwidth transmission technology. It is a LOS technology using invisible beams of light to provide optical bandwidth connections using laser or LED. While both millimeter-wave and free-space optical backhauls can provide gigabit transmission rate, the requirement of high-stability mounting for free-space optics and the potential of beam obstruction have rendered free-space optics less attractive for large-scale mobile backhaul deployment [43].

1.5.1 60 GHz: V-band

The 57–64 GHz band (best known as 60 GHz) is located in the millimeter-wave portion of the electromagnetic spectrum. The advantages of using this band include interference mitigation, strong security, traffic prioritization (QoS), frequency re-use, and rain fade mitigation. In the United State, the 60 GHz band is unlicensed.

1.5.2 80 GHz: E-Band

The E-Band refers to 10 GHz of licensed-band spectrum allocated by the U.S. FCC, split between 71–76 GHz and 81–86 GHz. With more spectrally efficient modulations, full duplex data rates of 10 Gbps (OC-192 or 10GigE) can be reached. Diffraction in the 80-GHz band is not as severe as it is at 60 GHz, thus allowing greater range at 80 GHz compared to 60 GHz. E-Band products are similar to millimeter-wave products, except that in the United States, for example, E-Band frequencies are licensed by the FCC in a streamlined licensing process. Licensed spectrum products are allowed more power than unlicensed spectrum products, further enhancing the range and throughput of 80-GHz products.

1.5.3 Very Narrow Beamwidth

One advantage of the 60- and 80-GHz bands is that they overcome the fallibilities of lower-frequency backhaul options. Antenna directivity (beamwidth) is limited by the physical principle of diffraction, wherein the beamwidth is inversely proportional to the operating frequency. So, at 60 GHz, for example, the beamwidth is far narrower than at lower frequencies (such as 5.8 GHz). This very narrow beam gives a backhaul solution the following advantages:

1. It avoids interference from other emitters in the same band, since the beam is so sharp that the potential for interfering with another such beam on the same frequency is very remote.
2. It offers superior security: the beam is so narrow, it is difficult to intercept or otherwise exploit.
3. It offers a high rate of frequency reuse in a backhaul network.
4. It has the power to overcome rain fading.

When engineering a backhaul network to support a millimeter wave/WiMAX network, the service provider should consider the following elements in backhaul planning:

1. Range and throughput demands of the market
2. Security
3. Quality of service (QoS)
4. Interference mitigation
5. Frequency reuse
6. Rain fade
7. Ease of licensing: E-Band in the United States

The selection of backhaul solutions, like the WiMAX platforms they support, should be driven by the business plan. That is, the appropriate platform should be selected for the market. More specifically, we consider an enterprise/urban market.

1.5.4 Range and Throughput

At first look, a backhaul solution offering a range of a couple of kilometers, as seen in Table 1.6, would not seem to fit the notion of backhaul. There are wireless backhaul solutions on the market that offer ranges of up to 160 km. For enterprise or dense urban

TABLE 1.6. Range and Throughput Parameters for 60- and 80-GHz Backhaul Solutions

Technology	V-Band	E-Band
Range	3 km	5 km
Throughput	1 Gbps	10 Gbps
Frequency	57–64 GHz	71–84 GHz
Licensed	No	Yes

markets, a long-range backhaul is not required in most cases. Millimeter wave and E-Band backhaul solutions are best suited to the enterprise and urban markets.

A well-known saying in the telecom world goes, "Bandwidth is the answer now, what was the question?" The 60- and 80-GHz products are also known as gigabit radios, indicating their high throughput. A WiMAX service provider in an enterprise/urban market must plan for gigabit speeds through their base stations. These speeds are not available through other wireless backhaul solutions.

1.5.5 Security

Given that a backhaul solution will need to support a base station that might have dozens of WiMAX radios servicing thousands of enterprise subscribers, the security of the wireless backhaul solution should be of paramount concern to the WiMAX service provider. In the service provider market, casual hackers are less of a concern than the wholesale theft of services via rogue base stations. Figure 1.18 illustrates the narrow beamwidths of 60- and 80-GHz solutions and demonstrates the difficulty of intercepting a point-to-point backhaul signal.

A very narrow beamwidth is not enough to ensure good security on any wireless networks. Most backhaul solutions can be engineered for rigorous authentication processes, followed by equally demanding encryption programs for the data stream. Figure 1.19 shows a possible 60 GHz/WiMAX OFDM mapping solution.

1.5.6 Quality of Service

When millimeter wave radio works with a WiMAX system, one of the major selling points of WiMAX is its ability to prioritize traffic to deliver the best possible quality of service (QoS) relative to the traffic (for example; VoIP and video can be assigned top priority). Most 60 and 80 GHz products offer traffic prioritization schemes to ensure VoIP and video over the backhaul link. In addition, most of these products also offer

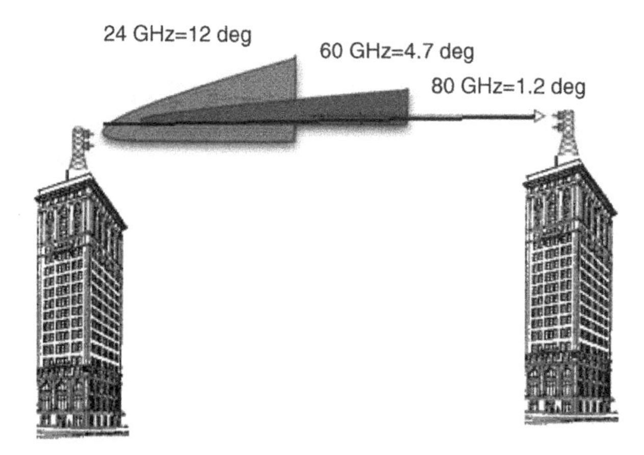

Figure 1.18. Beamwidth comparisons for wireless backhaul solutions

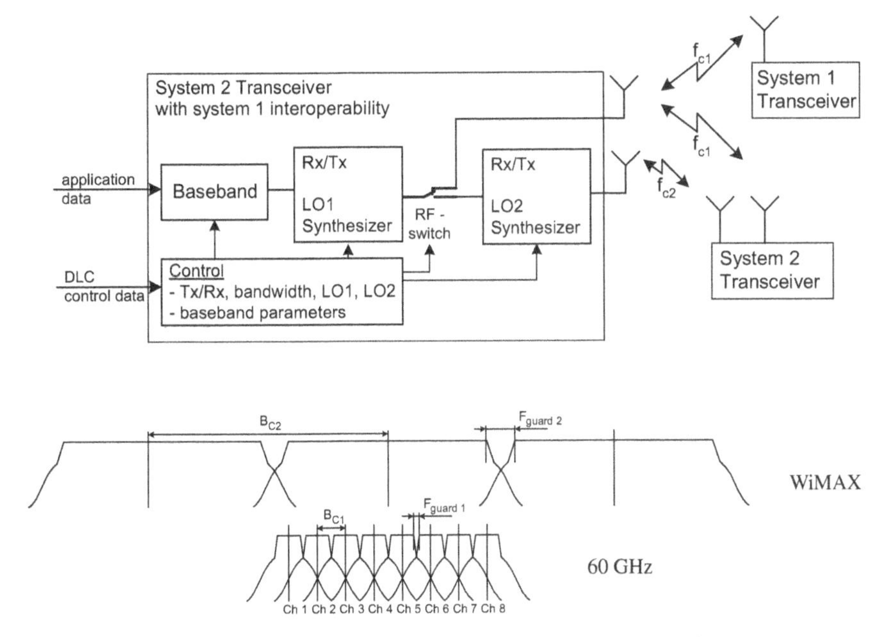

Figure 1.19. Possible 60 GHz/WiMAX OFDM mapping solution

sophisticated modulation schemes to ensure good QoS with up to 99.999% availability. forward error correction (FEC) and frequency division duplexing are features offered on some products to ensure QoS.

Potentially, the number one detractor to QoS in wireless links is latency. Millimeter wave V-band and E-Band products keep latency over their respective wireless links at single digit milliseconds (<10 ms). This ensures good QoS for time-sensitive applications, such as VoIP and video.

1.5.7 Interference Mitigation

Excessive interference can take down a backhaul link. It is imperative that the backhaul solution be engineered to mitigate interference as much as possible.

As illustrated in Figure 1.20, not only does the narrow beamwidths of 60 and 80 GHz products alleviate the possibilities of interference, they also focus the power of the beam, making for a strong link budget over its short transmission range, which further mitigates interference.

1.5.8 Frequency Reuse

One advantage of using the 60- and 80-GHz bands is that, owing to their short range propagation, the backhaul frequencies across the backhaul network can be reused. Figure 1.21 shows an example of frequency reuse in a 60 GHz communication network.

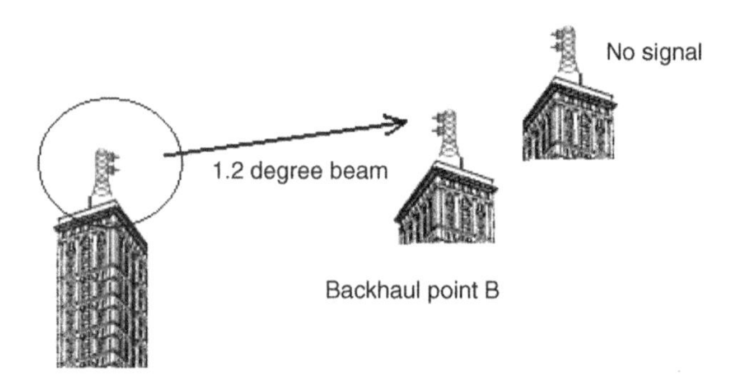

Figure 1.20. The narrow beamwidths of 60- and 80-GHz solutions mitigate interference while enhancing security

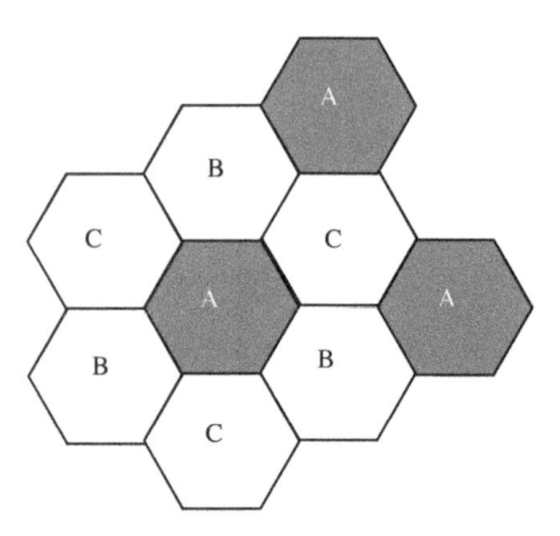

Figure 1.21. The 60- and 80-GHz bands offer dense frequency reuse

In the figure, frequencies *A*, *B*, and *C* are used three times to cover the communication area.

1.5.9 Rain Fade

Just as with excessive interference, service providers must engineer their backhaul solutions to deal with local meteorological conditions that can degrade the performance of their networks. Especially a sufficient rain margin will be needed. Data shows that the currently available commercial equipment can achieve gigabit speeds at 99.999% availability (uptime) with links of approximately one mile, regardless of rain. For a lower 99.9% availability, distances approaching three miles are routine.

1.5.10 Ease of Licensing: E-Band in the United States

An advantage to E-Band solutions is the relative ease of obtaining a license for the spectrum in the United States. Given its short range and narrow beamwidth, and its minimal propensity to interfere with other broadcasters, the FCC provides a streamlined licensing process for the E-Band spectrum. The price for a nationwide, nonexclusive, one-time license is about $1,500. Additional links thereafter are about $250.

E-Band link registration is normally completed using an automated online registration database system. The system checks for possible interference between a proposed E-Band installation and all the existing registered E-Band links based on the GPS coordinates of the E-Band radio locations and the operating parameters of the proposed radio. Registrants have 12 months to complete construction of the link, and the license is valid for 10 years, at which time it can be renewed.

Researchers believe that the advanced architectures used by 3G and 4G cell systems will drive network design and implementation so that cell site connectivity will demand 1-Gbps data rates and the aggregated cell site backbone cores will see data rate demands growing to 2.5 Gbps and even 10 Gbps [44]. Where fiber optic connections prove impractical, the millimeter wave radio community will stand ready to provide a cost-effective solution. There is a trend and new challenge in operating millimeter-wave technology with other mobile/wireless standards. The study of interoperability and convergence among these wireless standards is continuing.

Despite its large potential, development and applications of millimeter wave technology at 60 GHz band have not yet been promoted as expected. Short wavelength allows use of compact, low-profile devices. However, it suffers from high costs of hardware and high-speed signal processing compared with low-frequency bands below microwave.

REFERENCES

[1] H. T. Friis, "A note on a simple transmission formula." *Proc. IRE*, Vol. 34, No. 5, pp. 254–256, May 1946.

[2] M. Marcus, B. Pattan, "Millimeter wave propagation: spectrum management implications." *IEEE Micro. Mag.*, Vol. 6, No. 2, pp. 54–62, Jun. 2005.

[3] D. Sobel, "60 GHz wireless system design: towards a 1 Gbps wireless link." Berkeley Wireless Research Centre, Jun. 2003.

[4] D. Sobel, R. W. Brodersen, "60GHz CMOS system design: challenges, opportunities, and next steps." Research Retreat at Berkeley Wireless Research Centre, Jan. 2003.

[5] M. K. Simon and M. S. Alouini, *Digital Communication over Fading Channels*, 2nd edition, New York; Wiley-IEEE Press, 2004.

[6] K. Sato, T. Manabe, T. Ihara, H. Saito, S. Ito, T. Tanaka, K. Sugai, N. Ohmi, Y. Murakami, M. Shibayama, Y. Konishi, and T. Kimura, "Measurements of reflection and transmission characteristics of interior structures of office building in the 60GHz band." *IEEE Trans. Antennas Propag.*, Vol. 45, No. 12, pp. 1783–1792, 1997.

[7] J. Caruso, "Copper 10 gigabit ethernet NICs unveiled." *Network World*, Jan. 2007.

[8] W.I. Way, "Spectrally efficient parallel PHY for 100GbE MAN and WAN." *IEEE Commun. Mag.*, Vol. 45, No. 12, pp. 20–23, 2007.

[9] R. Merritt,"New tech breaks into network specs war." *EE Times,* Dec. 2006.

[10] IEEE 802.15 WPAN Task Group 3c (TG3c) Millimeter Wave Alternative PHY, available at http://www.ieee802.org/15/pub/TG3c.html

[11] K. Kimyacioglu, "WiMedia next gen UWB and 60 GHz considerations." *WiMedia Conference*, Mar. 2006.

[12] S. K. Yong and C. C. Chong, "An overview of multigigabit wireless through millimeter wave technology: potentials and technical challenges." *EURASIP Journal on Wireless Communications and Networking*, Vol. 2007 (2007), Article ID 78907, 10 pages doi:10.1155/2007/78907 .

[13] D. C. O'Brien, G. E. Faulkner, K. Jim, and D. J. Edwards, "Experimental characterization of integrated optical wireless components." *IEEE Photon. Technol. Lett.*, Vol. 18, No. 8, pp. 977–979, 2006.

[14] D. C. O'Brien, G. E. Faulkner, K. Jim, E. B. Zyambo, and D. J. Edwards, "High-speed integrated transceivers for optical wireless." *IEEE Commun. Mag.*, Vol. 41, No. 3, pp. 58–62, 2003.

[15] A. Tsukioka,"JVC develops base technologies for next-generation optical wireless access system." *JCN Network*, Oct. 2005.

[16] A. M. Street, P. N. Stavrinou, D. C. O'Brien, and D. J. Edwards, "Indoor optical wireless systems: a review." *Optical and Quantum Electronics*, Vol. 29, No. 3, pp. 349–378, 1997.

[17] D. C. O'Brien, E. B. Zyambo, G. Faulkner, D. J. Edwards, D. M. Holburn, R. J. Mears, R. J. Samsudin, V. M. Joyner, V. A. Lalithambika, M. Whitehead, P. Stavrinou, G. Parry, J. Bellon, and M. J. Sibley, "High-speed optical wireless transceivers for in-building optical local area networks (LANs)." *Optical Wireless Communications III*, 4124, paper 4124-16, SPIE, Boston 2000.

[18] K. C. Huang and Z. C. Wang, "Millimeter-wave circular polarized beam-steering antenna array for gigabit wireless communications." *IEEE Trans. Antennas Propag.*, Vol. 54, No. 2, Part 2, pp. 743–746, 2006.

[19] R. C. Qiu, H. Liu, and X. Shen, "Ultra-wideband for multiple access communications." *IEEE Commun. Mag.*, Vol. 43, No. 2, pp. 80–87, 2005.

[20] JVC Products, VIPSLAN OA-301, JVC Corporation Japan. Available at http://www.jvc.co.jp/

[21] A. Medeisis,"SE19 drafting group meeting on MGWS at 60 GHz, ERO, Copenhagen, 26 March 2007." ERO SE19 Broadband Applications in Fixed Service, Available at http://www.ero.dk/

[22] European Radiocommunications Committee (ERC), T/R 22-03E, "Provisional Recommended Use of the Frequency Range GHz by Terrestrial Fixed and Mobile Systems." pp. 3, 1990.

[23] CEPT, ERO, "The European table of frequency allocations and utilizations covering the frequency range 9 kHz to 275 GHz." Lisboa January 2002, Dublin 2003, Turkey 2004, Copenhagen 2004.

[24] ERC Recommendation 12-09, "Radio frequency channel arrangement for fixed service systems operating in the band 57.0–59.0 GHz which do not require frequency planning, the Hague 1998 revised Stockholm." Oct. 2004.

[25] ECC Recommendation (05)02, "Use of the 64–66 GHz frequency band for fixed services." Jun. 2005.

[26] ETSI DTR/ERM-RM-049, "Electromagnetic compatibility and radio spectrum matters (ERM); system reference document; technical characteristics of multiple gigabit wireless systems in the 60 GHz range." Mar. 2006.

[27] ECC Recommendation (05)02, "Use of the 64–66 GHz frequency band for fixed service." Revised ECC/REC/(05)02 Edition, Feb. 2009.

[28] IEEE 802.15-15-06-0044-00-003c document, "60 GHz regulation in Germany." Jan. 2006.

[29] Japan Regulations for enforcement of the radio law 6-4-2 specified low power radio station (11) 59–66 GHz band, Ministry of Public Management, Home Affairs, Posts, and Telecommunications, Japan, 2000.

[30] Ministry of Information Communication of Korea, "Frequency allocation comment of 60 GHz band." Apr. 2006.

[31] FCC, "Code of Federal Regulation, Title 47 Telecommunication, Chapter 1, Part 15.255." Oct. 2004.

[32] S. K. Yong and C. C. Chong, "An overview of multigigabit wireless through millimeter wave technology: potentials and technical challenges." *EURASIP Journal on Wireless Communications and Networking*, Volume 2007, Article ID 78907, 10 pages doi: 10.1155/2007/78907 (2007).

[33] Spectrum Management Telecommunications, "Radio Standard Specification-210, Issue 6, Low-Power Licensed-Exempt Radio Communication Devices (All Frequency Bands): Category 1 Equipment." Sep. 2005.

[34] FCC document, OMB 3060-1070, "Allocations and Service Rules for the 71–76 GHz, 81–86 GHz, and 92–95 GHz Bands."

[35] J. Wells,"Multigigabit wireless connectivity at 70, 80 and 90 GHz." RF Design, pp. 50–54, May 2006.

[36] ERC Recommendation 12-09, "Radio Frequency Channel Arrangement for Fixed Service Systems Operating in the Band 57.0–59.0 GHz Which Do Not Require Frequency Planning." The Hague, 1998; revised Stockholm, Oct. 2004.

[37] ARIB STD-T69, "Millimeter-Wave Video Transmission Equipment for Specified Low Power Radio Station." Jul. 2004.

[38] ARIB STD-T74, "Millimeter-Wave Data Transmission Equipment for Specified Low Power Radio Station (Ultra High Speed Wireless LAN System)." May 2001.

[39] IEEE 802.15 WPAN Task Group 3C (TG3c) Millimeter Wave Alternative PHY, available at http://www.ieee802.org/15/pub/TG3c.html

[40] IEEE 802.15-07-0761-15-003c, "Unified and flexible millimeter wave WPAN systems supported by common mode." Nov. 2007.

[41] IEEE 802.15-05-0353-07-003c, "Working group for wireless personal area networks (WPANs), TG3c System Requirements." Jan. 2007.

[42] "IEEE Standard for Local and Metropolitan Area Networks. Part 16: Air Interface for Fixed and Mobile Broadband Wireless Access Systems Amendment 2: Physical and Medium Access Control Layers for Combined Fixed and Mobile Operation in Licensed Bands and Corrigendum 1." IEEE Std 802.16e-2005 and IEEE Std 802.16-2004/Cor 1-2005 (Amendment and Corrigendum to IEEE Std 802.16-2004), p. 3, 2006.

[43] S. Chia, M. Gasparroni, and P. Brick, "The next challenge for cellular network backhaul." *IEEE Microw. Mag.*, Vol. 10, No. 5, pp. 55–66, 2009.

[44] D. Lockie and D. Peck, "High-data-rate millimeter-wave radios." *IEEE Microw. Mag.*, Vol. 10, No. 5, pp. 75–83, 2009.

2

REVIEW OF MODULATIONS FOR MILLIMETER WAVE COMMUNICATIONS

This chapter summarizes modern digital communication modulation techniques for millimeter wave communication systems. Digital modulation schemes transform digital signals into millimeter wave signals that are compatible with the nature of the communication channels.

There are two major categories of digital modulations. The first category uses a constant amplitude carrier to carry the information in phase or frequency variations, such as frequency shift keying (FSK) and phase shift keying (PSK). The second category conveys the information in carrier amplitude variations, such as amplitude shift keying (ASK) and quadrature amplitude modulation (QAM).

Millimeter wave radios require high power efficiency with low bit error rate (BER). Power efficiency is the ability of a modulation technique to preserve the fidelity of the digital message at low power levels. As millimeter wave power has high cost, either 64 or 256 QAM is not preferable due to the effect of phase noise and power consumption of the power amplifier (PA). In addition, receiver cost or complexity should also be considered, especially in multiple-input-multiple-output systems as mentioned in Chapters 5 and 6.

This chapter is organized as follows. Section 2.1 introduces on/off keying modulation scheme. The coherent and noncoherent demodulations are analyzed. Section 2.2

Millimeter Wave Communication Systems, by Kao-Cheng Huang and Zhaocheng Wang

describes various PSK schemes, including binary phase shift keying and quadrature phase shift keying. Section 2.3 discusses the concept of FSK scheme. Section 2.4 summarizes QAM modulation technology. Section 2.5 analyzes orthogonal frequency division multiplexing, which is used widely in the current broadband wireless communication systems.

2.1 ON/OFF KEYING (OOK)

On/off keying (OOK) modulation is a modulation scheme used in control applications. This is in part due to its simplicity and low implementation costs. OOK consists of keying a sinusoidal carrier signal on and off with a unipolar binary signal. OOK is equivalent to two-level ASK. The system diagram of OOK is shown in Figure 2.1. OOK modulation has the advantage of allowing the transmitter to idle during the transmission of a "0", therefore conserving power. Here input signal has two states ("1" and "0") and modulation factor is 100% (from full power to no transmitted power).

The disadvantage of OOK modulation arises in the presence of an undesired signal. The modulated signals can be graphically represented on a two-dimensional orthogonal plot, sometimes referred to as a signal diagram. Consider a set of two basis vectors φ_1 and φ_2. The signal diagrams for OOK are shown in Figure 2.2.

The idea of OOK is that the transmitter is on when logic "1" is transmitted and the transmitter is off when logic "0" is transmitted. OOK receivers require an adaptable threshold and automatic gain controller (AGC) in order to ensure an optimal threshold setting. A logarithm amplifier detector with an averaging bit slicer is employed, as shown in Figure 2.3. This circuit will ensure that the threshold is set between the signal levels of a "0" and a "1" transmission. The above circuit works well as long as the data received is effectively D.C. balanced.

There are two types of demodulation method, namely synchronous demodulation and envelope demodulation.

Synchronous demodulation is also known as coherent demodulation, and the block diagram is shown in Figure 2.4. As shown in Figure 2.4, the coherent carrier for demodulation is $2\cos(2\pi f_c t)$, where the amplitude factor of "2" used here is for calculation convenience, and f_c is the carrier frequency used for the generation of OOK signal. During the following analysis, the carrier phases of the transmitter and receiver are assumed to be the same and they are dropped for notational convenience.

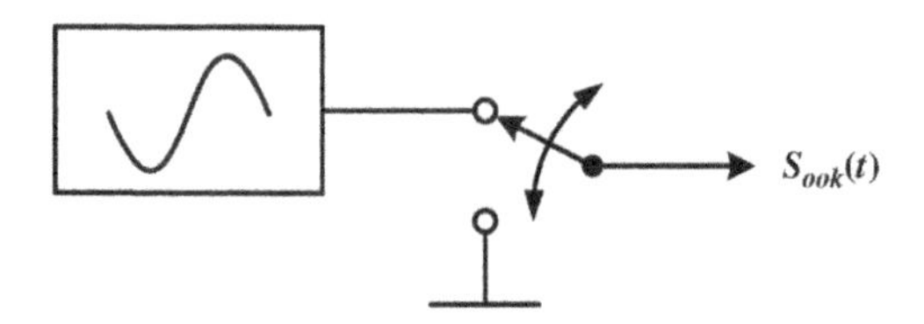

Figure 2.1. System diagram of OOK

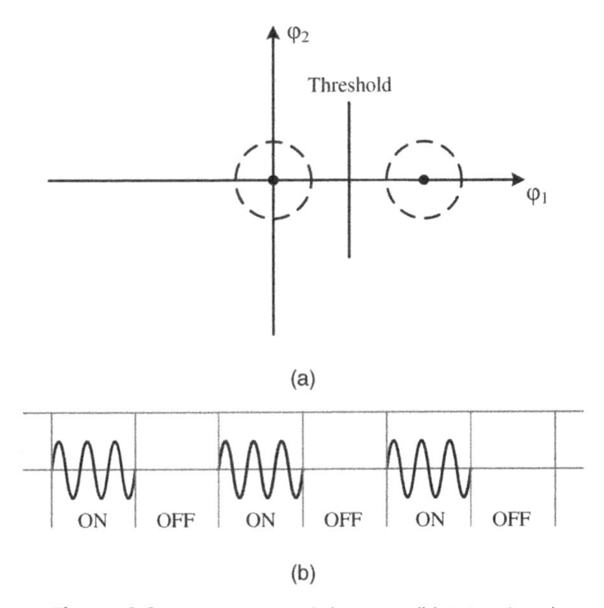

Figure 2.2. (a) OOK signal diagram, (b) OOK signal

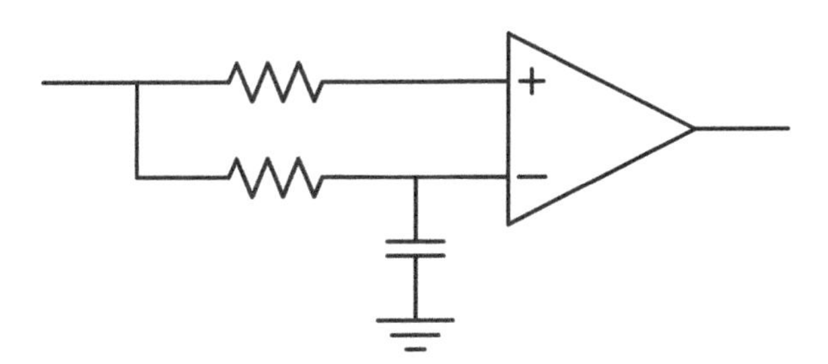

Figure 2.3. OOK receiver

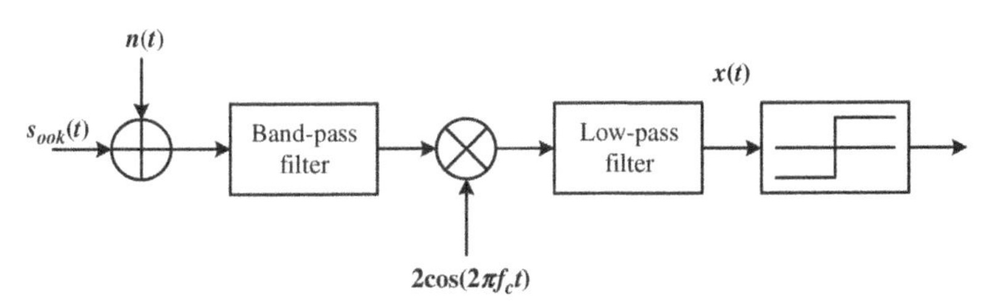

Figure 2.4. Block diagram of synchronous demodulation

When "1" is transmitted, the received OOK signal is $a \cdot \cos(2\pi f_c t)$ and can pass through the band-pass filter smoothly, where a is the amplitude of the received signal. $n(t)$ is the additive white Gaussian noise (AWGN). After the band-pass filter, $n(t)$ is converted to a narrow-band Gaussian noise, which is given as

$$n_i(t) = n_c(t)\cos(2\pi f_c t) - n_s(t)\sin(2\pi f_c t) \tag{2.1}$$

where $n_c(t)$ and $n_s(t)$ are the in-phase and quadrature components, respectively.

So when code "1" is transmitted, the signal after the band-pass filter is

$$a\cos(2\pi f_c t) + n_i(t) = [a + n_c(t)]\cos(2\pi f_c t) - n_s(t)\sin(2\pi f_c t) \tag{2.2}$$

The multiplier output is

$$\begin{aligned} &\{[a + n_c(t)]\cos(2\pi f_c t) - n_s(t)\sin(2\pi f_c t)\} \cdot 2\cos(2\pi f_c t) \\ &= [a + n_c(t)] + [a + n_c(t)]\cos(4\pi f_c t) - n_s(t)\sin(4\pi f_c t) \end{aligned} \tag{2.3}$$

After the low-pass filter, the later two items are filtered out. So the output signal is

$$x(t) = a + n_c(t) \tag{2.4}$$

$x(t)$ is the input signal to the decision device.

When code "0" is transmitted, the OOK signal is zero, but the noise still exists. The input signal to the decision device is

$$x(t) = n_c(t) \tag{2.5}$$

From (2.4) and (2.5), we can get

$$x(t) = \begin{cases} a + n_c(t) & \text{when code"1" is transmitted} \\ n_c(t) & \text{when code"0" is transmitted} \end{cases} \tag{2.6}$$

where $n_c(t)$ is a low-pass Gaussian noise with zero mean and variance of σ_n^2, and $a + n_c(t)$ is a low-pass Gaussian noise with mean of a. The two conditional probability density curves of the $x(t)$ are shown in Figure 2.5.

The total probability of error is determined by two possible error conditions, the probability of a "1" being sent and the receiver mistaking it for a "0" (a miss) and the probability of a "0" being sent and the receiver detecting a "1" (false alarm). The total probability of error is defined as follows:

$$P_e = \frac{1}{2}\int_{-\infty}^{V_t} p_0(r)\mathrm{d}r + \frac{1}{2}\int_{V_t}^{\infty} p_1(r)\mathrm{d}r \tag{2.7}$$

where V_t is the decision threshold, and $p_1(r)$ and $p_0(r)$ are the conditional probability density functions given "1" and "0" being sent, respectively.

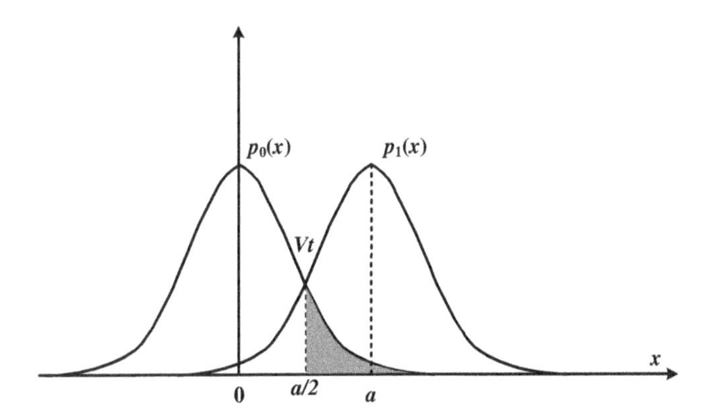

Figure 2.5. Joint probability density curve of the synchronous demodulation signal

When $V_t = a/2$ is chosen, P_e is given by

$$P_e = \frac{1}{2}\text{erfc}\left(\frac{a}{2\sqrt{2}\sigma_n}\right) \tag{2.8}$$

where erfc() is the complementary error function [1], which is defined as

$$\text{erfc}(z) = \frac{2}{\sqrt{\pi}}\int_z^{\infty} \exp(-t^2)\text{d}t \tag{2.9}$$

If we define the *SNR* to be the signal to noise ratio when code "1" is sent, then

$$SNR = \frac{a^2}{2\sigma_n^2} \tag{2.10}$$

and (2.8) can be rewritten as

$$P_e = \frac{1}{2}\text{erfc}\left(\frac{\sqrt{SNR}}{2}\right) \tag{2.11}$$

Envelope demodulation for OOK signal is a noncoherent demodulation method, as shown in Figure 2.6.

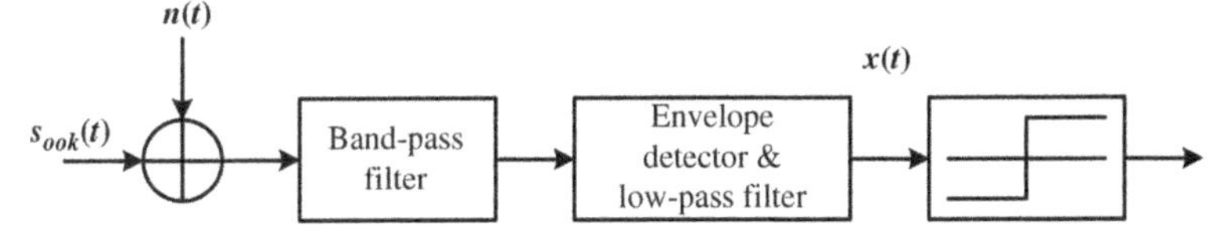

Figure 2.6. Envelope demodulation of OOK system

For the noncoherent demodulation, when code "1" is sent, the input signal to the envelope detector is $a \cdot \cos(2\pi f_c t) + n_i(t)$. The output of the envelope detector is the envelope of the sum of the useful signal and the narrow band Gaussian noise, which meets the Ricean distribution [2],

$$p_1(x) = \frac{x}{\sigma_n^2} \mathrm{I}_0\left(\frac{ax}{\sigma_n^2}\right) e^{-(x^2+a^2)/2\sigma_n^2} \tag{2.12}$$

where $\mathrm{I}_0(x)$ is the first zero-order modified Bessel function,

$$\mathrm{I}_0(x) = \frac{1}{2\pi} \int_0^{2\pi} \exp(x \cos\theta) \mathrm{d}\theta \tag{2.13}$$

When code "0" is sent, the input signal to the envelope detector is $n_i(t)$, and the output is the envelope of the narrow-band Gaussian noise, which meets the Rayleigh distribution,

$$p_0(x) = \frac{x}{\sigma_n^2} e^{-x^2/2\sigma_n^2} \tag{2.14}$$

The two conditional probability density functions of the envelope detector output are shown in Figure 2.7.

The threshold voltage Vt is the point where the two probability density functions intersect. Assuming that a "1" or a "0" is transmitted with equal probability, and the evaluation of Vt according to (2.12) and (2.14) yields the approximate probability of error for OOK modulation using envelope detection, which is expressed as

$$P_e = \frac{1}{2}\left(1 + \sqrt{\frac{1}{\pi \cdot SNR}}\right) \cdot \exp\left(-\frac{SNR}{4}\right) \approx \frac{1}{2} \cdot e^{-\frac{SNR}{4}} \tag{2.15}$$

The SNR is a physical quantity that can be easily measured, but it does not explicitly state the power efficiency. To evaluate the power efficiency, one must know the average

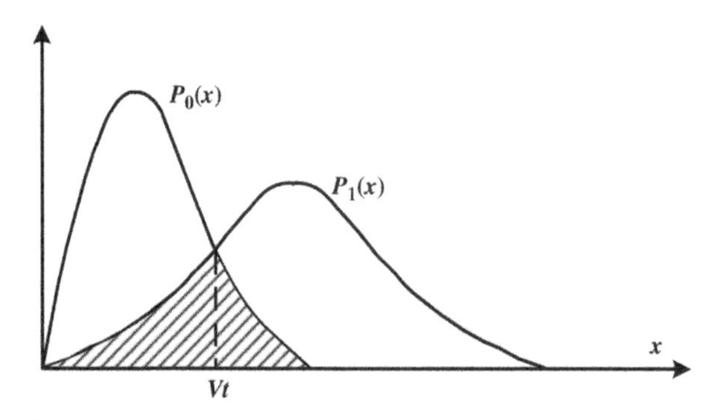

Figure 2.7. The probability density function of the envelope detector output

energy E_b per useful bit at the receiver that is needed for a reliable recovery of the information. If $\log_2(M)$ useful bits are transmitted by each symbol s_k, the relationship between the average energy per symbol E_S, and E_b

$$E_S = \log_2(M)E_b \tag{2.16}$$

holds, and the SNR is related to E_b by

$$SNR = \log_2(M)\frac{E_b}{N_0} \tag{2.17}$$

where E_b/N_0 is the signal energy per bit/noise power spectral density.

We note the important fact that E_b is just the average signal power needed per bit. Therefore, a modulation that needs less E_b/N_0 is more power efficient to achieve a reliable transmission.

2.2 PHASE SHIFT KEYING (PSK)

Phase shift keying (PSK) is a large class of digital modulation schemes. PSK is widely used in the communication industry. The simplest form of phase modulation is binary (two-level) phase modulation. For binary phase shift keying (BPSK) the carrier phase has only two states, 0 and π (see Figure 2.8). Obviously the transition from a "1" to a "0", or vice versa, will result in the modulated signal crossing the origin of the constellation diagram, resulting in 100% amplitude modulation.

Quadrature phase shift keying (QPSK) devices modulate input signals by 0°, 90°, 180°, and 270° phase shifts. Figure 2.9 shows three types of constellation diagram for QPSK modulations, (a) conventional QPSK, (b) offset QPSK (OQPSK), and (c) $\pi/4$ QPSK.

Conventional QPSK has transitions through zero (i.e., 180° phase transition). In OQPSK, the transitions on the I and Q channels are staggered. Phase transitions are therefore limited to 90°. In $\pi/4$-QPSK the set of constellation points are toggled each symbol, so transitions through the origin can be avoided. This scheme produces the lowest envelope variations. All these QPSK schemes require linear power amplifiers. In particular, a highly linear amplifier is required for the conventional QPSK.

Both QPSK and BPSK modulators are used in conjunction with demodulators that extract information from the modulated signal. Some QPSK and BPSK modulators

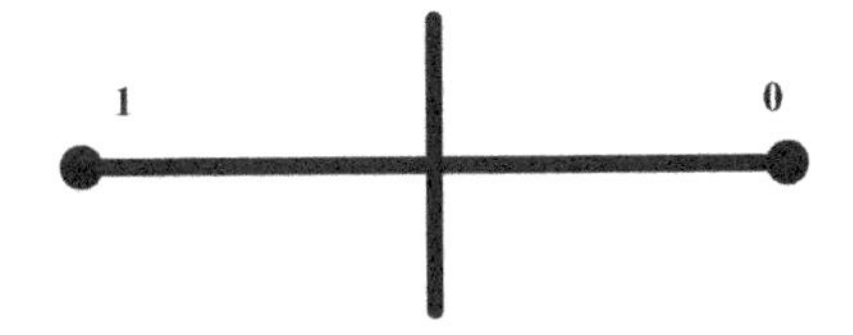

Figure 2.8. Constellation diagram for BPSK

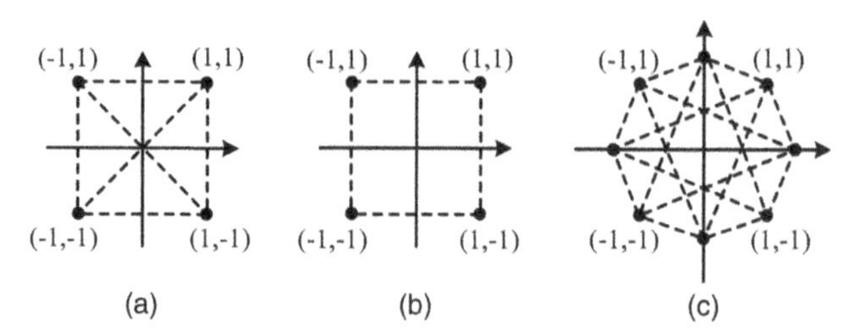

Figure 2.9. Three types of constellation diagram for QPSK: (a) conventional QPSK, (b) OQPSK, (c) $\pi/4$ QPSK

include an integral dielectric resonator oscillator. QPSK and BPSK modulators with root raised cosine (RRC) and Butterworth filters are also available.

Performance specifications for QPSK and BPSK modulators include input carrier frequency, insertion loss, amplitude unbalance, phase unbalance, and voltage standing wave ratio (VSWR). Insertion loss is the total RF power transmission loss through the device. Amplitude unbalance is the difference in power between the I output signal and the Q output signal. Phase unbalance is the deviation from 90° of the phase angle difference of the I and Q output signals. VSWR is a unitless ratio ranging from 1 to infinity that expresses the amount of reflected energy at the input of the device. A value of 1 indicates that all of the energy passes through. Any other value indicates that a portion of the energy is reflected. Other performance specifications for QPSK and BPSK modulators include frequency range, return loss, and reflected power.

If a sinusoidal carrier is modulated by a bipolar bit stream $a(t)$ according to the scheme illustrated in Figure 2.10, its polarity will be reversed every time the bit stream changes polarity. This, for a sine wave, is equivalent to a phase reversal (shift). The multiplier output is a BPSK signal.

In the example of Figure 2.11, the upper trace is the binary message sequence and the appearance of a BPSK signal in the time domain is shown in the lower trace. The bipolar data stream $a(t)$ from the binary data stream is

$$a(t) = \sum_{k=-\infty}^{\infty} a_k p(t-kT) \tag{2.18}$$

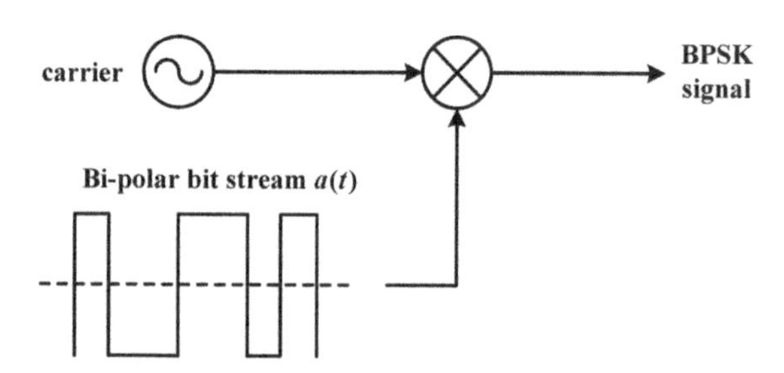

Figure 2.10. BPSK signal generation

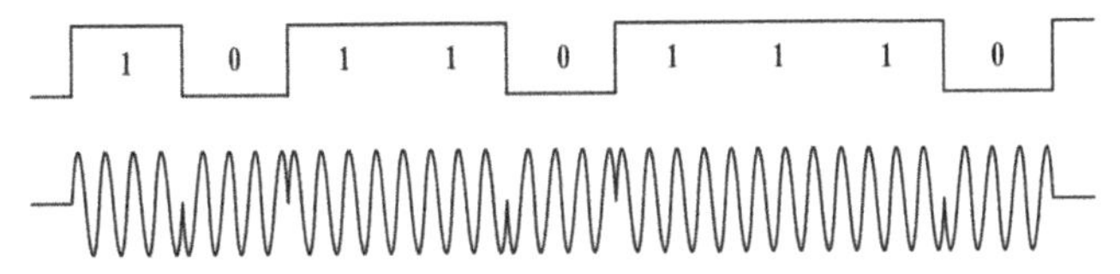

Figure 2.11. BPSK signals in the time domain

where a_k is the bipolar data symbol, T is the bit interval, and $p(t)$ is the rectangular pulse with unit amplitude defined on $[0,T]$. Then the BPSK signal can be expressed as

$$s(t) = Aa(t)\cos(2\pi f_c t) \tag{2.19}$$

where f_c is the carrier frequency and A is the amplitude.

The coherent demodulator for BPSK signal is shown in Figure 2.12.

The BPSK coherent demodulator is one type of binary coherent detectors. The coherent detector could be in the form of a correlator or matched filter with the reference signal of $\cos(2\pi f_c t)$. The frequency and phase between the reference signal and the received signal have to be synchronous. The synchronous reference signal can be generated by the carrier recovery circuit.

In the absence of noise, assuming that $A = 1$, the output of the correlator at time instance $t = (k+1)T$ is

$$\begin{aligned}
&\int_{kT}^{(k+1)T} r(t)\cos(2\pi f_c t)dt \\
&= \int_{kT}^{(k+1)T} a_k \cos^2(2\pi f_c t)dt \\
&= \frac{1}{2}\int_{kT}^{(k+1)T} a_k(1+\cos(4\pi f_c t))dt \\
&= \frac{T}{2}a_k + \frac{a_k}{8\pi f_c}[\sin(4\pi f_c(k+1)T) - \sin(4\pi f_c kT)]
\end{aligned} \tag{2.20}$$

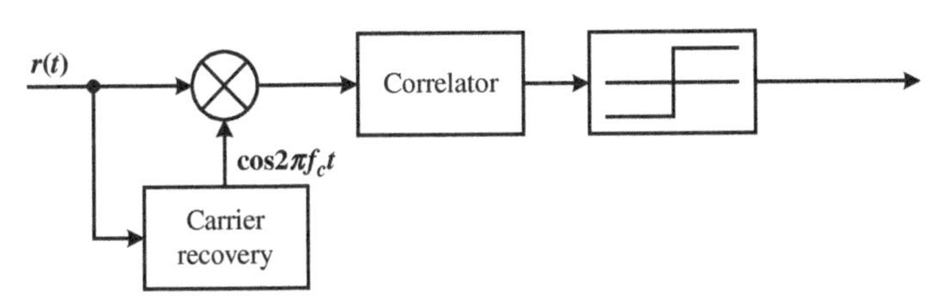

Figure 2.12. Coherent demodulator for BPSK signal

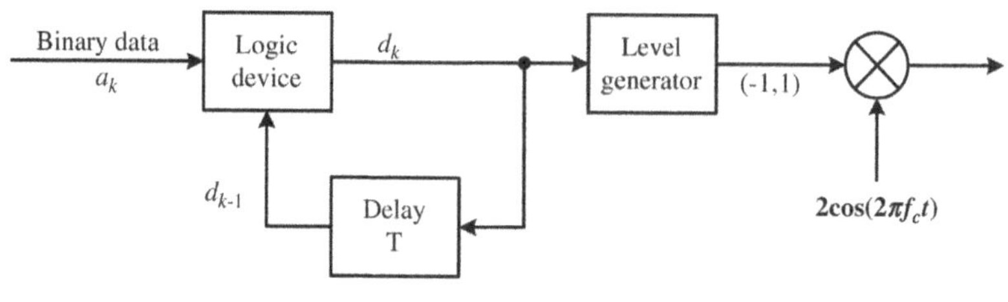

Figure 2.13. DBPSK modulator

Similar to the case of OOK coherent demodulation, the probability of error for a coherent receiver system with BPSK modulation can be derived as

$$P_e = \frac{1}{2}\text{erfc}\left(\sqrt{\frac{E_b}{N_0}}\right) \tag{2.21}$$

The difficulty with the coherent PSK receiver is that the receiver cannot know the exact phase of the transmitted signal. It is not possible even if the transmitter and the receiver clocks were accurately linked because the path length would determine the exact phase of the received signal. To overcome this problem, PSK systems can use a differential method for encoding the data onto the carrier. This is accomplished, for example, by making a change in phase equal to a "1", and no phase change equal to a "0". Further improvements can be made upon this basic system and a number of other types of PSK have been developed. One simple improvement can be made by making a change in phase of 90° in one direction for a "1", and 90° in the other way for a "0". This retains the 180° phase reversal between "1" and "0" states, but gives a distinct change for a "0". In a basic system not using this process it may be possible to loose synchronization if a long series of "0" is sent. This is because the phase will not change state for this occurrence.

Figure 2.13 shows the structure of the differential BPSK (DBPSK) modulator. The logic device could be binary exclusive-OR (XOR) operator. DBPSK signal can be coherently demodulated or differentially demodulated, which does not require a coherent reference signal. Figure 2.14 is a simple differential demodulator where the previous symbol is used as the reference for demodulating the current symbol.

In Figure 2.14 the band-pass filter can reduce noise power but preserve the signal. The low-pass filter can also be used instead of the integrator.

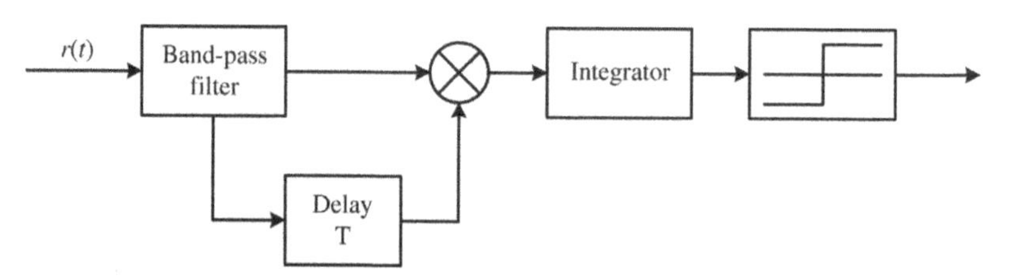

Figure 2.14. DBPSK differential demodulator

The output of the integrator is

$$x(t) = \int_{kT}^{(k+1)T} r(t)r(t-T)dt \tag{2.22}$$

In the absence of noise,

$$x(t) = \int_{kT}^{(k+1)T} s_k(t)s_{k-1}(t)dt = \begin{cases} E_b & \text{if } s_k(t) = s_{k-1}(t) \\ -E_b & \text{if } s_k(t) = -s_{k-1}(t) \end{cases} \tag{2.23}$$

where $s_k(t)$ and $s_{k-1}(t)$ are the current and the previous symbols, respectively. The integrator output is positive if the current symbol is the same as the previous one; otherwise the output is negative. This means that the decisions are based on the difference between the two adjacent symbols. Thus, information data must be encoded according to the difference between adjacent signals.

One of the differential encoding rules for DBPSK signal generation is

$$d_k = a_k \oplus d_{k-1} \tag{2.24}$$

where $\oplus$ is the operation of binary XOR. The recovery of a_k from d_k is

$$a_k = d_k \oplus d_{k-1} \tag{2.25}$$

According to (2.24) and (2.25), Figure 2.15 gives an example of the DBPSK signal generation and demodulation. During the encoding process, when bit "1" is transmitted, the encoded signal will change the polarity, and otherwise the encoded signal will hold the previous state. At the receiver, the opposite operation is applied. The demapping rules used in Figure 2.15 are $+1$ to "0" and -1 to "1".

The demodulator structure shown in Figure 2.14 is nonoptimal. In order to improve the performance of DBPSK systems, an optimized demodulator, as shown in Figure 2.16, can be applied. The module named $-\pi/2$ is a $-\pi/2$ phase shifter. Note that the optimum demodulator shown in Figure 2.16 does not require phase synchronization between the reference signal and the received signal. But the reference frequency must be the same as the received signal. Therefore, the differential detector in Figure 2.14 is more practical even though its error performance is slightly inferior to that of the optimum demodulator.

Message a_k		1	0	1	1	0	0	0	1	1
Encoding d_k	0	1	1	0	1	1	1	1	0	1
Signal phase θ	π	0	0	π	0	0	0	0	π	0
Demodulation		-1	1	-1	-1	1	1	1	-1	-1
Demodulator output		1	0	1	1	0	0	0	1	1

Figure 2.15. Example of the DBPSK signal

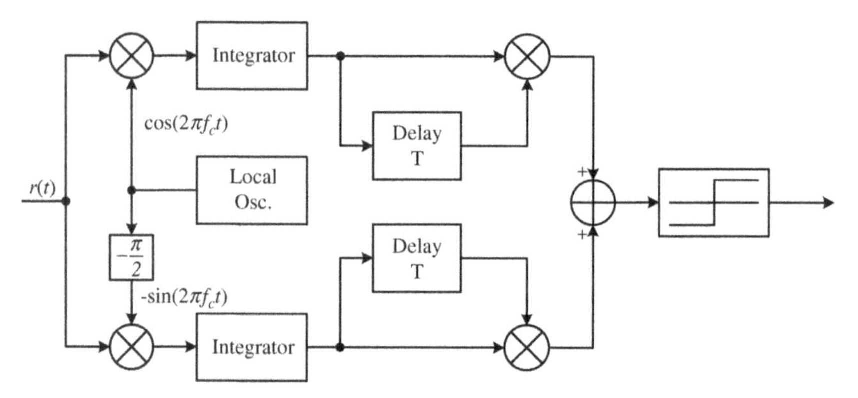

Figure 2.16. Optimum demodulator of DBPSK signal

The bit error probability of the optimum demodulator is [6]

$$P_e = \frac{1}{2} e^{-\frac{E_b}{N_0}} \tag{2.26}$$

The performance of the differential demodulator is given in [3]. When an ideal narrow-band intermediate frequency (IF) filter with bandwidth W is used before the integrator, the bit error probability is

$$P_e = \frac{1}{2} e^{-0.76\frac{E_b}{N_0}}, \quad \text{for } W = 0.5/T \tag{2.27}$$

The performance loss is about 1.2 dB compared with the optimum demodulator. If an ideal wide-band IF filter is used, the approximate bit error probability is

$$P_e \approx \frac{1}{2\sqrt{\pi}\sqrt{E_b/2N_0}} e^{-\frac{E_b}{2N_0}}, \quad \text{for } W > 1/T \tag{2.28}$$

The signal of QPSK is expressed as

$$s_i(t) = A\cos\left[2\pi f_c t + (2i-1)\frac{\pi}{4}\right], \quad 0 \le t \le T, \quad i = 1,2,3,4 \tag{2.29}$$

In Figure 2.17(a), QPSK is effectively two independent BPSK systems (I and Q), and therefore exhibits twice bandwidth efficiency. The two signal components with their bit assignments and the total combined signals are shown in Figure 2.17(b). The phase of I or Q signal changes abruptly at some of the bit-period boundaries. QPSK can be filtered using raised cosine filters to achieve excellent out-of-band suppression. Large envelope variations occur during phase transitions, thus requiring linear amplification.

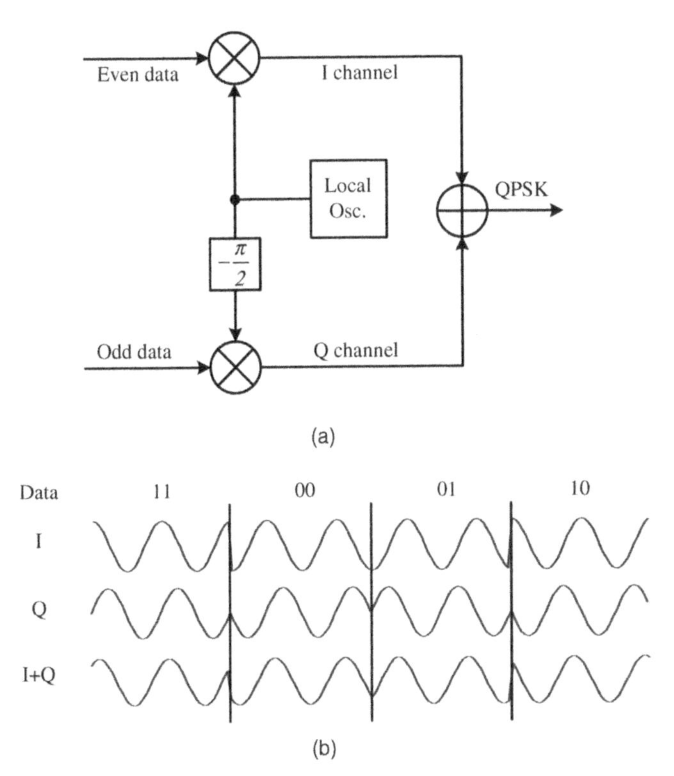

Figure 2.17. (a) Conventional QPSK signal generation, (b) QPSK signals in the time domain

The coherent demodulator for QPSK signal is shown in Figure 2.18. It consists of two individual BPSK demodulators for both I and Q channels. The two demodulated signals are converted into one data sequence by the parallel to serial converter (P/S). This is possible due to the correspondence and orthogonality between data bits from I and Q channels.

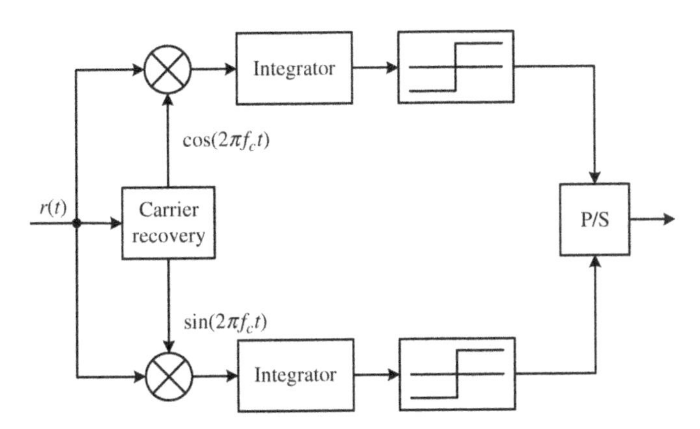

Figure 2.18. QPSK demodulator

Data	$\Delta\theta_i$	$\cos\Delta\theta_i$	$\sin\Delta\theta_i$
00	0	1	0
01	$\pi/2$	0	1
10	$-\pi/2$	0	−1
11	π	−1	0

Figure 2.19. DQPSK phase assignment

For a coherent QPSK receiver system, the probability of error is about [6]

$$P_e = Q\left(\sqrt{\frac{2E_b}{N_0}}\right) \tag{2.30}$$

where

$$Q(x) = \int_x^{\infty} \frac{1}{\sqrt{2\pi}} e^{-u^2} du \tag{2.31}$$

Differential QPSK (DQPSK) is an important special case of QPSK system in which the phase ambiguity can be eliminated. In the DQPSK system the information bits are represented by the phase differences $\Delta\theta_i$ from symbol to symbol, where $i = 0, 1, 2, 3$. There are different phase assignments between $\Delta\theta_i$ and logic bits. A possible phase assignment is shown in Figure 2.19.

The differential coding rules are given as follows:

$$\begin{aligned} u_k &= \overline{(I_k \oplus Q_k)}(I_k \oplus u_{k-1}) + (I_k \oplus Q_k)(Q_k \oplus v_{k-1}) \\ v_k &= \overline{(I_k \oplus Q_k)}(Q_k \oplus v_{k-1}) + (I_k \oplus Q_k)(I_k \oplus u_{k-1}) \end{aligned} \tag{2.32}$$

where $\oplus$ is the XOR operation, I_k and Q_k are the original information bits, and u_k and v_k are coded I and Q channel bits, respectively. An example of the phase of the DQPSK is shown in Figure 2.20. The structure of DQPSK modulator is shown in Figure 2.21. The DQPSK modulator is basically the same as the QPSK modulator, except that the differential encoder must be inserted before the mixer in each channel.

Message I_k		1	0	1	0	1	1	0	1
Q_k		0	1	0	1	1	0	0	1
Encoded u_k	1	1	1	1	1	0	0	0	1
v_k	1	0	1	0	1	0	1	1	0
Signal phase θ	$\frac{\pi}{4}$	$\frac{7\pi}{4}$	$\frac{\pi}{4}$	$\frac{7\pi}{4}$	$\frac{\pi}{4}$	$\frac{5\pi}{4}$	$\frac{3\pi}{4}$	$\frac{3\pi}{4}$	$\frac{7\pi}{4}$

Figure 2.20. Example of DQPSK signal

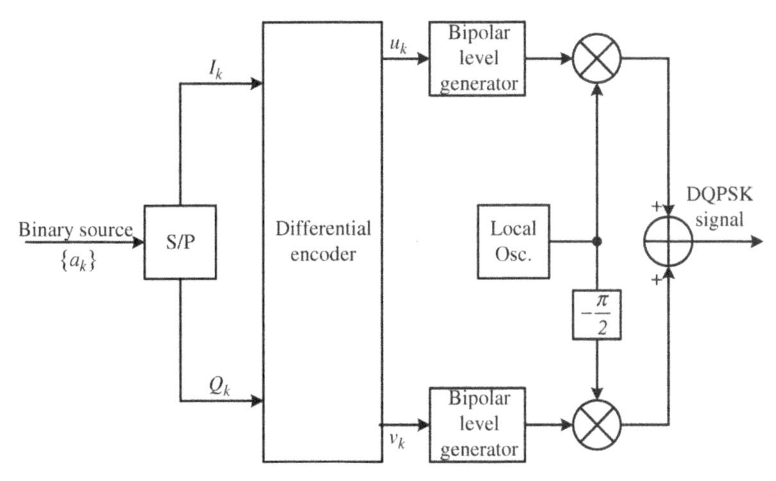

Figure 2.21. DQPSK modulator

A differential demodulator similar to DBPSK demodulator is shown in Figure 2.22, where the integrator can be replaced by a low-pass filter [4]. The bit error probability of the differential demodulator is given as [5]

$$P_e \approx e^{-\left(A^2/2\sigma^2\right)\left(1-1/\sqrt{2}\right)} = e^{-0.59\frac{E_b}{N_0}} \tag{2.33}$$

where $A^2/2\sigma^2$ is the SNR.

The coherent demodulator for the DQPSK signal is basically the same as the coherent QPSK demodulator except that a differential decoder detecting phase differences has to be used. However, because the demodulated signals from both I and Q channel are digital values with "0" and "1", the differential decoding can be simplified.

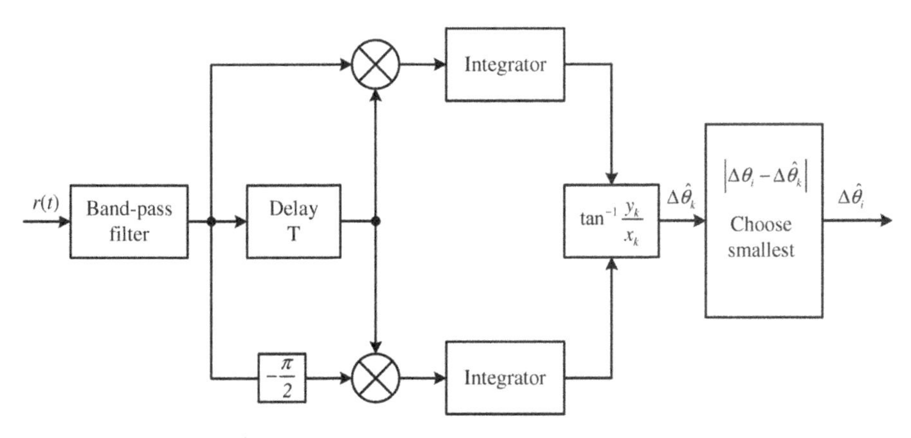

Figure 2.22. Differential DQPSK demodulator

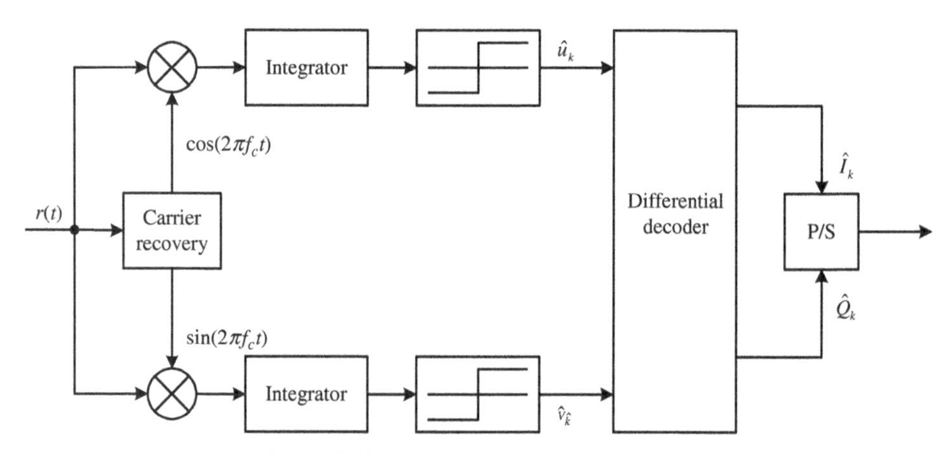

Figure 2.23. Coherent DQPSK demodulator

The demodulator using the simple decoder is depicted in Figure 2.23, where the decoding rules are

$$\begin{aligned} \hat{I}_k &= \overline{(\hat{u}_k \oplus \hat{v}_k)}(\hat{u}_k \oplus \hat{u}_{k-1}) + (\hat{u}_k \oplus \hat{v}_k)(\hat{v}_k \oplus \hat{v}_{k-1}) \\ \hat{Q}_k &= \overline{(\hat{u}_k \oplus \hat{v}_k)}(\hat{v}_k \oplus \hat{v}_{k-1}) + (\hat{u}_k \oplus \hat{v}_k)(\hat{u}_k \oplus \hat{u}_{k-1}) \end{aligned} \tag{2.34}$$

OQPSK is a variant of QPSK modulation. The modulator and demodulator of OQPSK are shown in Figure 2.24, which differ from the QPSK only by an extra $T/2$ delay in the Q-channel.

Because of the offset between I and Q channels, the OQPSK signal has a symbol period of $T/2$. At any symbol boundary, only one of the two bits can change the sign. Thus the phase changes between adjacent symbols can only be 0° and 90° whereas the phase of QPSK signal can jump as much as 180° at a time. Since the 180° phase shifts no longer exist in OQPSK signal, the amplitude fluctuations is less severe compared with the conventional QPSK.

One problem of OQPSK systems is that the differential encoding cannot be used. $\pi/4$-QPSK is an improved scheme compared with OQPSK, because it not only has no 180° phase shifts, but also can be differentially encoded. Therefore, $\pi/4$-QPSK is more suitable for mobile communication systems. It has been adopted in the digital cellular telephone systems in the United States and Japan.

Figure 2.25 illustrates the structure of the $\pi/4$-QPSK modulator, where $(I(t), Q(t))$ and $(u(t), v(t))$ are the uncoded and coded I and Q bits, respectively. The relationship between $(I(t), Q(t))$ and $(u(t), v(t))$ is as follows:

$$\begin{cases} u_k = \dfrac{1}{\sqrt{2}}(u_{k-1}I_k - v_{k-1}Q_k) \\ v_k = \dfrac{1}{\sqrt{2}}(u_{k-1}Q_k + v_{k-1}I_k) \end{cases} \tag{2.35}$$

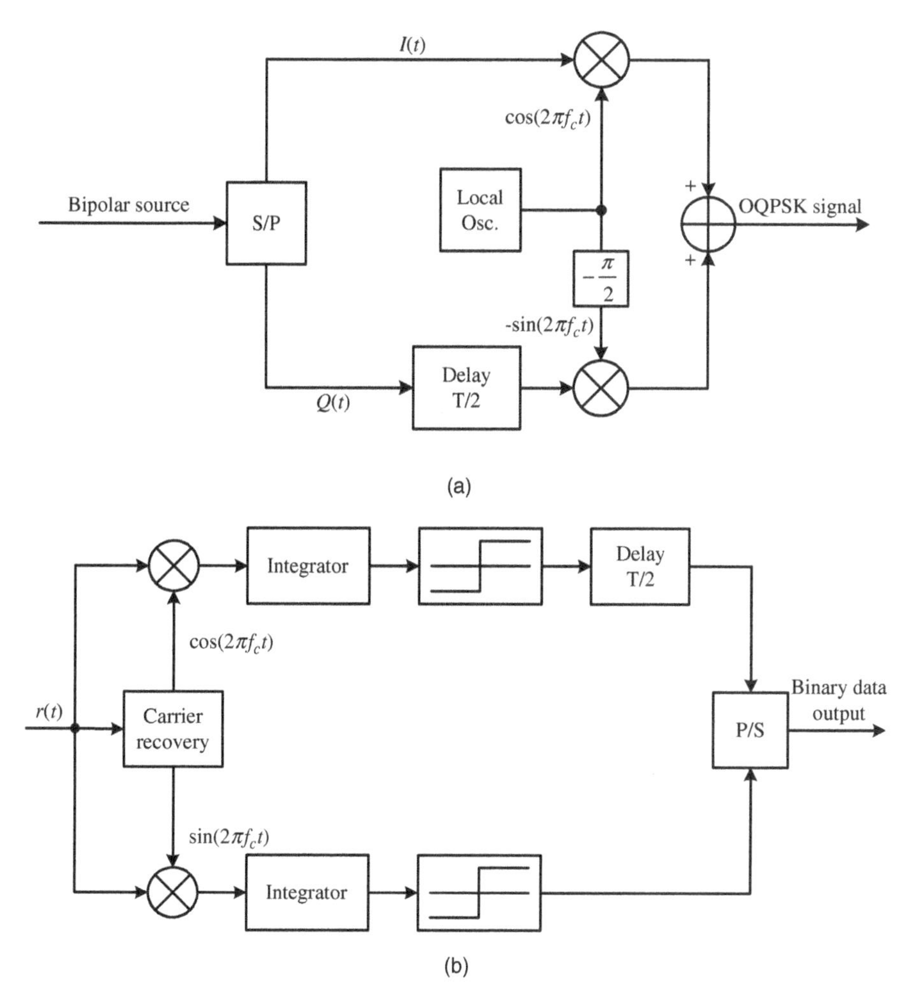

Figure 2.24. (a) OQPSK modulator and (b) demodulator

where u_k is the amplitude of $u(t)$ in the kth symbol duration and so on. u_k and v_k can take the values of ±1, 0, and $\pm1/\sqrt{2}$. The output signal of the modulator is

$$\begin{aligned} s(t) &= u_k \cos 2\pi f_c t - v_k \sin 2\pi f_c t \\ &= A \cos(2\pi f_c t + \Phi_k) \end{aligned} \tag{2.36}$$

where

$$\Phi_k = \tan^{-1}\frac{v_k}{u_k}, \quad A_k = \sqrt{u_k^2 + v_k^2} \tag{2.37}$$

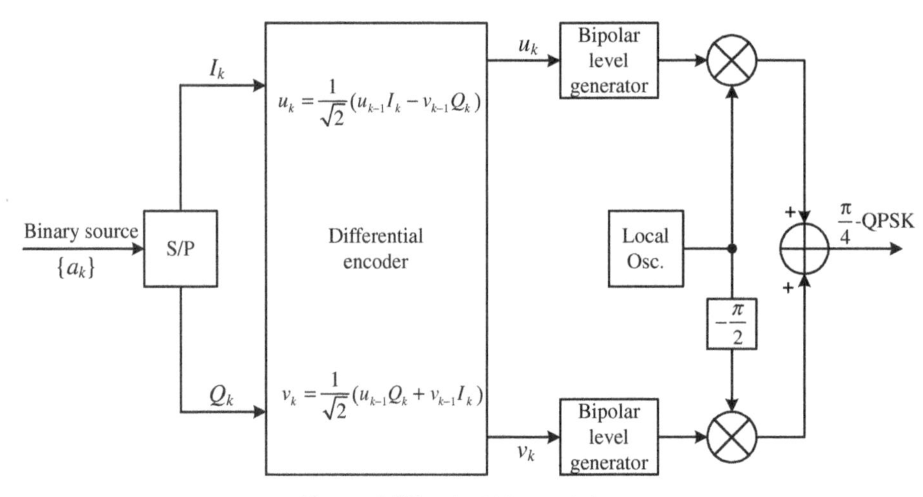

Figure 2.25. π/4-QPSK modulator

It can be proved that the phase relationship between two consecutive symbols is

$$\begin{cases} \Phi_k = \Phi_{k-1} + \Delta\theta_k \\ \Delta\theta_k = \tan^{-1}\dfrac{Q_k}{I_k} \end{cases} \tag{2.38}$$

and (2.35) can be rewritten as

$$\begin{aligned} u_k &= u_{k-1}\cos\Delta\theta_k - v_{k-1}\sin\Delta\theta_k \\ v_k &= u_{k-1}\sin\Delta\theta_k + v_{k-1}\cos\Delta\theta_k \end{aligned} \tag{2.39}$$

where $\Delta\theta_k$ is the phase difference determined by input data.

Since information is carried by the phase changes $\Delta\theta_k$, differentially coherent demodulation can be used. However, coherent demodulation is desirable when higher power efficiency is required. There are four ways to demodulate a π/4-QPSK signal: (1) baseband differential detection, (2) IF differential detection, (3) frequency modulation (FM) discriminator detection, and (4) coherent detection.

The diagrams of these four differential demodulators are shown in Figure 2.26(a), (b), (c) and (d), respectively [6].

For the same bandwidth, one can transmit more information with higher order modulation schemes (see Table 2.1). With two phases, BPSK can encode one bit per symbol, whereas QPSK with four phases can encode two bits per symbol, which is twice the rate of BPSK. BPSK can be considered as 2QAM with two real symbols: -1 and $+1$. QPSK is therefore treated as 4QAM with 4 complex symbols: $-1-1$j, $-1+1$j, $+1-1$j and $+1+1$j. 16QAM has 16 symbols from $-3-3$j to $+3+3$j and 64QAM has 64 symbols from $-7-7$j to $+7+7$j. Quadrature amplitude modulation (QAM) is discussed in Section 2.4.

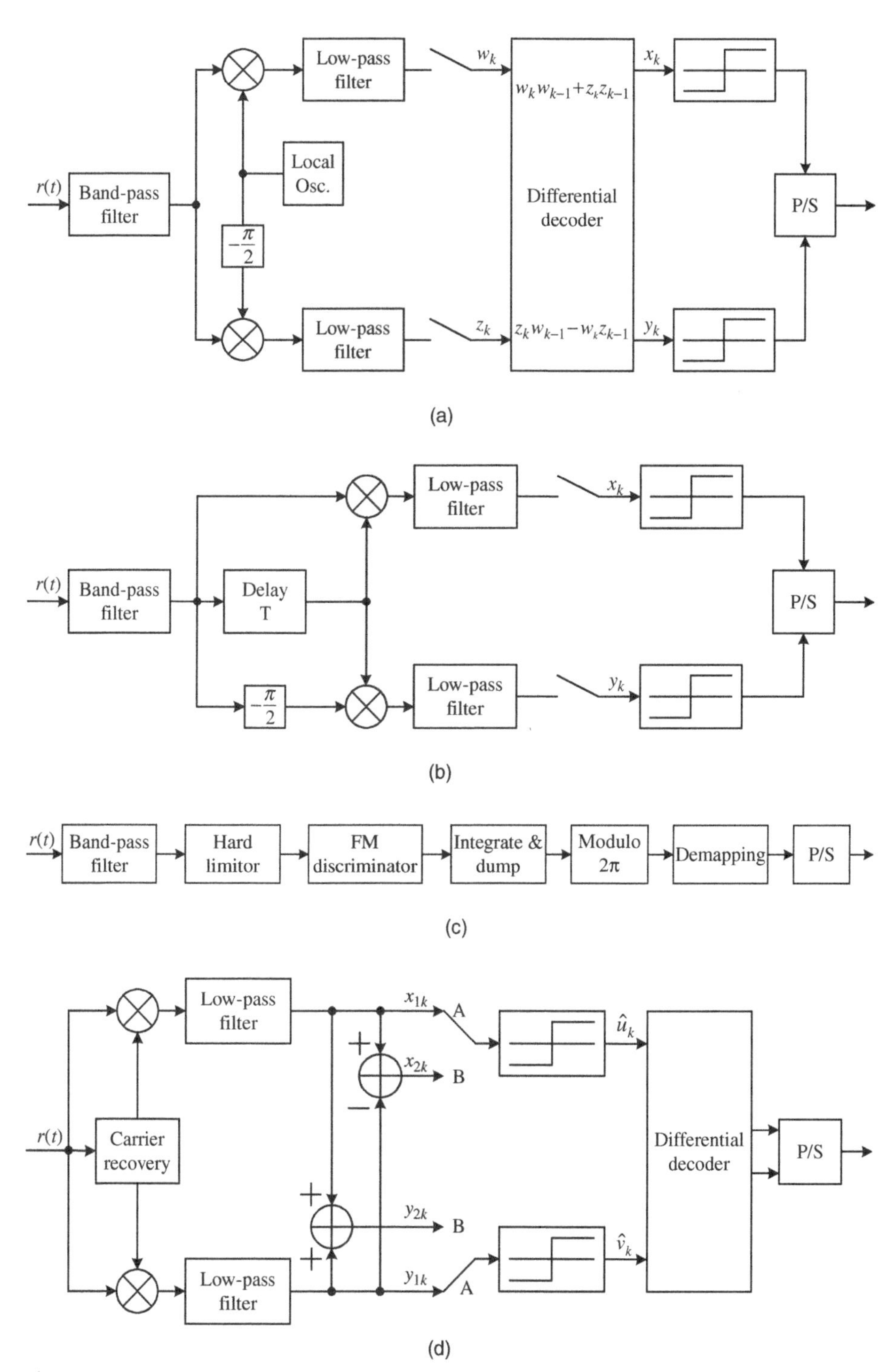

Figure 2.26. Demodulator for π/4-QPSK: (a) baseband differential detection, (b) IF band differential detection, (c) FM-discriminator detection, (d) coherent detection

TABLE 2.1. Bit Rate and Complexity for Different Modulations

Modulation	Bit Rate	Complexity
BPSK	1 bit per symbol	Low
QPSK	2 bits per symbol	Medium
16 QAM	4 bits per symbol	High
64QAM	6 bits per symbol	Very high

2.3 FREQUENCY SHIFT KEYING (FSK)

Digital information can also be transmitted by modulating the frequency of the carrier. If we use a binary signal, two different frequencies, f_1 and f_2, are used to transmit the information. The two resulting signal waveforms, s_1 and s_0, can be expressed as

$$s_i(t) = \begin{cases} A\cos(2\pi f_i t) & 0 \leq t \leq T \\ 0, & elsewhere \end{cases} \tag{2.40}$$

where $i = 0, 1$.

If the bit "1" is emitted, then the signal waveform $s_1(t)$ is transmitted. If the bit "0" is emitted, then the signal waveform $s_0(t)$ is transmitted. Note that $f_1 - f_0$ is so chosen such that $s_1(t)$ and $s_0(t)$ are orthogonal (see Figure 2.27).

In Figure 2.28(a), the upper trace is the baseband data and the appearance of a FSK signal in the time domain is shown in the lower trace. Bandwidth occupancy of FSK is dependent on the spacing of the two frequencies. A frequency spacing of 0.5 times the symbol period is typically used. FSK can be expanded to an M-ary scheme, employing multiple frequencies as different states, as shown in Figure 2.28(b).

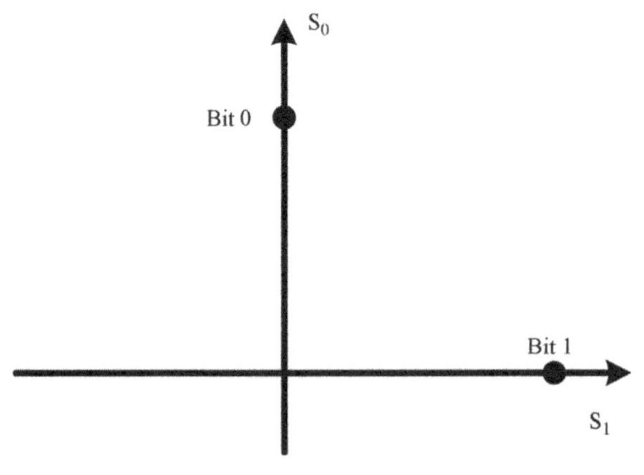

Figure 2.27. FSK constellation

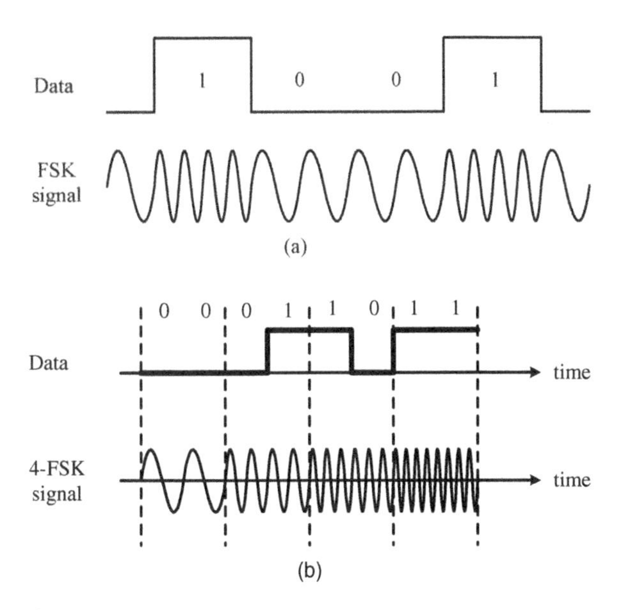

Figure 2.28. (a) BFSK, (b) 4FSK signals in the time domain

FSK signal can be generated both coherently with an IQ modulator and noncoherently with simply a voltage controlled oscillator (VCO) and a digital voltage source, as shown in Figure 2.29. The terms coherent and noncoherent are frequently used when discussing the generation and reception of digital modulation. When linked to the process of modulation the term coherence relates to the ability of the modulator to control the phase of the signal, not just the frequency.

With the system in Figure 2.29(a) the instantaneous frequency of the output waveform is determined by the modulator (within a tolerance set by the VCO and data amplitude and etc.) but the instantaneous phase of the signal is not controlled and can have any value. Alternatively coherent generation of modulation is achieved as shown in Figure 2.29(b). Here the phase of the signal is controlled besides the frequency.

When a coherent modulator is used to generate the FSK signal, both frequency and phase are controlled. The modulator offers the possibility to shape the resultant carrier phase trajectory at baseband either with analogue filtering or digital signal processing and a digital-to-analog converter (DAC). This can be used to generate both constant amplitude and amplitude-modulated signals.

There are two types of demodulation schemes in FSK, that is, synchronous demodulation and envelope demodulation, just like the OOK system.

The block diagram of synchronous demodulation is shown in Figure 2.30.

From Figure 2.30 we can observe that there are two branches in the synchronous demodulation scheme. After the two band-pass filters with center frequency of f_1 and f_2, the input FSK signal is split into two OOK signals. Then the demodulation is similar to

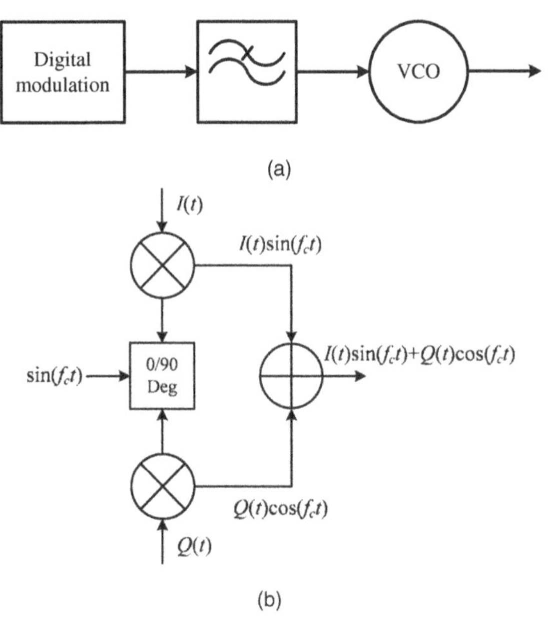

Figure 2.29. FSK signal generators: (a) noncoherent, (b) coherent

the OOK. Assuming that the output signal of the upper low-pass filter is x_1 and the lower low-pass filter output is x_2, the decision criterion is

$$\begin{cases} x_1 - x_2 > 0 & \text{signal } f_1 \text{ is present} \\ x_1 - x_2 < 0 & \text{signal } f_2 \text{ is present} \end{cases} \tag{2.41}$$

When noise is considered, the performance of the synchronous demodulation is analyzed as follows. If the input signal is f_1, the output of the upper band-pass

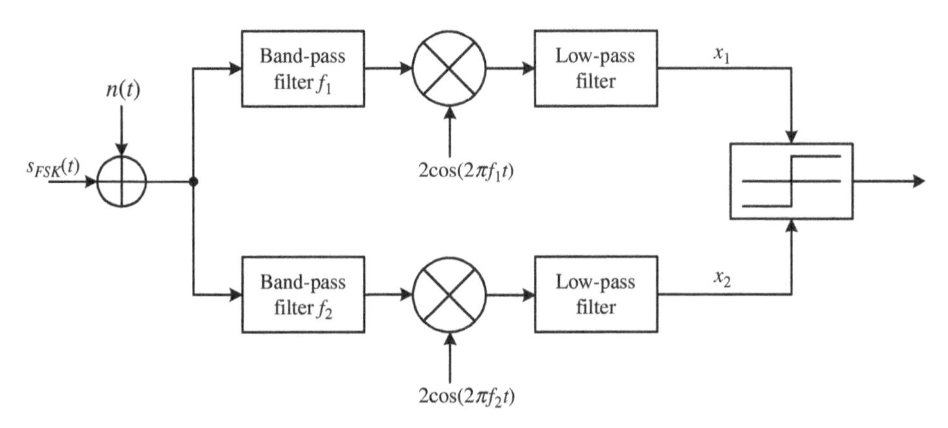

Figure 2.30. Synchronous demodulation of FSK

filter can be expressed as $x_1 = A + n_{1,c}(t)$, where $n_{1,c}(t)$ is the in-phase component of the narrow-band Gaussian noise $n_{i,1}(t)$ with the center frequency of $f_1 \cdot n_{i,1}(t)$ is given by

$$n_{i,1}(t) = n_{1,c}(t)\cos 2\pi f_1 t - n_{1,s}(t)\sin 2\pi f_1 t \tag{2.42}$$

The output of the lower branch is $x_2 = n_{2,c}(t)$, where $n_{2,c}(t)$ is the in-phase component of the narrow-band Gaussian noise $n_{i,2}(t)$ with the center frequency of f_2. $n_{1,c}(t)$ and $n_{2,c}(t)$ are independent and there is no overlap between them. The variances of $n_{1,c}(t)$ and $n_{2,c}(t)$ are the same as σ_n^2. Based on the above analysis, when input signal is f_1, we have

$$\begin{cases} x_1 = A + n_{1,c}(t) \\ x_2 = n_{2,c}(t) \end{cases} \tag{2.43}$$

According to (2.41), when $x_1 - x_2 > 0$,

$$\begin{aligned} & A + n_{1,c}(t) - n_{2,c}(t) > 0 \\ & \text{or } A + \left[n_{1,c}(t) - n_{2,c}(t)\right] > 0 \end{aligned} \tag{2.44}$$

Let $n_z(t) = n_{1,c}(t) - n_{2,c}(t)$, $n_z(t)$ is a subtraction of the two independent zero-mean low-pass Gaussian noise. So $n_z(t)$ is also a Gaussian stochastic process with zero-mean and variance of $2\sigma_n^2$. Equation (2.44) can be rewritten as

$$z_1(t) = A + n_z(t) > 0 \tag{2.45}$$

The probability distribution of $z_1(t)$, which is denoted as $p_1(z)$, is Gaussian distribution with mean of A and variance of $2\sigma_n^2$. The distribution curve of $p_1(z)$ is shown in Figure 2.31. The shadow area 1 is the probability of decision error when f_1 is sent and f_2 is decided, which is denoted as $P_e(f_2/f_1)$,

$$\begin{aligned} P_e(f_2/f_1) &= \int_{-\infty}^{0} p_1(z)\mathrm{d}z = \int_{-\infty}^{0} \frac{1}{\sqrt{2\pi\sigma_z^2}} e^{-(z-A)^2/2\sigma_z^2}\mathrm{d}z \\ &= \frac{1}{2}\operatorname{erfc}\left(\frac{A}{\sqrt{2}\sigma_n}\right) \end{aligned} \tag{2.46}$$

Similarly, when f_2 is inputted, we have

$$\begin{cases} x_1 = n_{1,c}(t) \\ x_2 = A + n_{2,c}(t) \end{cases} \tag{2.47}$$

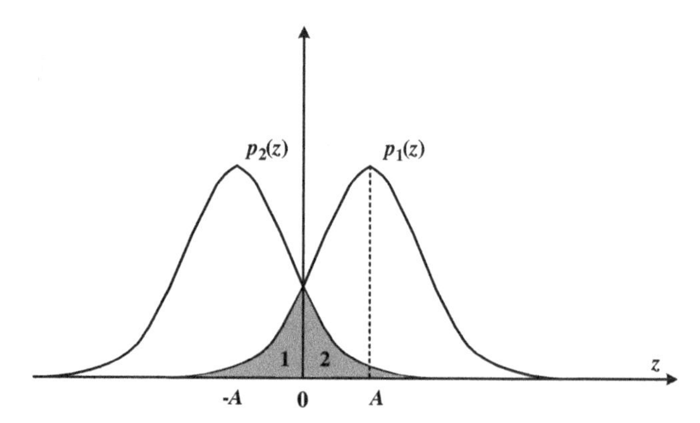

Figure 2.31. Probability distribution curve of FSK coherent demodulation

According to (2.41), when $x_1 - x_2 < 0$, f_2 is decided, we have

$$-A + n_z(t) < 0 \tag{2.48}$$

Let $z_2(t) = -A + n_z(t)$, the corresponding probability distribution $p_2(z)$ is also shown in Figure 2.31. The shaded area 2 is the decision error probability when f_2 is sent and f_1 is decided, which is denoted as $P_e(f_1/f_2)$,

$$P_e(f_1/f_2) = P_e(f_2/f_1) = \frac{1}{2}\mathrm{erfc}\left(\frac{A}{\sqrt{2}\sigma_n}\right) \tag{2.49}$$

Then the total probability of decision error for noncoherent FSK demodulation is

$$P_e = P(f_1)P_e(f_2/f_1) + P(f_2)P_e(f_1/f_2) \tag{2.50}$$

where $P(f_1)$ and $P(f_2)$ are the priori probabilities of f_1 and f_2, respectively. When $P(f_1) = P(f_2) = 1/2$, then

$$P_e = \frac{1}{2}\mathrm{erfc}\left(\sqrt{\frac{SNR}{2}}\right) \tag{2.51}$$

where $SNR = A^2/2\sigma_n^2$.

The envelope demodulation for FSK signal is shown in Figure 2.32. It consists of two branches of ASK envelope demodulators. When f_1 is sent, the upper envelope detector's output x_1 obeys Ricean distribution, and the lower envelope detector's output is Rayleigh distribution. When f_2 is sent, the distribution of x_1 becomes Rayleigh while x_2 has a Ricean distribution.

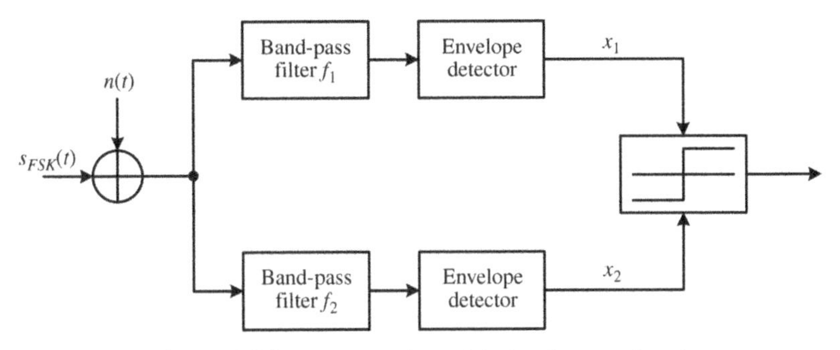

Figure 2.32. Envelope demodulation for FSK signal

Assume that f_2 is sent. Let $v_1(t)$ be the random voltage of Rayleigh distribution and $v_2(t)$ the random voltage obeying Ricean distribution. The probability density functions of $v_1(t)$ and $v_2(t)$ are

$$p(v_1|f_2) = \begin{cases} \dfrac{v_1}{\sigma_n^2}\exp\left(-\dfrac{v_1^2}{2\sigma_n^2}\right) & v_1 \geq 0 \\ 0 & v_1 < 0 \end{cases} \tag{2.52}$$

$$p(v_2|f_2) = \begin{cases} \dfrac{v_2}{\sigma_n^2}\exp\left[-\dfrac{(v_2^2+A^2)}{2\sigma_n^2}\right] I_0\left(\dfrac{Av_2}{\sigma_n^2}\right) & v_2 \geq 0 \\ 0 & v_2 < 0 \end{cases} \tag{2.53}$$

Then the probability of error conditioned on f_2 is

$$\begin{aligned} P_e &= P(v_1 > v_2|f_2) \\ &= \int_0^{\infty} p(v_2|f_2)\left[\int_{v_2}^{\infty} p(v_1|f_2)\mathrm{d}v_1\right]\mathrm{d}v_2 \\ &= \int_0^{\infty} \frac{v_2}{\sigma_n^2}\exp\left[-\frac{(v_2^2+A^2)}{2\sigma_n^2}\right] I_0\left(\frac{Av_2}{\sigma_n^2}\right)\left[\int_{v_2}^{\infty}\frac{v_1}{\sigma_n^2}\exp\left(-\frac{v_1^2}{2\sigma_n^2}\right)\mathrm{d}v_1\right]\mathrm{d}v_2 \\ &= \frac{1}{2}\exp\left(-\frac{SNR}{2}\right) \end{aligned} \tag{2.54}$$

where $SNR = A^2/2\sigma_n^2$.

For a coherent demodulation, system makes a demodulation decision based on the received signal phase, not frequency. The additional "information" available results in an improved BER performance. The high level of digital integration in semiconductor devices has made digitally based coherent demodulators common in mobile communications systems.

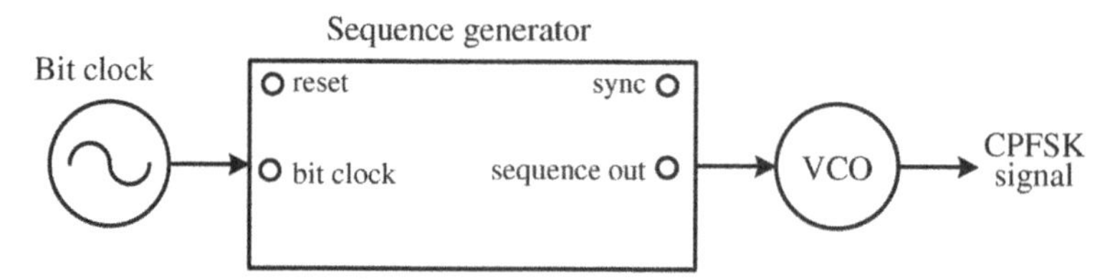

Figure 2.33. A CPFSK signal generator

When the noncoherent FSK transmitted signal switches from one frequency to another one following the symbol stream, there is an abrupt switching at the beginning of each waveform. This results in some sharp discontinuities in the phase of the transmitted signal, giving rise to some high-frequency components in the spectrum. Therefore, this method does not have high bandwidth efficiency. One way to avoid wasting bandwidth is to modulate the signal such that the phase of the transmitted signal changes continuously. The resulting FSK signal is referred to as continuous phase FSK (CPFSK). CPFSK is a modulation method with memory (see Figure 2.33).

A CPFSK signal can be expressed as

$$s(t) = \begin{cases} A\cos[2\pi f_1 t + \theta(0)] & \text{for symbol "1"} \\ A\cos[2\pi f_2 t + \theta(0)] & \text{for symbol "0"} \end{cases} \tag{2.55}$$

Until now, we have reviewed the modulation techniques that are referred to as memoryless. Either OOK or PSK modulations are described as memoryless because the amplitude or the phase of the transmitted signal is statistically independent over successive signal intervals. It means that the value of, say, the amplitude of the carrier for OOK modulation at the instant t is independent of its value at the instant $t+T_s$, where T_s is the symbol duration. In other words, the transmitted signal has no memory of its previous values. This is not the case with CPFSK modulation because the memory results from the phase continuity of the transmitter carrier phase from one signal interval to the next. The analysis of signals with memory, in particular CPFSK signals, is not a trivial exercise. Reader is referred to the literature (e.g., [6]) for further information.

2.4 QUADRATURE AMPLITUDE MODULATION (QAM)

Quadrature amplitude modulation (QAM) is a complicated name for a simple technique. In the simplest terms, QAM is the combination of amplitude modulation and phase shift keying. A QAM signal can be expressed as

$$\begin{aligned} s_m(t) &= \mathrm{Re}\left[(A_{mc} + jA_{ms})g(t)e^{j2\pi f_c t}\right] \\ &= A_{mc}g(t)\cos(2\pi f_c t) - A_{ms}g(t)\sin(2\pi f_c t) \\ &= V_m g(t)\cos(2\pi f_c t + \theta_m) \\ m &= 1, 2, \cdots, M \qquad 0 \le t \le T \end{aligned} \tag{2.56}$$

where g(t) is a pulse waveform to control the spectrum, for example, raised cosine, A_{mc} and A_{ms} are the in-phase and quadrature components of the modulating signal, respectively, and V_m and θ_m denote the signal amplitude and phase, given, respectively, by,

$$V_m = \sqrt{A_{mc}^2 + A_{ms}^2}$$
$$\theta_m = \tan^{-1}\frac{A_{ms}}{A_{mc}} \tag{2.57}$$

We can write $s_m(t)$ as a linear combination of two orthogonal waveforms

$$s_m(t) = s_{m1}f_1(t) + s_{m2}f_2(t) \tag{2.58}$$

From (2.58), it can be seen that QAM is a modulation scheme in which data is transferred by modulating the amplitude of two separate orthogonal carrier waves, which are out of phase by 90° (sine and cosine). Due to their 90° phase difference, they are called quadrature carriers.

As stated previously, QAM involves sending digital information by periodically adjusting the phase and amplitude of a sinusoidal electromagnetic wave. 4-QAM uses four combinations of phase and amplitude of a sinusoidal electromagnetic wave, and each combination is assigned a 2-bit digital pattern. Each unique 2-bit digital pattern is called a symbol. When the bit stream (0,0,0,1,1,0,1,1) is generated, for example, they are mapped or grouped into the corresponding four symbols (00, 01, 10, 11). These combinations or symbols are shown in the constellation plot of Figure 2.34. The digital bit patterns that the four symbols represent are labelled on the constellation plot of Figure 2.34, where the lines represent the phase and amplitude transitions from one symbol to another.

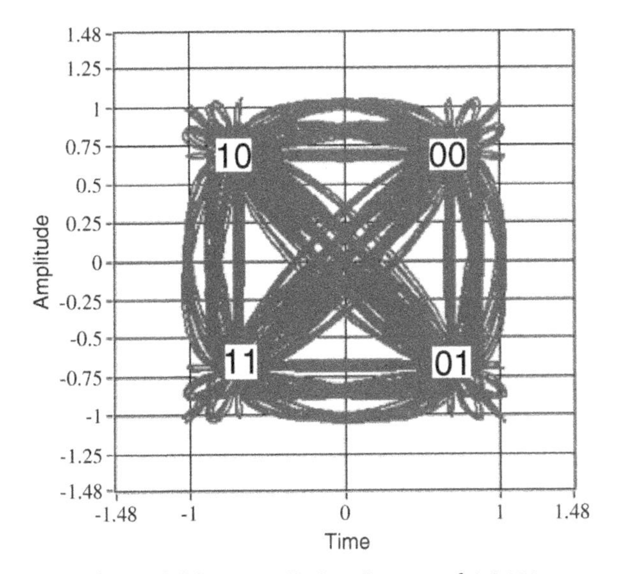

Figure 2.34. Constellation diagram of 4-QAM

As just stated, a digital bit pattern or symbol can be sent over a carrier signal by generating a unique combination of phase and amplitude. Since multiple discrete amplitudes and phases can be used, it is possible to convey multiple bits per symbol in QAM. Thus, QAM has a tremendous advantage in capable of achieving high bandwidth efficiency.

Unmodulated signals exhibit only two positions enabling a transfer of either a "0" or "1". In QAM, it is possible to transfer more bits per symbol as there are multiple points of transfer. In QAM, a signal obtained by summing the amplitude and phase modulation of a carrier signal (a modulated sine and cosine wave or quadrature waves) is used for the data transfer. Since the number of transfer points remains high, it is possible to convey more bits per symbol.

The possible states of a particular configuration can be represented using a constellation diagram. In a constellation diagram, constellation points or symbols are arranged in a regular grid with equal horizontal and vertical spacing. In digital communication, as data are binary, it follows that the number of points in the grid is a function of the power of 2 (2, 4, 8, etc.).

While two bits per symbol can be sent using 4-QAM, it is possible to send data at higher rate by increasing the number of symbols, M, in the symbol map, where M is a function of the power of 2. Thus, the number of bits that can be represented by a symbol in M-QAM is $\log_2(M)$. For example, two bits can be represented by each symbol in 4-QAM, while each symbol in 256-QAM can be used to represent an 8-bit digital pattern ($\log_2(256) = 8$). The higher the value M is, the higher the actual data transmission rate.

The most common QAM constellations, shown in Figure 2.35 include the squared QAMs, 4-QAM, 16-QAM, 64-QAM, and 256-QAM, as well as the odd-bit QAMs, 32-QAM (5-bit per symbol) and 128-QAM (7-bit per symbol).

Although it is possible to transfer more bits per symbol with higher-order M-QAM constellations to achieve a higher bandwidth efficiency, an inherent technical problem exists. In order to maintain the mean energy of a higher order constellation at the same level of a lower order one, the constellation points in the higher-order scheme have to be packed much closer to each other. However, such a configuration dramatically reduces the noise immunity of the modulation scheme. Thus, in practical implementation, a higher-order QAM delivers data with lower reliability than a lower-order one. Furthermore, for a higher-order QAM, a highly linear PA with a large dynamic range is required at transmitter, so that edge constellation symbol points are not distorted by the nonlinearity of the PA. Therefore, the higher the value of M is, the lower power efficiency will be.

The block diagrams of a QAM modulator and demodulator are shown in Figure 2.36(a) and (b), respectively. A highly stable local oscillator is normally required for a QAM system.

The 64-QAM and 256-QAM are often used in cable modem and digital cable television applications. In millimeter wave communications, neither 64-QAM nor 256-QAM is preferable due to the effect of phase noise and power consumption of PA.

Phase shift keying in Section 2.2 can be regarded as special cases of quadrature amplitude modulation where the amplitude of the modulating signal is constant and the phase is changing. The same theory can further be extended to frequency shift keying and frequency modulation. Both are the special cases of phase modulation.

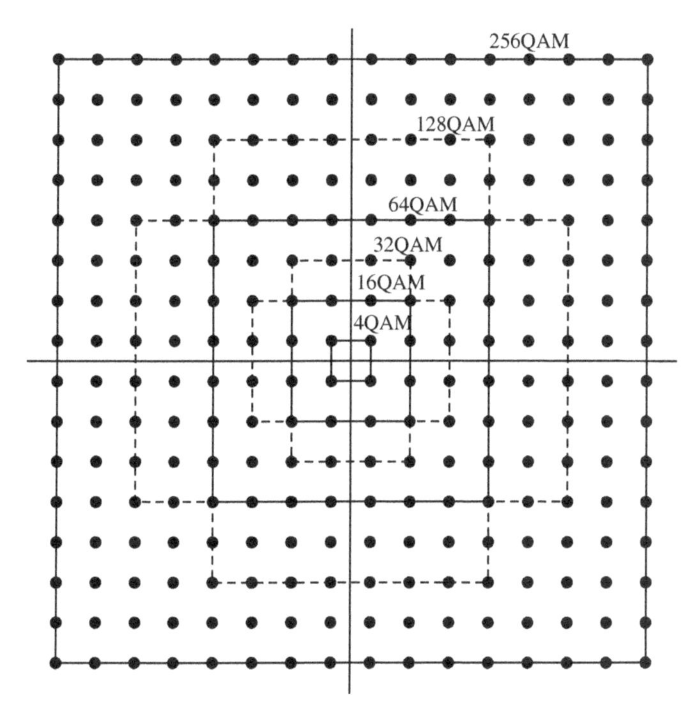

Figure 2.35. QAM constellation

Owing to the symmetry of the signal space and the orthogonality of the I and Q branches, error probability in a coherent detection of M-QAM system can be derived below.

$$P_e(M) = 1 - P_{Correct}(M) = 1 - P_r(\text{no error detected on I and Q branches}) \quad (2.59)$$

where

$$\begin{aligned} &P_r(\text{no error detected on I and Q branches}) \\ &= P_r(\text{no error on } I) \cdot P_r(\text{no error on Q}) \\ &= P_r(\text{no error on I})^2 = \left(1 - P_e(\sqrt{M})\right)^2 \end{aligned} \quad (2.60)$$

with $P_e(\sqrt{M})$ being the error probability of a coherent $\sqrt{M}$-QAM system.
Finally, the error probability can be expressed as [6, 7]

$$P_e(M) = 4\left(1 - \frac{1}{\sqrt{M}}\right) Q\left(\sqrt{\frac{3\log_2 M}{M-1}\frac{E_b}{N_0}}\right) \quad (2.61)$$

where E_b is the energy per bit, and N_0 is the noise power spectral density.

This error probability is derived when the channel is assumed to be AWGN, and the transmitting and receiving filters are the square root of a Nyquist filter. In this case we have zero inter-symbol interference (ISI) and matched receive filter. The error probability of QAM signal with different M value is shown in Figure 2.37.

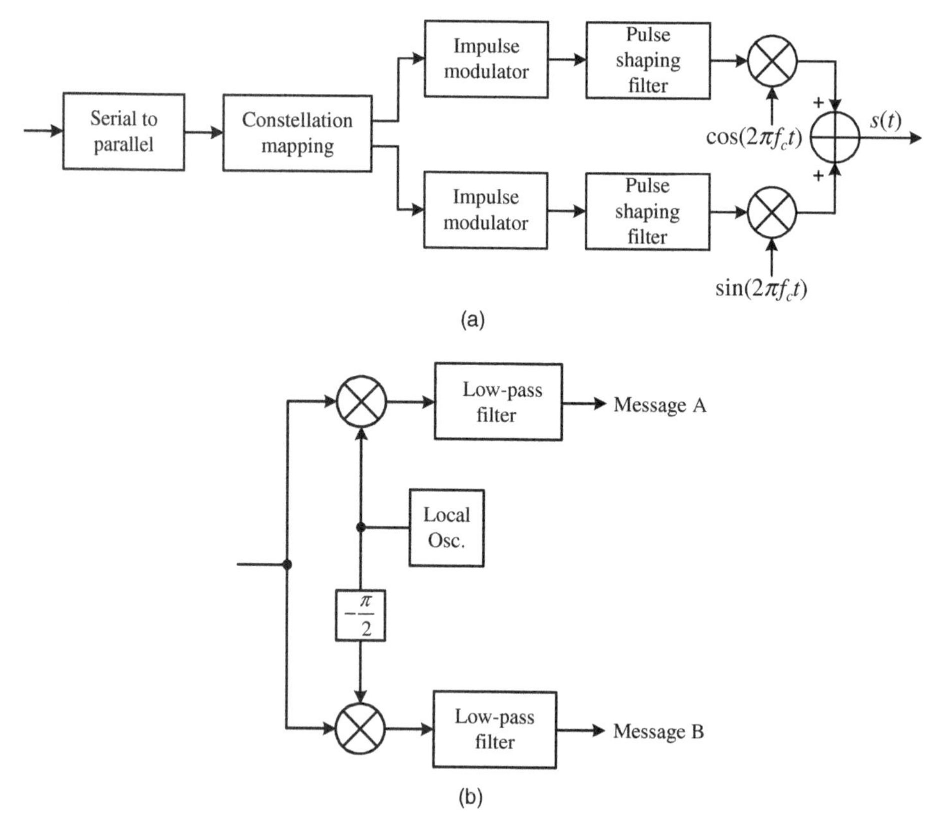

Figure 2.36. (a) QAM modulator, (b) QAM demodulator

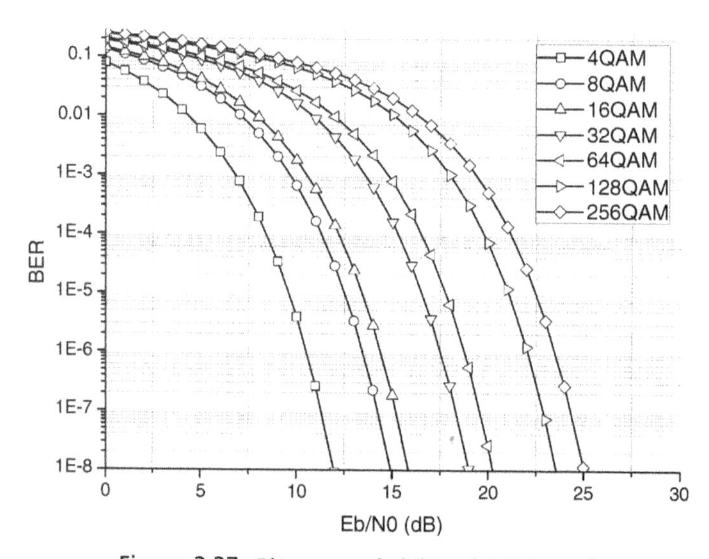

Figure 2.37. Bit error probability of QAM signal

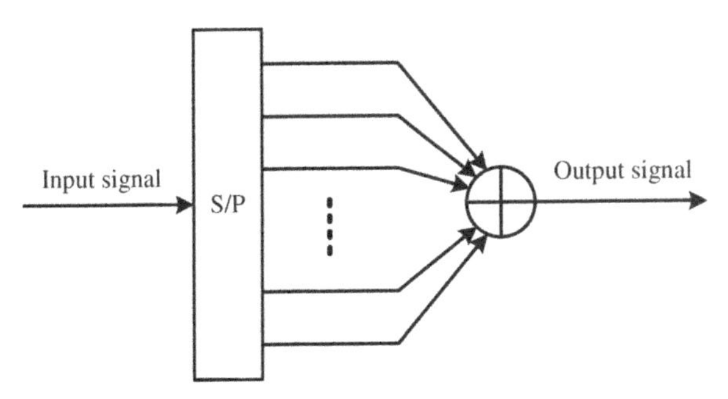

Figure 2.38. OFDM configuration

2.5 ORTHOGONAL FREQUENCY DIVISION MULTIPLEXING (OFDM)

Orthogonal frequency division multiplexing (OFDM) is used widely in the current broadband wireless communication system because of its high data rate transmission and the robustness against frequency selective fading. OFDM technology is to split a high-rate data stream into a number of lower rate streams that are transmitted simultaneously over a number of subcarriers as shown in Figure 2.38 [8]. Because the symbol duration increases with the lower rate parallel subcarrier, the relative amount of dispersion in time caused by multi-path delay spread is decreased. OFDM can be viewed as a multiplexing technique and the output signal is the linear sum of the modulated subcarrier signals. In other words, the radio is transmitting multiple RF subcarriers instead of a single RF carrier. Each of those OFDM subcarriers will still be modulated exactly the same way as using, for example, BPSK, QPSK, or QAM.

At time instant t, the baseband OFDM signal can be expressed as

$$x(t) = \frac{1}{\sqrt{N_c}} \sum_{k=0}^{\infty} p(t-kT_s) \sum_{m=0}^{N_c-1} X_m(k) \exp[\, j2\pi f_m(t-kT_s)] \tag{2.62}$$

where $X_m(k)$ is the modulating signal on the mth subcarrier of the kth OFDM symbol. T_s is the duration of one OFDM symbol, $p(t)$ ($p(t) = 1$ when $0 \le t < T_s$, otherwise $p(t) = 0$) is the time domain rectangle window function, $f_m = m/T_s$ is the mth subcarrier frequency, and N_c is the total number of subcarriers. $X_m(k)$ can take different values according to the modulation type, such as QPSK, 16-QAM and 64-QAM. Furthermore, $p(t)$ can be any other window functions that obey Nyquist criterion.

The discrete expression of (2.62) is

$$x(n) = \frac{1}{\sqrt{N_c}} \sum_{k=0}^{\infty} p(n-kN_c) \sum_{m=0}^{N_c-1} X_m(k) \exp\left[j\frac{2\pi m(n-kN_c)}{N_c} \right] \tag{2.63}$$

The implementation complexity of OFDM modems can be reduced significantly by employing inverse discrete Fourier transform (IDFT) to replace the bank of sinusoidal

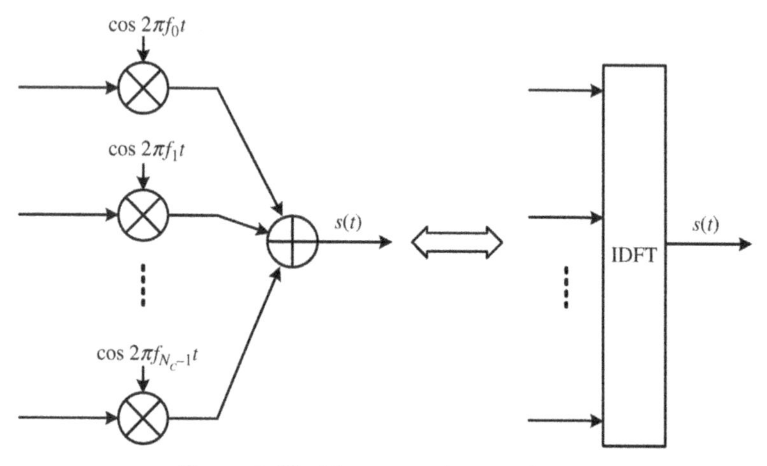

Figure 2.39. Discrete Fourier transformer

generators at transmitter (see Figure 2.39) and using discrete Fourier transform (DFT) to replace the bank of local oscillators at receiver.

With OFDM systems, a sinc-shaped pulse is applied in the frequency domain of each channel. As a result, each subcarrier remains orthogonal to one another. The orthogonality of subcarriers is manifested in both time and frequency domains by the mathematical expressions for subcarrier i and subcarrier l as

$$\int_{-\infty}^{\infty} x_i(t)x_l^*(t)dt = 0 \quad \text{time domain} \tag{2.64}$$

$$\int_{-\infty}^{\infty} X_i(f)X_l^*(f)df = 0 \quad \text{frequency domain} \tag{2.65}$$

where $i \neq l$, and $X_i(f)$ and $X_l(f)$ are the Fourier transforms of $x_i(t)$ and $x_l(t)$, respectively.

In OFDM systems, the spectrum of individual subcarrier is overlapped with minimum frequency spacing, which is carefully designed so that each subcarrier is orthogonal to the other subcarriers. The bandwidth efficiency of OFDM is another advantage.

Figure 2.40 shows an example of one OFDM signal with four subcarriers. In this example, all subcarriers are with the same phase and amplitude. However, in the real system, all the amplitudes and phases could be different according to the modulated symbol for each subcarrier. During the symbol interval T_s, all the subcarriers have integer number of cycles, and the number of cycles between adjacent subcarriers just differs by one. This accounts for the orthogonality between subcarriers.

The spectral shapes for the OFDM signal are shown in Figure 2.41. This figure is based on the rectangle window function. When $p(t)$ are other window functions, the spectral shape will be changed. Therefore, we can use different window functions to adjust the shapes of the subcarriers to reduce the inter-carrier interference (ICI). The

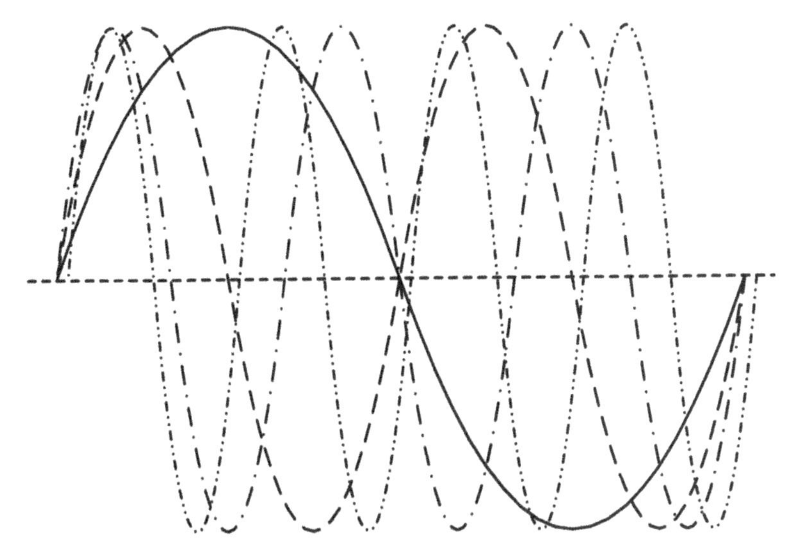

Figure 2.40. Example of four subcarriers within one OFDM symbol

subchannel spacing in OFDM signal is $1/T_s$, and the transmission rate of each subchannel is also $1/T_s$, so the subchannels are overlapped. Because of this property, the spectral efficiency of OFDM system is better than the traditional frequency division multiplexing (FDM) system.

One of the main advantages of OFDM is the ability of dealing with the multi-path delay spread. The data to be transmitted over an OFDM signal are spread across N_c subcarriers of the signal, and each subcarrier carrying out the payload. This reduces the data rate for each subcarrier. The lower data rate makes it easy to handle the interference from reflections. This is achieved by adding a guard interval with zero padding between adjacent OFDM symbols. Figure 2.42 gives an example of OFDM signals with zero-padded guard interval. It ensures that the data is only sampled when the signal is stable and no new delayed signals arrive that would alter the timing and phase of the signal. Therefore, the ISI can be eliminated.

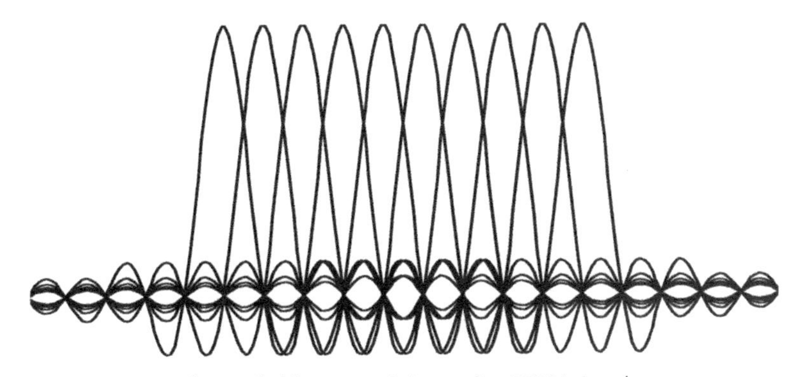

Figure 2.41. Spectral shapes for OFDM signal

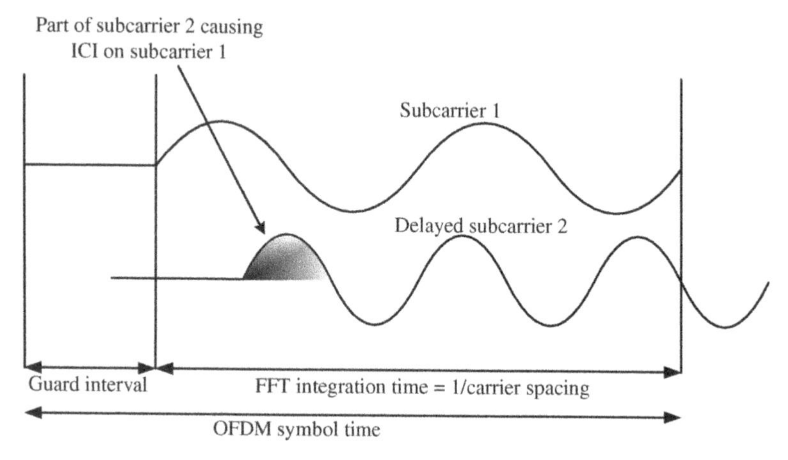

Figure 2.42. ISI elimination

With the guard interval, the OFDM symbol can be cyclically extended to avoid ICI, as shown in Figure 2.43. The cyclic prefix is a repetition of the last section of a symbol that is inserted just before the symbol. It enables multi-path representations of the original signal to fade so that they do not interfere with the subsequent symbol. In general, OFDM has the advantages of the immunity to multi-path delay spread, superior resistance to frequency selective fading, enabling simple equalization, and efficient bandwidth usage.

OFDM requires synchronization and FFT units at transmitter and receiver. It is very sensitive to carrier frequency offset. Also, the problem of high peak to average power ratio is particularly challenging in millimeter wave communication systems. The peak power of the OFDM signal might be minimized by employing different data encoding schemes before modulation [9].

High peak to average ratio of OFDM causes difficulties to power amplifiers. They generally have to be operated at a large backoff to avoid out-of-band interference. If this interference is required to be lower than 40 dB below the power density in the OFDM band, an input backoff of more than 7.5 dB is required [3]. Crest factor is defined as the ratio of peak amplitude to root mean square (RMS) amplitude. Designers can employ

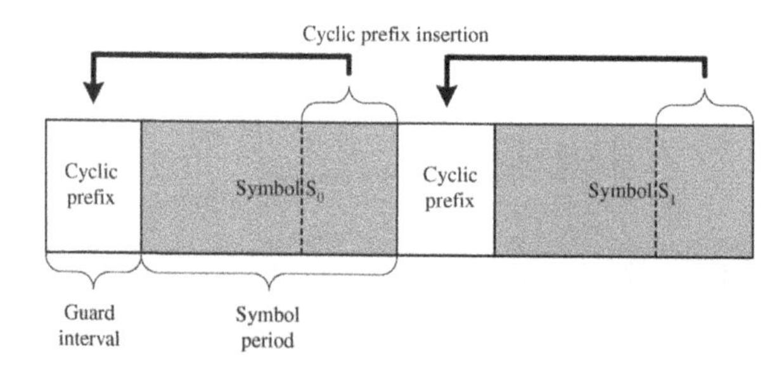

Figure 2.43. Guard interval with cyclic prefix

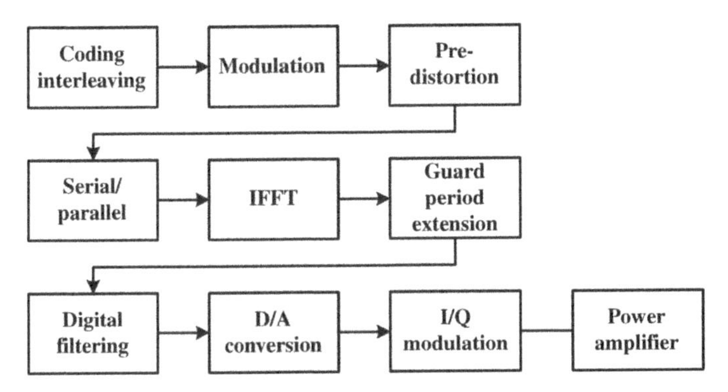

Figure 2.44. Typical OFDM transmitter chain

various crest factor reduction (CFR) techniques to help combating the problem of high peak to average ratio.

One of the problems of filtering an OFDM signal is the passband ripple. It is well-known in filter design theory that if we want to minimize this ripple, the number of taps on the filter should be increased. The trade-off is between performance and cost-complexity. A higher ripple leads to higher BER. Ripple has a worse effect in OFDM systems because some subcarriers get amplified and others get attenuated. One way to combat this issue is to equalize the SNR across all subcarriers using what is called digital predistortion (DPD). Applying DPD before filtering increases the signal power and hence out-of-band interference. The latter must be limited by using the bandpass filter having a higher attenuation outside the passband, as compared to the system without predistortion. The sequence of operations at the transmitter is represented in Figure 2.44.

A typical FFT-based OFDM communication system is described in Figure 2.45. At the transmitter, the serial to parallel converter converts a serial bit stream into several parallel bit streams to be divided among the individual subcarriers. Each subcarrier is modulated as an individual channel before all the subcarriers are combined back together and transmitted as a single signal. The parallel to serial conversion stage is the process of summing all the subcarriers together to form a single signal. The modulation of data into a complex waveform occurs at the IFFT stage of the transmitter. The role of the IFFT is to modulate each subchannel onto the appropriate subcarrier. Here, the modulation scheme can be chosen completely independently of the specific channel and can be chosen based on the channel requirements. In fact, it is possible for each individual subcarrier to use a different modulation scheme.

The receiver performs the reverse process to first divide the incoming signal into the appropriate subcarriers via the S/P converter and then to demodulate them individually via the FFT before reconstructing the original bit stream with the P/S converter [10].

A traffic performance comparison of QPSK and 16-QAM modulation techniques for OFDM system is studied in the context of BER for the same information rate and peak to average power ratio. It is found that spectral width of 16-QAM is narrower than that of QPSK for the same data rate. Each symbol of QPSK conveys 2 bits but for 16-QAM it is 4 bits/symbol. Therefore, time domain equivalent symbol period of 16-QAM is twice as

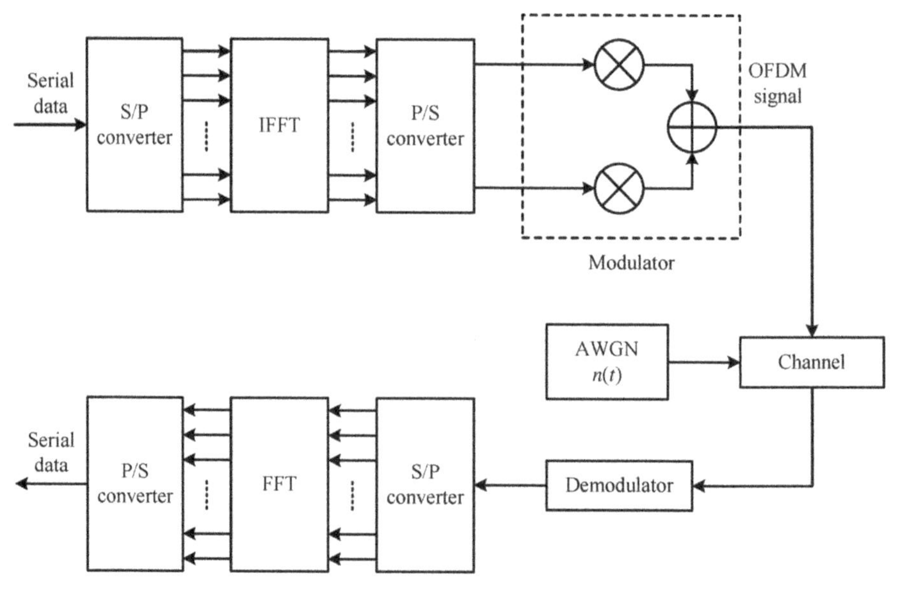

Figure 2.45. OFDM communication system

long. It is concluded that BER performance of QPSK is better than that of 16-QAM at the expense of large spectral width. Alternatively, 16-QAM can carry more traffic than QPSK at the expense of higher BER [11].

Recently, OFDM has been used by a plurality of standardization bodies for a wide range of wireless and wire line systems. Applications range from digital video/audio broadcasting to power-line communications. The attractive features of OFDM technology are summarized as follows:

1. The possibility of achieving channel capacity if the transmitted signal is adapted to the state of the wireless channel (i.e., if energy and adaptive bit-loading procedures are adopted),
2. The robustness to multi-path propagation providing a viable low-complexity and an optimal (in the maximum likelihood sense) solution for inter-symbol interference (ISI) mitigation, and
3. The availability of strategies for frequency diversity scheduling in multi-user environment.

REFERENCES

[1] S. Benedetto, E. Biglieri, and V. Castellani, *Digital Transmission Theory*, Englewood Cliffs, NJ: Prentice Hall, 1987.

[2] T. S. Rappaport, *Wireless Communications: Principles and Practice*, 2nd edition, Englewood Cliffs, NJ: Prentice Hall PTR, 2002.

[3] J. H. Park, Jr "On binary DPSK detection." *IEEE Trans. Commun.*, Vol. 26, No. 4, pp. 484–486, Apr. 1978.

[4] K. Feher, *Digital Communications: Satellite/Earth Station Engineering*, Englewood Cliffs, NJ: Prentice Hall, 1983.

[5] R. Lucky, J. Salz, and J. Weldon, *Principles of Data Communications*, New York: McGraw-Hill, 1968.

[6] F. Xiong, *Digital Modulation Techniques*, Boston, London: Artech House, 2000.

[7] J. Proakis and M. Salehi, *Communication Systems Engineering*, Englewood Cliffs, NJ: Prentice Hall, 1994.

[8] J. G. Proakis, *Digital Communications*, 5th edition, McGraw Hill Higher Education, New York, 2008.

[9] I. Islam and S. Hossain, "Comparison of traffic performance of QPSK and 16-QAM modulation techniques for OFDM system." *J. Telecommun. Information Technol.*, pp. 147–152, 2005.

[10] R. Prasad, *OFDM for Wireless Communications Systems*, Boston, London: Artech House Publishers, 2004.

[11] D. Wulich, "Peak factor in orthogonal multicarrier modulation with variable levels." *IEEE Electron. Lett.*, Vol 32, No. 20, pp. 1859–1861, 1996.

3

MILLIMETER WAVE TRANSCEIVERS

A transceiver is the main component of a wireless communication system. This chapter reviews classical and modern transceiver architectures. First, Section 3.1 discusses a system link budget to calculate the signal power and noise figure for a cascaded system. Section 3.2 describes the conventional transceiver architecture. Section 3.3 discusses novel transceivers with no mixer, whereas Section 3.4 introduces a special receiver with no local oscillator. Both configurations can be useful in a low power communication system. Finally, millimeter wave calibration methods are reviewed in Section 3.5. Readers can apply appropriate calibration method prior to their millimeter wave measurement.

3.1 MILLIMETER WAVE LINK BUDGET

A link budget is a signal-power plan for a radio system. It is used to determine a proposal's capabilities under specific operating conditions for the standard specified data rates, ranges, and bit error rates. The formula below identifies the necessary parameters

Millimeter Wave Communication Systems, by Kao-Cheng Huang and Zhaocheng Wang

that should be used to compute the final link margin:

1. Path loss at 1 m ($PL_0 = 20\log_{10}(4\pi f_c/c)) = 68.00$ dB
   where center frequency $f_c = 60$ GHz, light speed $c = 3 \times 10^8$ m/s
2. Average noise power per bit (dB) $N = -174 + 10\log_{10}(R_b)$
   where R_b (Gbps) is system payload bit rate
3. Average noise power per bit (dBm) $P_N = N$ + Rx noise figure in reference to the antenna terminal (dB)
4. Tolerable path loss (dB): $PL = P_T + G_T + G_R - P_N - S - M_{shadowing} - I - PL_0$

   where

   P_T is the average Tx power (dBm)

   G_T is the Tx antenna gain (dBi)

   G_R is the Rx antenna gain (dBi)

   S is the minimum signal-to-noise ratio E_b/N_0 for the additive white Gaussian noise (AWGN) channel (dB)

   $M_{shadowing}$ is the shadowing link margin (dB)

   I is the implementation loss (dB), including filter distortion, phase noise, and frequency errors
5. Maximum operating range $d = 10^{PL/10n}$ (m)

   where

   n is the path loss exponent whose value is subject to scenarios. The following path loss parameters are considered in the IEEE 802.15.3c standard [1].

For line-of-sight scenarios:

- Path loss at 1 m: $PL_0 = 68$ dB
- Path loss exponent: $n = 2$
- Shadowing link margin: $M_{shadowing} = 1$ dB

For non-line-of-sight scenarios:

- Path loss at 1 m: $PL_0 = 68$ dB
- Path loss exponent: $n = 2.5$
- Shadowing link margin: $M_{shadowing} = 5$ dB

A simple millimeter wave link can be drawn as shown in Figure 3.1.

From this perspective, we can now calculate the signal-to-noise ratio for the system. In Table 3.1, the example of case 4 of Figure 1.11 is set up to calculate the signal and noise of a millimeter wave system with two 12 dBi directional antennas in a 5-m wireless link. Both the free space loss and reflection loss are taken into account.

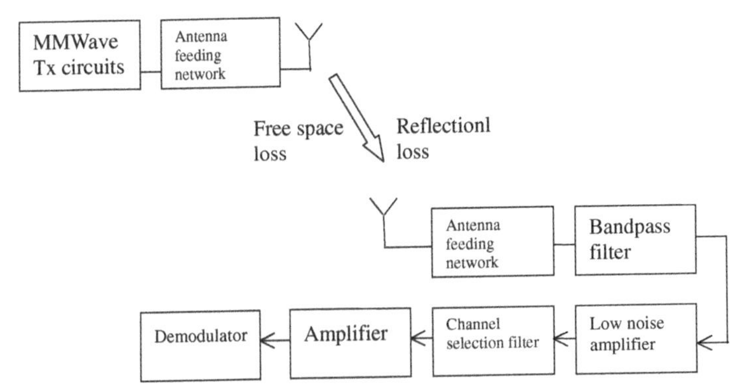

Figure 3.1. 60 GHz transmitter, receiver, and wireless link

When we set the transmitted signal level to 15 dBm, we can budget a 5-dB loss due to the feeding network of the transmitting antennas. The power delivered to the antenna is therefore 10 dBm. The effective isotropic radiated power (EIRP) is then effectively increased to 22 dBm when the transmitting antenna has a 12 dBi gain. In a 5-m link, the 60 GHz signal suffers a free space loss of approximately 81.98 dB. The signal is therefore attenuated to −59.98 dBm due to this free space loss. A further 15 dB attenuation due to the reflection loss gives rise to the final EIRP of −74.98 dBm at the receiving antenna input.

The input noise of the converter is the theoretical thermal noise floor limit, $kT\bar{B}$, which is calculated as follows:

$$kT\bar{B} = 4.002 \times 10^{-21} \text{ Watts (or in log form} = -174 \text{ dBm)},$$

where

$k =$ Boltzmann's constant $= 1.381 \times 10^{-23}$ W/Hz/K,
$T = 290$ K at room temperature,
$\bar{B} =$ a normalized bandwidth of 1 Hz.

TABLE 3.1. An Example of a Millimeter Wave Scenario

Transmission power to Tx antenna (dBm)	10
Bandwidth (GHz)	2
Distance (m)	5
Free space loss (dB)	81.98
Tx antenna gain (dBi)	12
Rx antenna gain (dBi)	12
Reflection loss (dB)	15
Input signal level to Rx antenna (dBm)	−74.98
Input noise level (dBm)	−81

TABLE 3.2. Components and Their Gain/Noise Figures

	Feeding Network	Bandpass Filter	LNA	Switch and Channel Selection Filter	Amplifier
Gain (dB)	−5	−1	20	−5	30
Cumulative Gain (dB)	−5	−6	14	9	39
Noise Figure (dB)	5	1	3	5	10
Noise Figure (real)	3.16	1.26	2.00	3.16	10.00

When we take the bandwidth into account, the input noise level is calculated as given by

$$\text{Input noise level} = 10\log_{10}(kT\bar{B}\cdot B) = 10\log_{10}(B) - 174\ (\text{dBm})$$

where B = bandwidth (Hz). From the above, for a 2-GHz bandwidth, we have −81 dBm of noise at the receiver.

The millimeter wave communication link budget for this example is summarized in Table 3.1. Based on this table, we can calculate the signal-to-noise ratio of the system.

Table 3.2 is an example of a cascaded millimeter wave receiver, which includes a feeding network, a bandpass filter, a low noise amplifier (LNA), a switch and channel selection filter, and an amplifier. The gain and noise figure for each component are provided and the cumulative gain and noise figure are calculated.

This cascade millimeter wave system is demonstrated below. The transmission power is assumed to be 10 dBm and the feeding network loss for the transmitting antenna is assumed to be 5 dB. We should have 15 dBm of power before the signal enters the feeding network. The transmitting antenna has a 12 dBi gain, so the signal is increased to 22 dBm. During the propagation, the signal suffers from free space loss and reflection loss, reducing it to −75 dBm. After the 12 dBi gain of the receiving antenna and the 5 dB loss from the feeding network, the signal is increased to −68 dBm. After this, the signal goes through a filter with a −1 dB loss, a low noise amplifier with a 20 dB gain, a selection filter with a −5 dB loss, and an amplifier with a 30 dB gain. The final signal power is therefore −24 dBm. The power level is plotted in Figure 3.2.

Commercial softwares such as *SysCalc* are available to calculate system gain, noise figure and link budget in a cascade microwave/millimeter-wave system [3]. More details of link budget calculation can be found in [4].

3.2 TRANSCEIVER ARCHITECTURE

Cost-effective millimeter wave solutions for high data rate transmission at 60 GHz still have to be determined. In this respect, some important selections have to be made that might be crucial to its commercial success:

- Design of a 60 GHz radio front-end architecture.
- Design of antennas (see Chapter 4).

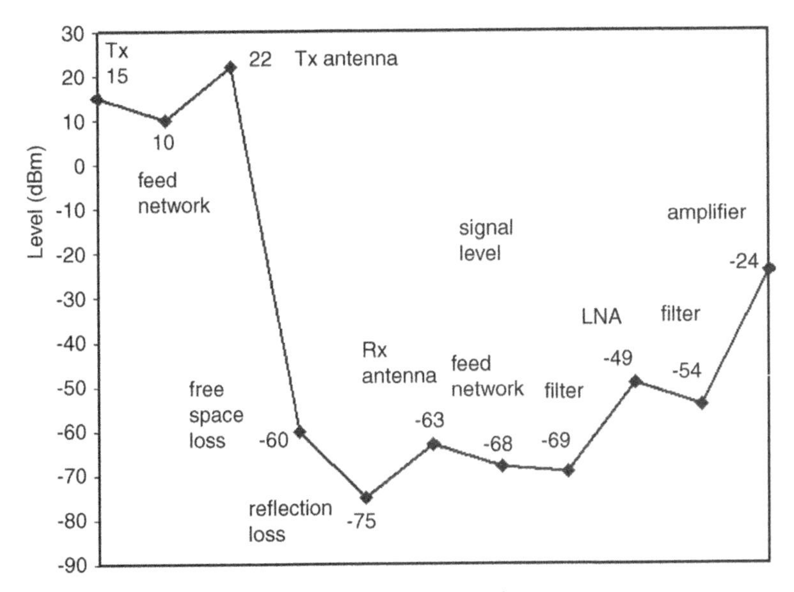

Figure 3.2. 60 GHz link budget from Tx to Rx

With respect to the choice of architecture for the front-end of a 60 GHz radio there are, in principle, four options:

1. Employing a superheterodyning architecture
2. Employing a direct conversion architecture
3. Employing a software radio architecture
4. Employing a six-port technology (Section 3.3)

3.2.1 Superheterodyning Architecture

Let us consider a simple architecture as depicted in Figure 3.3(a). This figure shows a basic 60 GHz RF front-end architecture for application at a portable station (PS). Ideally, it should be an integrated on-chip solution consisting of a receiving branch, a transmitting branch, and a frequency generation function. The receiving branch consists of the receiving antenna, a low noise amplifier, and a mixer that down converts the signal from millimeter wave range to intermediate frequency (IF) range. The transmitting branch consists of a mixer, a power amplifier (PA), and the transmitting antenna. The antennas are (integrated) patch antennas. The mixers are image rejecting mixers. They need not to be in-phase/quadrature (IQ) mixers. The IF in this example is considered to be 5 GHz, with the idea that, with appropriate modifications, an IEEE 802.11a RF chip set can serve as the IF to allow dual mode operation and interoperability. A superheterodyning architecture requires more components and more DC power, so it is not preferred for mobile devices.

A radio receiver generally includes an antenna section filter, a low noise amplifier, a down conversion mixer, an intermediate frequency stage, and a demodulator. In

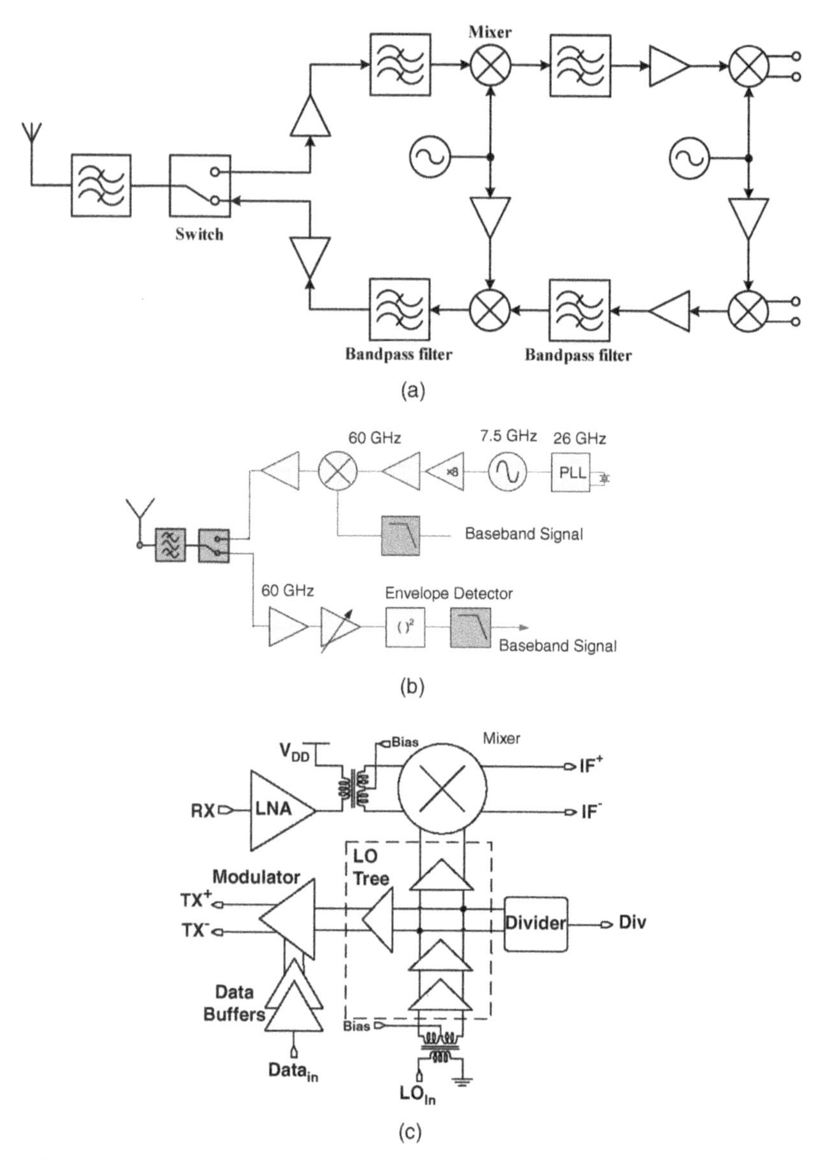

Figure 3.3. (a) Block diagram for millimeter wave/microwave superheterodyning system. (b) Block diagram for 60 GHz direct conversion system with noncoherent on/off keying (OOK) modulation. (c) Block diagram for 65 nm CMOS 60 GHz direct conversion system with coherent BPSK/QPSK modulation [11] (©2008 IEEE)

operation, the antenna section filter receives RF-modulated signals and provides them to the low noise amplifier, which amplifies the received RF signals and provides the amplified RF signals of interest to the down conversion mixer, which down converts the frequency of the RF signals to an intermediate frequency using a local oscillator. The IF stage includes one or more local oscillators, one or more mixers, and one or more adders to step-down the frequency of the intermediate frequency signals to a base-band frequency. The IF stage provides the base-band signal to the demodulator which, based on the modulation/demodulation protocol, recaptures the data.

3.2.2 Direct Conversion Architecture

The advantages of direct conversion are that it is well suited to monolithic integration, due to the lack of image filtering, and is an intrinsically simple architecture [5, 6]. The challenge of this configuration is that a 60 GHz local oscillator is required. Frequency shift keying (FSK)-modulated signals are especially well-suited to direct conversion, due to their low-signal energy at DC. However, the direct conversion receiver has not gained widespread acceptance to date, especially in high-performance wireless transceivers, due to its intrinsic sensitivity to the problems of DC offset, even harmonics of the input signal, and local oscillator (LO) leakage back to the antenna. The last problem may be considered to be the most serious. Offset arises from three sources [7]:

1. Transistor mismatch in the signal path
2. LO signal leakage to the antenna because of poor reverse isolation through the mixer and RF amplifier, which then reflects off the antenna and self-down-converts to DC through the mixer
3. A large near-channel interferer leaking into the LO part of the mixer, then self-down-converting to DC

These effects may be reduced to a certain extent through circuit design, but cannot be eliminated completely, particularly if quadrature phase shift keying (QPSK) or Gaussian minimum shift keying (GMSK) is used, since the spectra of these schemes exhibit a peak at DC. But when orthogonal frequency division multiplexing (OFDM) is applied, a solution may be chosen to avoid the use of those subcarriers that, after conversion, correspond with, or will be close to, the DC component. This is just an example of a possible solution. There exist also other solutions that exploit the characteristics of the 60 GHz physical layer.

A block diagram of conventional millimeter wave direct conversion system with noncoherent on/off keying (OOK, see Section 2.1) is shown in Figure 3.3(b). OOK enables the simplest radio architecture to realize the lowest cost and fastest time to market. It supports high rate applications (beyond Gbps) with very low power consumption due to its noncoherent architecture. It has a sufficient performance over AWGN channels. Its phase noise of local oscillator has no effects on the detection performance

The block diagram in Figure 3.3(c) shows a millimeter wave direct conversion system with binary phase shift keying (BPSK)/QPSK. It has been implemented in 65 nm CMOS technology [11]. For CMOS technology, one should note that available gain and power decrease as frequency increases.

The modulation of QPSK/BPSK can be an upper compatible system for OOK to support high-end applications. It has no major hardware impacts for BPSK transceiver to adopt OOK mode. The systems consist of transmitting and receiving paths, combined with a 60 GHz switch at the antenna side.

In Figure 3.3(b), the voltage controlled oscillator (VCO) is realized in the 7–8 GHz range. This VCO is modulated with the data stream (>1 Gbps), which will not affect the low bandwidth phase-locked loop (PLL) circuitry. The modulated signal is multiplied (8 times for the transmitting path) and filtered several times, before being transmitted or used to drive the receiver's subharmonic mixer.

To support the appropriate output power, two amplifier MMICs (monolithic microwave integrated circuits) are cascaded in series. At the same time, an LNA is added in the receiving chain to guarantee low noise figure values for the receiver. The most critical issues for the functionality of the demonstrator are the filtering networks placed behind each multiplication stage. Each filter must be specified in detail and checked against the simulation to avoid spurious emissions in the Tx and Rx-band.

The oscillator circuit could be a VCO controlled by an (off-chip) frequency synthesizer. In traditional designs, the VCO is generally implemented off-chip because providing sufficient performance would require it to occupy too much space on the chip. However, at frequencies as high as 60 GHz it becomes feasible to implement the VCO directly on the chip because the minimum dimensions to achieve the required performance become much smaller. The advantage of this approach is a reduction in the number of components that have to be mounted on an external circuit board and the avoidance of on-chip frequency multiplication circuits, thus saving space on the chip and reducing VCO performance degradation, which would arise from phase noise and frequency offsets. It is important to note that an on-chip VCO that directly generates a reference frequency close to 60 GHz may have a relatively relaxed performance when compared with the requirements of a VCO that operates on a much lower frequency in combination with a couple of frequency multipliers.

Millimeter wave transceivers can be implemented using expensive processes such as silicon germanium (SiGe), gallium nitride (GaN), and gallium arsenide (GaAs) [8]. Given that antenna size is presently limited by typical installation issues and that noise figure is essentially as good as technology allows, the next variable that requires attention is transmitter output power. A power amplifier made of GaN that operates at 70 and 80 GHz has its S parameter [9] shown in Figure 3.4.

As one looks forward, technical progress demands that millimeter wave applications in silicon be considered, particularly standard complementary metal-oxide-semiconductor (CMOS) technology. CMOS provides the most cost-effective and the highest digital integration solution even though it has high process variability, low carrier mobility constants and small device breakdown voltages. This makes millimeter wave design, especially the design of LNAs, transmit PAS, and low phase noise VCOs particularly challenging [10]. Thus, the integration of these building blocks to construct a transceiver with suitable performance remains a challenge.

A 1.2-V transceiver chip has been produced in a 65-nm CMOS technology [11]. It employs direct BPSK modulation, a 60 GHz LO distribution tree, a 64-GHz static frequency divider, and zero-IF down-conversion (see Figure 3.3(c)). This transceiver represents a compact architecture appropriate for millimeter wave communication system. The receiver has 14.7-dB gain, a low 5.6-dB noise figure, both occurring at

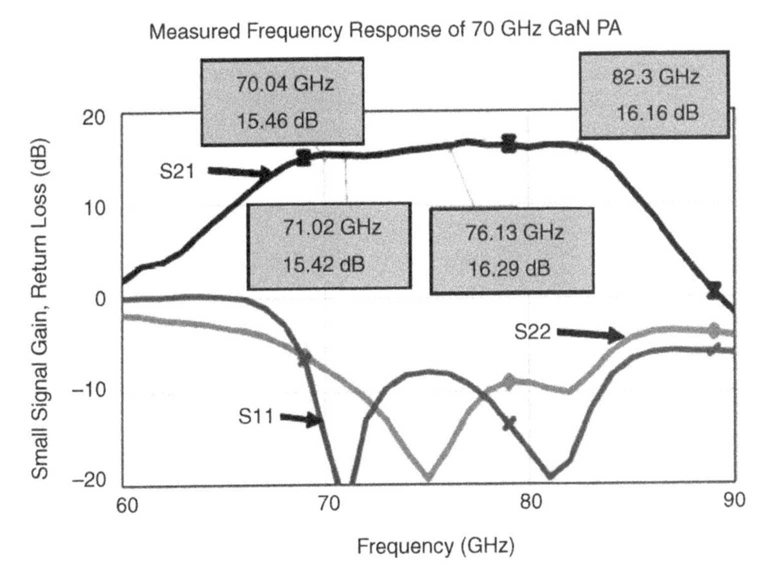

Figure 3.4. S parameter of a GaN MMIC power amplifier manufactured by HRL Laboratories [9] (© 2008 IEEE)

60 GHz. The integrated circuit consumes 374 mW (232 mW) from 1.2 V (1.0 V) and occupies $1.28 \times 0.81\ mm^2$. A feasibility study of 60 GHz radio circuits is presented with measurements of transistors, the low-noise amplifier, and the receiver on typical and fast process splits. The transceiver performance is demonstrated using a 3.5 Gbps 2-m wireless transmit-receive link over the 55–64 GHz range. Figure 3.5 shows the measured receiver gain and noise figure [11].

3.2.3 Software Radio

Employing analog-to-digital conversion (ADC) and digital-to-analog conversion (DAC) directly at the antennas would appear to make the complete RF and IF portions of the

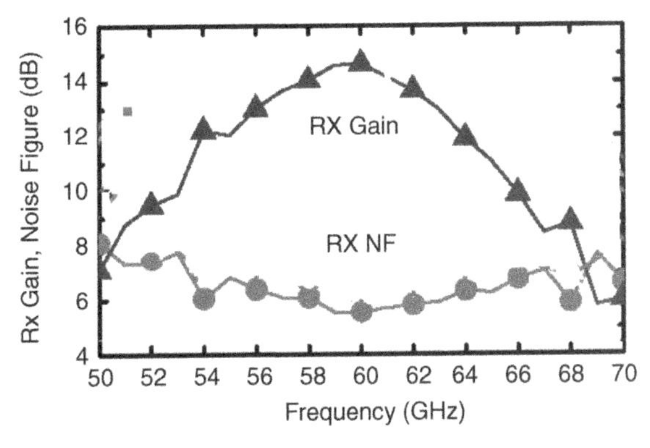

Figure 3.5. Measured receiver gain, noise figure [11] (©2008 IEEE)

transceiver chain obsolete. However, this technology is extremely challenging because it would require ultra-high speed ADC and DAC devices having a 60 GHz bandwidth. A low-cost implementation of this in the medium term is not feasible. An alternative approach, the subsampling receiver, might represent the "ultimate" solution for simple, low power, down conversion. This essentially consists of a sampling switch, clocked at a much lower frequency, and an analog-to-digital (A/D) converter. The limitations of the subsampling approach, however, demonstrate some of the inherent problems in low power receiver implementations. In a subsampling receiver, image frequencies exist at integral multiples of the sampling rate and can alias into the band of interest. As a result, careful filtering is required prior to down conversion. For example, the down conversion of an RF signal with a bandwidth of 500 MHz would require a sample rate of at least 1 GHz, assuming the use of a "brick wall filter." In practice, the sampling rate will have to be much higher, for example, at least 2 GHz, in order to minimize the finite bandwidth effects of the filter. It is questionable whether 2 GHz A/D conversion, with, let us say, 10 bit quantization, will become feasible and low cost in the medium term. In addition, the problem exists that the resulting signal-to-noise ratio of the down-sampled signal will inevitably be poorer than that of an equivalent system employing a mixer for down conversion. This is due to the noise aliased from the bands between DC and the passband [12].

3.3 TRANSCEIVER WITHOUT MIXER

No mixer transceiver refers to a six-port radio that uses no mixers in transceivers. Six-port technology (or five port technology) is a passive linear device, with two input ports and three outputs (see Figure 3.6) [13]. A phase shifter is used to adjust the phase between the RF and LO. Diode detectors are used on the output ports as frequency converters, instead of mixers. The five-port technology has been extended to be a direct digital transmitter and can be used for software-defined radio applications, since it can accommodate different wireless modulation standards without requiring hardware modification.

In 1994, a six-port radio was used, for the first time, in a six-port interferometer (SPI) radio receiver operated at millimeter wave frequencies [14]. The modifications required to transform a six-port instrument function into a useful radio function were not obvious in 1994, and it was necessary to introduce different approaches to previous six-port designs. The first reported changes involved the narrow-band single-carrier demodulation of digital data. These changes were reported in 1994/1995 publications and in subsequent publications discussing new phase spectrum modulation/demodulation schemes (PSMS/PSDS) using SPI modulators/demodulators [14, 15].

A six-port radio's modulator performs analog signal processing (vector divisions and additions) on reflected spectrum phase-modulated pulse waves using the phase spectrum of reference pulse waves. The six-port radio's demodulator does the reverse analog signal processing to directly obtain data with a decoder (analog or digital) from interferometer output signals. The changes made in the six-port radio are summarized below:

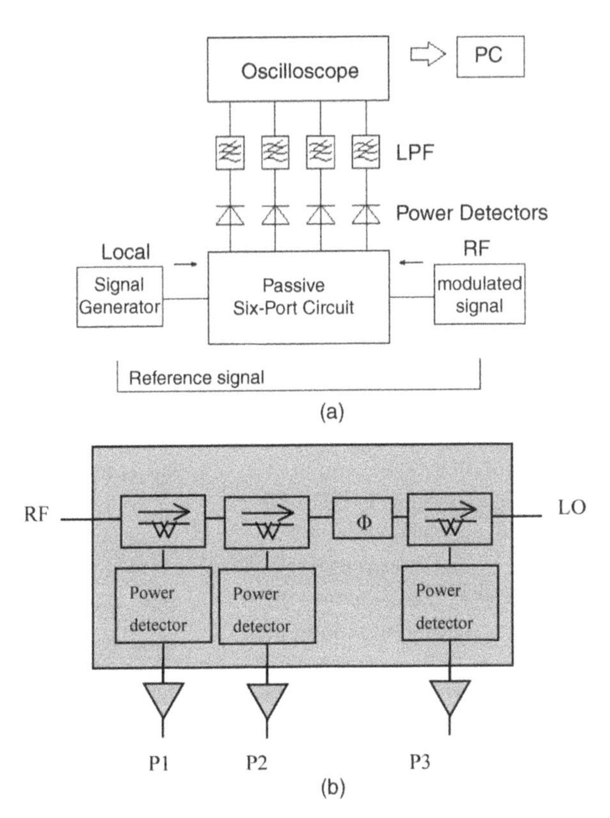

Figure 3.6. (a) Block diagram of six-port radio with four interferometer output power signals. (b) Block diagram of a five-port radio [16] (©2003 IEEE)

- The reference signal and modulated signal (or unknown signal) are fed to different input ports in the six-port interferometer radio. The modulated signal can be a single carrier or multiple carriers signal.
- A new PSMS is introduced. The PSMS phase modulates digital data on the entire phase spectrum of a monocycle pulse, on a single frequency carrier, or on multiple carriers. This modulation allows the unique six-port radio hardware to be utilized for communications.

The following features can be tested in a six-port radio for quadrature phase shift keying (QPSK)/binary phase shift keying (BPSK):

- Development of a new modulator circuit to realize PSMS
- Development of a simple, fast decoder (analog or digital) for the receiver to obtain output signals from the interferometer's demodulation of digital data transmitted by wired or wireless means

- Development of digital signal processing (DSP) algorithms for PSMS modulation schemes and PSDS demodulation schemes.
- Demonstration of wired and wireless (with phase linear antennas) digital data transmission in a laboratory environment
- Unique six-port radio (hardware and software) is used for broadband communications

The test data rate can be limited by the available rectangular pulse generator. The digital or analog decoders in the demodulator, the speed of the switch matrix in the modulator, and the speed of the DSP in the radio platform are the ultimate data rate limiting factors for digital data transmissions in the ideal propagation channel. The speed of a six-port radio can be increased by upgrading the present technology with meta-materials, fast acting switch matrices, rapid signal processing algorithms, and coding. Six-port radios can work with QPSK/BPSK modulation (see Section 2.2). On the other hand, for quadrature amplitude modulation (QAM)/multiple-phase shift keying (MPSK) data, more research is needed to develop advanced algorithms, new interferometer designs, and related software implementations in DSP [17]. Six-port radio platforms are also developed for software-defined radios (SDRs) to be used in modern communications [18, 19]. CMOS transceiver chips with SPI modulators and demodulators that utilize the 60 GHz band were fabricated for commercial applications [20].

Research results were obtained using digital QPSK/BPSK data [21]. Six-port radio research and development is ongoing, and includes QAM/MPSK data transmissions in addition to QPSK/BPSK transmissions. Six-port radios mainly use PSMS modulation. Six-port radios use interferometer functions [13, 15] to directly modulate and demodulate digital data. Other radios use nonlinear mixing (heterodyne or superheterodyne) to lower the modulated carrier frequency to an intermediate frequency or zero frequency before recovering digital data. It has been found that six-port radios using PSMS/PSDS can transmit digital data in wireless propagation channels via Tx/Rx antennas with linear phase features [22]. A six-port radio platform and linear phase antennas are sufficient for use in radio tests.

The six-port radio platform contains two six-port interferometers (one for modulation and the other for demodulation), a four-channel digital signal processor, four single-pole double-throw (SPDT) switches, two antennas (Rx/Tx), and various minor components, such as wideband, short, and open circuit terminations. Six-port radio technology provides cognitive radio hardware and software at a low cost and with wideband/narrow-band performance, along with integrated chips for QPSK/BPSK data transmissions. Promising means are also available to extend the present six-port radio to include QAM/MPSK data and other digital formats.

Six-port radios can operate in multimedia user environments with a central radio provider to serve military, commercial, and consumer software needs. The calibration process plays a key role in the six-port communication system. It is done periodically online to avoid tuning procedures and to decrease quality control costs. An efficient calibration process can improve the speed of six-port radios [23]. Six-port radios are designed to operate in wideband channels (over 500 MHz per channel), and in wideband single/multiple carrier communication systems, for example, orthogonal frequency division multiplexing (OFDM, see Section 2.5).

3.3.1 Demodulators

The six-port interferometer demodulator was mounted in a black box (Figure 3.6), with the reference signal input and a second input for the received unknown or modulated signal. The six-port radio performs linear analog vector addition and division operations on the two input waves (signals). The input signal can be either a continuous wave (CW) or pulsed, with a wide operating bandwidth determined by the pulse shape and bandwidth limitations of the linear passive components of the interferometer circuits. These circuits perform simple analog functions, such as vector division and addition on the two input waves. The results of the above linear vector operations are routed to the four output ports of the demodulation interferometer. It is easy to understand that the addition of vectors can result in one output port having a signal with a zero amplitude vector. At the same time, the remaining three output ports of the interferometer can have nonzero vector signal amplitudes. This feature of the interferometer's design is very useful for realizing rapid digital or analog QPSK/BPSK demodulation when the output ports of the demodulation interferometer are terminated with four signal amplitude or power sensors.

If the phase of the unknown input signal is changed by 90°, the interferometer design is such that the next neighbor of the four interferometer output ports has zero amplitude, while the remaining three ports have nonzero amplitudes. Thus, if the four output signals from the demodulator interferometer are simultaneously fed to the four inputs of the DSP, a simple DSP algorithm can rapidly determine the modulation state of the QPSK or BPSK data. More simply, if the four output signals are fed simultaneously to a four-channel analog amplitude comparator, the modulation state of the QPSK or BPSK data is detected very rapidly.

The amplitudes of the four RF output signals from a demodulation interferometer are easily detected with Schottky diodes or other high-speed amplitude sensors to rapidly determine the modulation state of QPSK/BPSK data. It is possible to calculate the QAM/MPSK modulation state with DSP circuitry using four nonzero output power levels, if the DSP is programmed with QAM/MPSK demodulation algorithms. Planar SPI circuits integrated with planar antennas permit the mass fabrication of low-cost and miniature radio chips [20]. The synchronization of the reference signals with the received signals presents the same challenge in the new radio as in any radio. The correct solution in each case is best determined by the given application. A number of interferometer circuit architectures are shown in Figure 3.7. The SPI radio interferometers contain simple planar passive linear components, such as power dividers, hybrid couplers, and transmission lines (f), to provide the desired vector addition of the reference and unknown waves and enable the fast decoding of BPSK/QPSK data.

Attention is continuously being given to the development of DSP algorithms for fast QAM/MPSK digital signal decoding based on existing SPI hardware. Decisions must inevitably be made on the relative merits of the different SPI architectures available to satisfy given communication system requirements. The passive circuits of Figure 3.7 are planar and use microstrip or coplanar waveguide lines. They have the advantage of

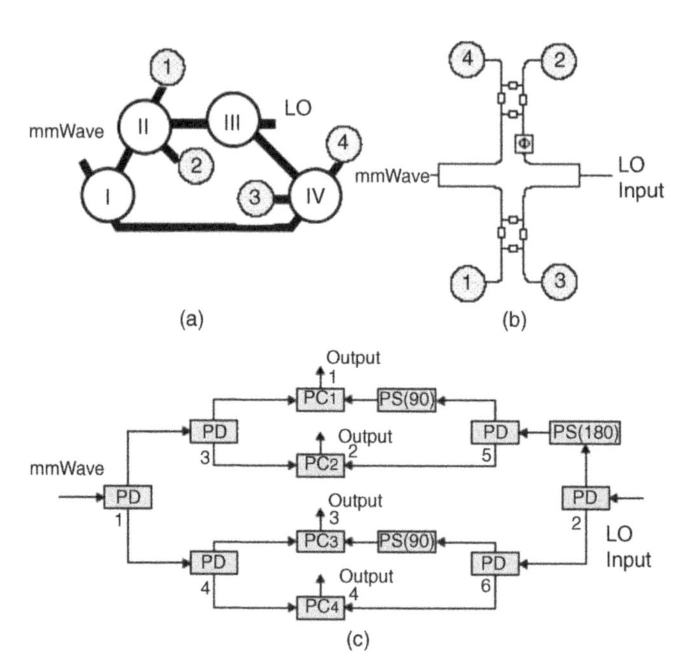

Figure 3.7. (a) Six-port radio architecture using four Rat-Race couplers [24] (©2007 IEEE) (b) Six-port radio architecture using two power dividers, two hybrid couplers, and a delay line. (c) Six-port radio architecture using power dividers (PD), power combiners (PC), and 180° and 90° delay lines [25] (©2008 IEEE)

relatively low loss and the simplicity of its design and fabrication. Figure 3.7(a) shows an example of a six-port circuit composed of four Rat-Race couplers and connected by 50 Ω microstrip transmission lines. The Rat-Race couplers are four-port devices used to either equally split an input signal or to combine two signals. An additional benefit of the hybrid ring is to alternately provide equally split but 180° phase-shifted output signals. This four-coupler configuration creates a lossless device with low VSWR (voltage standing wave ratio), excellent phase and amplitude balance, high output isolation, and matched output impedances. The low-loss airline construction also makes the device a perfect choice for combining high-power mixed signals. The incident waves, millimeter wave input, and LO signals are applied to the couplers *I* and *III*, respectively. The couplers (*I, II*) and (*II, III*) are connected by a transmission line length of 0.25 λ, the couplers (*I, IV*) by λ, and the couplers (*III, IV*) by 0.5 λ [24].

In some millimeter wave applications, it is preferable to use rectangular waveguides, substrate integrated waveguides (SIWs), or dielectric waveguides. At submillimeter wave and optical frequencies, optical integrated waveguides can be used to implement SPI radios.

We define the millimeter wave port in Figure 3.7(a), as port 5. The phase S parameters are described in Figure 3.8. The phases of S_{51} and S_{52} are equal as well as phases of S_{53} and S_{54}. Ports (1, 2) and ports (3, 4) are 90° out of phase over a very wide band, suitable for a high-quality I/Q (in-phase/quadrature-phase) mixer [24]. Six-port

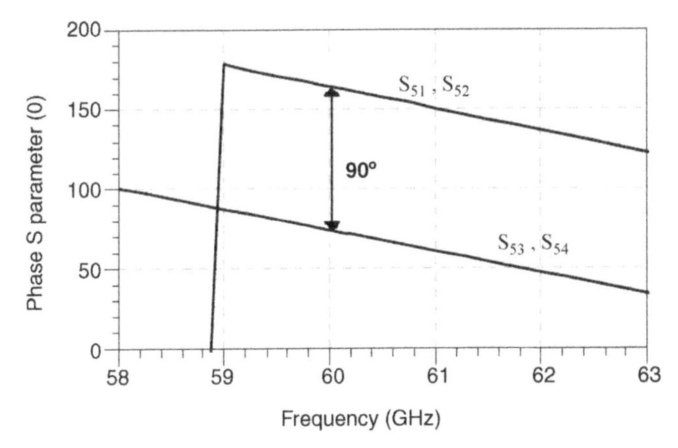

Figure 3.8. Simulation results of S parameters phase [24] (©2007 IEEE)

radio circuits can be implemented using soft substrates, alumina substrates, semiconductor substrates, and numerically controlled precision machined waveguides.

The six-port receiver has an advantage in the point that the user can select the frequency arbitrarily, but there is a challenge of nonlinear distortion in the diode detectors when the envelope detection is made. A nonlinearity compensation method can be used to reduce the distortion of system parameters [16].

3.3.2 Modulators

Figure 3.9 shows a drawing of a modulator with switches and output terminations (shorts or open circuits) for BPSK/QPSK modulation, as given in Table 3.3. The activation of the switches is done directly by DSP output signals. Terminal circuit shorts provide reflection coefficients ($\Gamma = -1$), and terminal open circuits also provide reflection coefficients ($\Gamma = +1$). For QAM/MPSK data, it is necessary to provide switch terminations with passive circuit reflection coefficients of less than unity. In such cases, the reflection coefficient Γ has an angle between zero and 360°. Lossless passive circuits, with the required reflection coefficients, can be designed as terminations, to act with fast switches and operate a modulator at the desired high rate for QPSK/BPSK data [26–28]. High data rates can be achieved with quality switches, but important choices regarding cost and performance need to be made for any given application. Figure 3.10 shows the simulated output signal constellations for two signal-to-noise ratios (SNRs). A white noise is added to the input QPSK signal and the output constellations are presented in Figure 3.10 (a) and (b) for 30 dB and 10 dB of SNR, respectively.

A comparison between six-port receiver and a homodyne receiver at 60 GHz has been made [29]. Simulation results demonstrate that the six-port architecture offers superior bit error rate (BER) performance. Also, the six-port receiver architecture is less sensitive to LO power variations and phase errors than the conventional homodyne architecture. In addition, by increasing the LO power, a larger bandwidth is obtained for both architectures, but considerably less LO power is needed for six-port receiver. Based

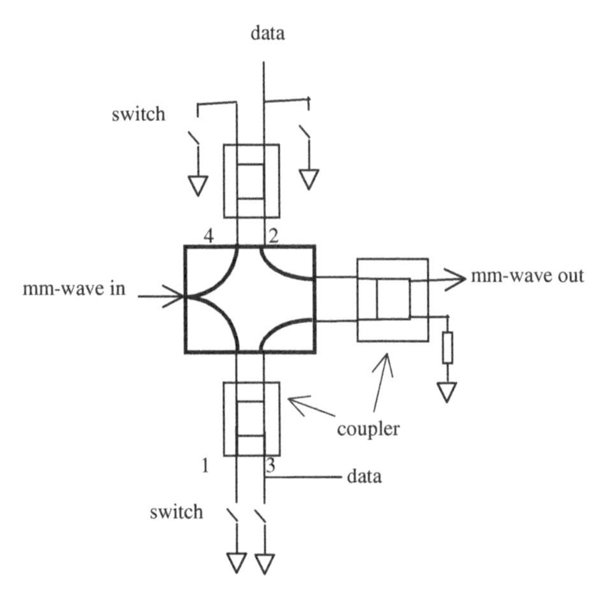

Figure 3.9. Configuration of six-port radio modulator with switches and short/open circuit terminations

on the results of [29], the six-port architecture enables the design of high-performance, compact, and low-cost wireless millimeter wave communication devices.

3.4 RECEIVER WITHOUT LOCAL OSCILLATOR

In a millimeter wave system, millimeter wave LOs are costly and consume much more DC power than other RF local oscillators. If a millimeter wave receiver can be built without using an LO, the system will save DC power and the cost for the system will be reduced. This section discusses a system configuration for frequency-shift-keying (FSK, see Section 2.3) demodulation. Such a method and apparatus include the process that begins by generating a charge signal, a data acquisition signal, and a reset signal from an I

TABLE 3.3. Switch Terminations for Six-Port Radio Operated with BPSK/QPSK Data Transmission. "OFF" is for Open and "ON" is for Short (see Figure 3.9)

Modulation State (i)	Port 1	Port 2	Port 3	Port 4	$\Delta\Psi$	I/Q
0	OFF	OFF	OFF	OFF	0	0 0
1	OFF	ON	OFF	ON	$\pi/2$	0 1
2	ON	OFF	ON	OFF	π	1 0
3	ON	ON	ON	ON	$3\pi/2$	1 1

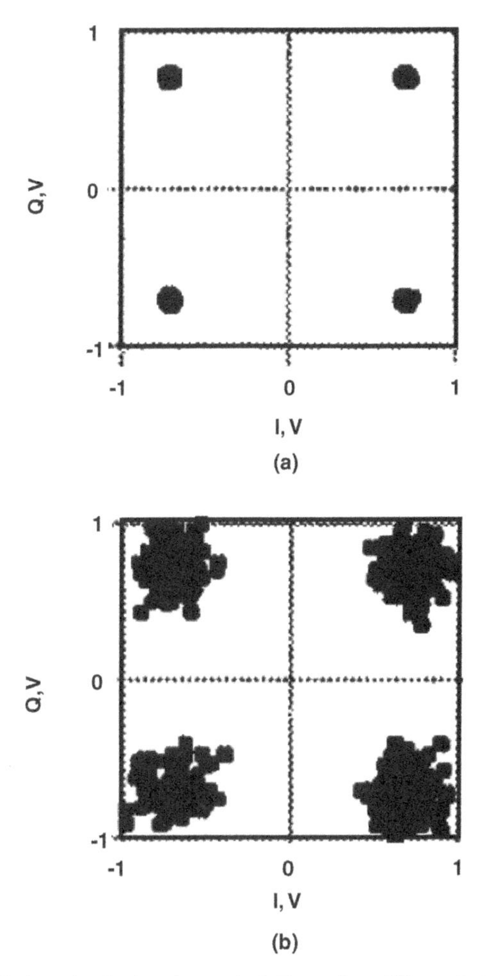

Figure 3.10. Simulated output signal constellations for different signal-to-noise ratios: (a) SNR = 30 dB, (b) SNR = 10 dB

component and a Q component of an FSK-modulated signal. This process continues by generating a delta frequency signal based on the charge signal, the data acquisition signal, and the reset signal. The delta frequency signal is representative of the frequency difference used within the FSK modulation to differentiate between a logic "*1*" and a logic "*0*". The process then demodulates the delta frequency signal to recapture a stream of data. A low-power, reliable, and low-profile FSK demodulator can be obtained.

A conventional demodulator can be implemented using analog and/or digital circuitry. While fully digital demodulators are better suited for use in an integrated circuit radio receiver than analog demodulators, due to their reduced sensitivity to noise, digital demodulators are complex, and relatively costly. Also, current digital demodulators consume a relatively large amount of power at millimeter wave frequency. For example, a demodulator that uses a local oscillator to convert signals to base-band

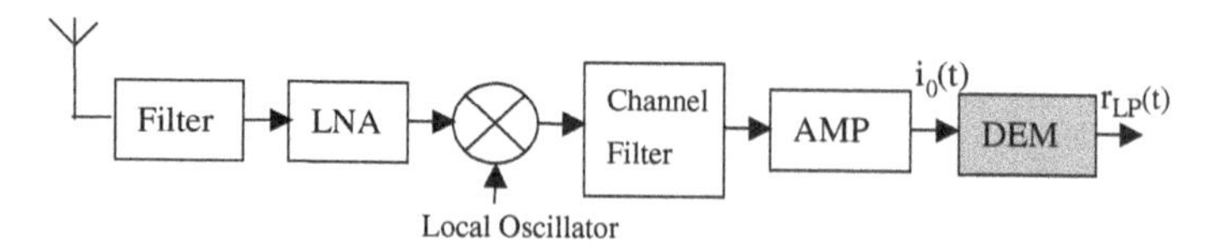

Figure 3.11. A block diagram of a conventional receiver architecture

requires up to 5 mA at 3 volts. If the demodulator processes data at the intermediate frequency, the power requirements are 3 mA at 3 volts.

Therefore, there is a demand for a mixed signal demodulator with low-power, high reliability, and reduced complexity [30].

In order to reduce the power consumption of the power amplifier on the transmitter side for wireless communication, and to realize bandwidth-efficient wireless transmission, various continuous phase FSK-type modulation schemes can be considered. With respect to the demodulation of a signal, frequency discriminators or FM-AM converters are the most popular forms of demodulators, due to their simplicity. Figures 3.11 and 3.12 show a receiver and a differential demodulator for FSK modulation, respectively.

Figure 3.11 shows a conventional receiver architecture, which includes an antenna for the reception of transmitted signals. The antenna is connected to a mixer via an RF-bandpass filter and a low-noise-amplifier (LNA). The mixer multiplies the signal received by the antenna using a signal generated by a local oscillator to down convert the signal to an intermediate frequency (IF). The IF signal travels from the mixer, via a channel selection filter and another amplifier, to a conventional FSK demodulator (DEM). The FSK demodulator receives the FSK input signal $i_O(t)$. The differential FSK demodulator demodulates the FSK input signal $i_O(t)$ and outputs a low-pass filtered signal $r_{LP}(t)$.

Figure 3.12 shows the internal structure of a differential FSK demodulator. It consists of a phase shifter, a mixer, and a low-pass filter. The input signal $i_O(t)$ is supplied to both the phase shifter and the mixer. The phase shifter shifts the phase of the input signal $i_O(t)$ by a predetermined number of degrees and outputs a shifted signal $i(t)$ to the mixer. The mixer multiplies the input signal $i_O(t)$ by the shifted signal $i(t)$ and outputs a mixed signal $r(t)$ to the low-pass filter.

The low-pass filter filters the mixed signal $r(t)$ and outputs a low-pass filtered signal $r_{LP}(t)$.

The FSK input signal $i_O(t)$ of the differential FSK demodulator can be written as

$$i_O(t) = s(t) + n(t) = A\cos[2\pi f_i t + 2\pi h \int_{-\infty}^{t} m(\tau)d\tau] + n(t) \tag{3.1}$$

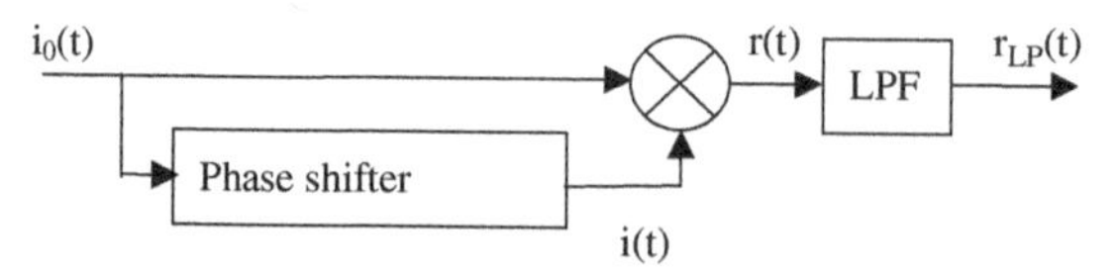

Figure 3.12. A block diagram of a differential frequency-shift-keying demodulator

where A is the amplitude, f_i is the carrier frequency, $m(\tau)$ is the filtered data signal, $n(t)$ is an additive white Gaussian noise (AWGN), and h is the modulation index of the signal $i_O(t)$.

The modulation index h of the signal $i_O(t)$ is defined as $h = 2\,f_d T_s$, where f_d is the frequency deviation and T_s is the symbol duration.

The function of the differential FSK demodulator is to provide an output signal $r_{LP}(t)$ that is proportional to the instantaneous frequency of the input signal $i_O(t)$.

The frequency response of the phase shifter of the FSK demodulator is

$$\varphi(f) = -\pi/2 + 2\pi K(f - f_i) \cdot (K = 1, 2, 3 \ldots)$$

Thus, the output $i(t)$ of the phase shifter is

$$\begin{aligned} i(t) &= s_d(t) + n_2(t) \\ &= A\cos\left[2\pi f_i t + 2\pi h \int_{-\infty}^{t} m(\tau)d\tau - \pi/2 + 2\pi K h m(t) + n_1(t)\right] + n_2(t) \end{aligned} \quad (3.2)$$

where $n_1(t)$ and $n_2(t)$ are the noise terms generated by the phase shifter due to the coexistence of $n(t)$ in the input signal $i_O(t)$.

The low-pass filtered signal output $r_{LP}(t)$ by the low-pass filter after eliminating the double frequency terms is

$$r_{LP}(t) = \left(\frac{A^2}{2}\sin[2\pi K h m(t) + n_1(t)] + [s(t) \times n_2(t) + s_d(t) \times n(t) + n(t) \times n_2(t)]\right)\Bigg|_{LP} \quad (3.3)$$

When the term $2\pi K h m$(t) is small and the noise terms are negligible, the resulting low-pass filtered signal $r_{LP}(t)$ is

$$r_{LP}(t) \approx \frac{A^2}{2} \cdot 2\pi K h m(t) \quad (3.4)$$

Thus, the transmitted data contained in the input signal $i_O(t)$ can be recovered correctly. Corresponding differential FSK demodulators are described in [31]. Another FSK demodulator based on a delay-locked loop with a digital frequency offset canceller is explained in [32].

One disadvantage of the differential frequency-shift-keying demodulators described above is that an active mixer is required to achieve sufficient signal strength in the FSK signal. The active mixer requires a local oscillator signal that consumes much power for millimeter wave frequency (e.g., 60 GHz) applications. Furthermore, mixers working at millimeter wave frequencies are very complicated and thus expensive.

Therefore, this section discusses an FSK demodulator and a demodulation method that has reasonable performance even at millimeter wave frequencies. Furthermore, the demodulator should be easy to implement using passive and cheap components.

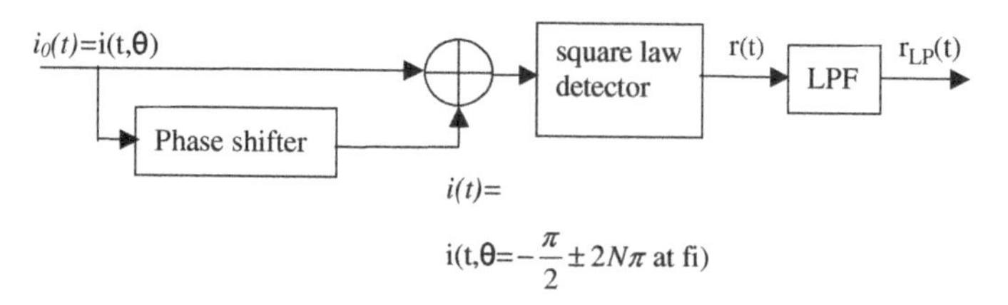

Figure 3.13. A block diagram of a frequency-shift-keying demodulator

Figure 3.13 shows a block diagram of an FSK demodulator. The FSK demodulator comprises a phase shifter, an adder, a square law detector, and a low-pass filter.

An input signal $i_0(t) = s(t) + n(t) = A\cos[2\pi f_i t + 2\pi h \int_{-\infty}^{t} m(\tau)d\tau] + n(t)$ is supplied to both the phase shifter and the adder. The phase shifter shifts the phase of the input signal $i_O(t)$ by a predetermined phase and outputs a shifted signal $i\,(t)$ to the adder. The adder adds the input signal $i_O(t)$ and the shifted signal $i(t)$ and outputs the resulting signal to the square law detector. The square law detector receives this resulting signal output from the adder and outputs a squared signal $r(t)$. The squared signal $r(t)$ is the square of the resulting signal output by the adder. Thus, the adder and the square law detector constitute a combining unit for combining the input signal $i_O(t)$ and the shifted signal $i(t)$ generated by the phase shifter to produce a corresponding signal $r(t)$. In the present configuration, the square law detector can be realized by using a diode, so it is a very cheap and simple passive element.

The squared signal $r(t)$ generated by the square law detector is provided to the low-pass filter, which filters the signal $r(t)$ to produce a low-pass filtered signal $r_{LP}(t)$. To allow for suitable filtering, the bandwidth of the low-pass filter is matched with the bandwidth of the data signal contained in the input signal $i_O(t)$.

The output signal $r_{LP}(t)$ is

$$r_{LP}(t) = [s(t) + n(t) + s_d(t) + n_2(t)]^2|_{LP} = \alpha + \beta + \gamma \tag{3.5}$$

where

$$\alpha = 2 \cdot \left(\frac{A^2}{2} \sin[2\pi Khm(t) + n_1(t)] + [s(t) \times n_2(t) + s_d(t) \times n(t) + n(t) \times n_2(t)] \right)\bigg|_{LP}$$

$$\beta = [s^2(t) + s_d^2(t)]|_{LP}$$

$$\gamma = [2 \cdot s(t) \times n(t) + 2 \cdot s_d(t) \times n_2(t) + n^2(t) + n_2^2(t)]|_{LP}$$

- α corresponds to the output of a conventional differential FSK demodulator.
- β indicates that after low-pass filtering a DC offset is introduced, since $i_O(t)$ and $i(t)$ are constant envelope FSK-modulated signals.
- γ is the enhanced noise (comprised in the output signal $r_{LP}(t)$).

To maintain its sensitivity, the differential FSK demodulator of Figure 3.13 uses the basic principle of frequency demodulation but no mixer or multiplier are required for differential detection. This novel differential FSK demodulator can be used directly in the RF portion of a receiver architecture. Thus, no signal conversion from radio frequency to baseband is required. As a consequence, the local oscillator and corresponding mixer of the receiver architecture can be eliminated, simplifying the entire receiver architecture.

Referring to Figure 3.13, after channel selection filtering there is a small amplitude fluctuation for $i_O(t)$ and thus a small AM modulation effect exists due to itemβ. The spectrum of the fluctuation signal cannot be eliminated because it is within the bandwidth of the modulated data.

To solve the issue, the simple FSK demodulator of Figure 3.13 is replaced by a balanced configuration, which is adopted to cancel the corresponding DC offset and to eliminate the corresponding enhanced noise.

Figure 3.14 shows a block diagram of a balanced FSK demodulator. This demodulator can be made using purely passive components. Its power consumption and complexity are very low. The FSK demodulator comprises three adders, two square law detectors, a phase shifter, and a low-pass filter. An input signal $i_O(t)$ is supplied to the phase shifter and the adders A and B.

The phase shifter shifts the phase of the input signal $i_O(t)$ by $-\frac{\pi}{2} \pm 2N\pi$, where N is an integer, at the carrier frequency of the input signal $i_O(t)$, and outputs a shifted signal $i(t)$ to the adders A and B. In the present configuration, the phase shifter operates as a delay line.

The adder A adds the input signal $i_O(t)$ and the shifted signal $i(t)$ generated by the phase shifter and outputs the resulting signal to the first square law detector. By squaring the output signal from the adder A, the first square law detector produces the first squared signal $r_1(t)$ to the adder C.

By contrast, the adder B subtracts the input signal $i_O(t)$ from the shifted signal $i(t)$ generated by the phase shifter and outputs the resulting signal to the second square law

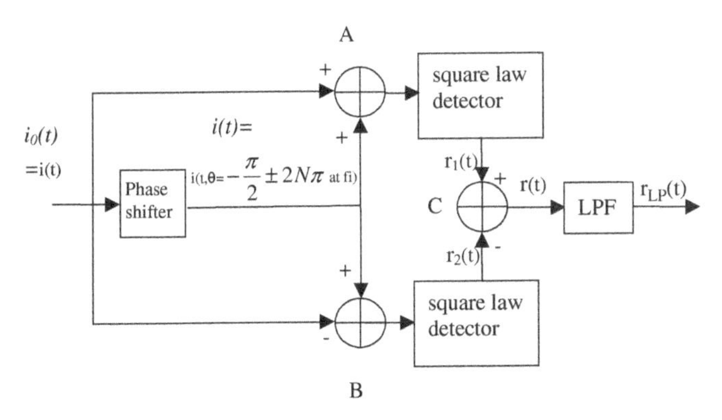

Figure 3.14. A block diagram of a balanced frequency-shift-keying demodulator

detector. Similarly, by squaring the output signal from the adder B, the second square law detector produces the second squared signal $r_2(t)$ to the adder C.

The adder C subtracts the second squared signal $r_2(t)$ from the first squared signal $r_1(t)$, which are received from the second and first square law detectors, respectively, and passes the combined signal $r(t)$ to the low-pass filter.

The low-pass filter filters the signal $r(t)$ received from the adder C to produce the low-pass filtered signal r_{LP}(t).

The novel FSK demodulator described in Figure 3.14 constitutes a balanced architecture. This balanced architecture is used to eliminate the second item β and the third item γ, contained in the output of the FSK demodulator of Figure 3.13, that degrade the demodulation performance.

From the above described embodiment it is clear that

$$r_1(t)|_{LP} = [s(t) + n(t) + s_d(t) + n_2(t)])^2|_{LP} = \alpha + \beta + \gamma \tag{3.6}$$

and

$$r_2(t)|_{LP} = [-s(t) - n(t) + s_d(t) + n_2(t)])^2|_{LP} = -\alpha + \beta + \gamma \tag{3.7}$$

Consequently, the low-pass filtered signal r_{LP}(t) produced by the low-pass filter is

$$r_{LP}(t) = r_1(t)|_{LP} - r_2(t)|_{LP} = 2 \cdot \alpha \tag{3.8}$$

Thus,

$$r_{LP}(t) = 4\left(\frac{A^2}{2}\sin[2\pi Khm(t) + n_1(t)] + [s(t) \times n_2(t) + s_d(t) \times n(t) + n(t) \times n_2(t)]\right)\Big|_{LP} \tag{3.9}$$

When the term $2\pi Khm(t)$ is small and the noise terms are negligible, this low-pass filtered signal $r_{LP}(t)$ is

$$r_{LP}(t) \approx 4\frac{A^2}{2}2\pi Khm(t) = 2 \cdot A^2 \cdot 2\pi Khm(t) \tag{3.10}$$

Thus, the transmitted data contained in the input signal $i_0(t)$ can be recovered correctly by using this novel FSK demodulator.

It can be concluded that the novel FSK demodulator in Figure 3.14 has the same performance as the conventional differential FSK demodulator.

The above FSK demodulator can be applied to any constant envelope modulation scheme, such as Gaussian frequency-shift-keying (GFSK), minimum shift-keying (MSK), shaped offset quadrature phase-shift-keying (SOQPSK), or Gaussian minimum shift-keying (GMSK).

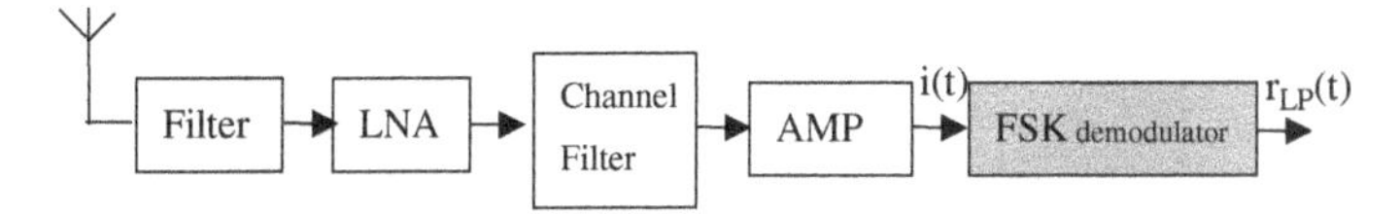

Figure 3.15. A block diagram of a direct conversion receiver using the novel frequency-shift-keying demodulator

In Figure 3.15, the direct conversion receiver comprises an antenna, an RF-bandpass filter, an LNA, a channel selection filter, an amplifier, and an FSK demodulator. The FSK demodulator of Figure 3.13 and Figure 3.14 can be used in this system.

Here, the FSK demodulator is used in the RF range. The LO and the corresponding mixer of the conventional receiver architecture shown in Figure 3.11 are eliminated and the entire architecture of the receiver becomes simpler and consumes less power. Nevertheless, the novel FSK demodulators can also be used in any conventional receiver architecture (such as the architecture shown in Figure 3.11).

In conclusion, it is possible to build an FSK demodulator at 60 GHz without using a complicated mixing/multiplying process. The demodulator has the following features:

1. A simple adder and a square law detector are used in an FSK demodulator to replace a mixer.
2. For a differential FSK demodulator, the basic principle of frequency discrimination is used instead of a mixer or multiplier for differential detection.
3. Only passive components are used in the FSK demodulator. The power consumption and complexity of the demodulator can be reduced consequently.
4. The structure of a differential FSK demodulator can be used in the RF portion of a 60 GHz receiver architecture. Thus, no signal conversion from intermediate frequency to baseband is required. In consequence, the local oscillator and corresponding mixer of the receiver architecture can be eliminated, simplifying the entire architecture of the receiver.

3.5 MILLIMETER WAVE CALIBRATION

There are several millimeter wave design challenges, such as cross-talk through substrate and chip interconnect issues. To characterize those issues, calibration is a significant step prior to millimeter wave measurement. There are numerous calibration methods available. Wafer probes are commonly used together with a vector network analyzer to measure S-parameters for millimeter wave devices and antennas. In this section we review three methods for two-port wafer-probe calibrations: SOLT, LRM/LRRM, and TRL. After calibration, the measurement reference plane is exactly at the probe tip (see Figure 3.16).

The SOLT (short, open, load, and thru) method is the most frequently used calibration where users must correctly define the reference planes and parasitic inductances and capacitances for the short, open, and thru. At higher frequencies, inaccuracies

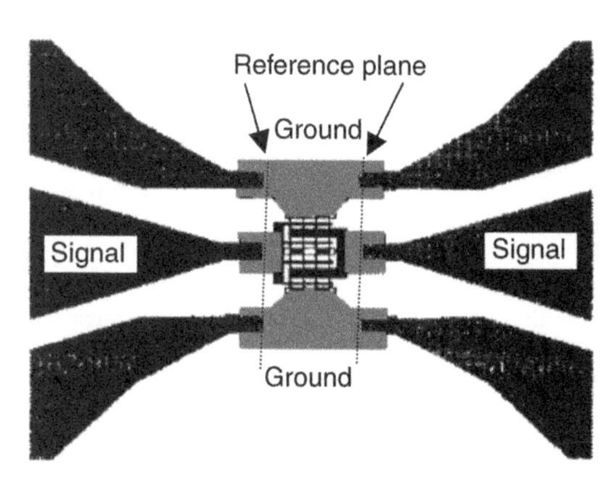

Figure 3.16. Two ground-signal-ground probes and reference planes after calibration

in the parasitic descriptions greatly impact the accuracy of subsequent measurements. After calibration, users should note that the open may not have perfectly zero return loss. The probe in air will actually have slightly less loss than the probe in contact with the wafer. Thus, a small amount of return gain from the calibrated open may normally be observed. In SOLT calibration, the placement of probes can affect the measurement result (Figure 3.17).

The LRM (line-reflect-match) method uses a line standard, a reflect standard, and two identical matched load standards. With an LRM calibration, the matched load determines the reference impedance. An advanced method is called LRRM (line-reflect-reflect-match) (see Figure 3.18). LRRM is a broadband calibration. The two reflects are undefined shorts and undefined opens. Users need to measure the match on only one of the two ports. With LRRM, you need these calibration standards: a line (or thru) standard, two different reflect standards, which do not require characterization, and a match

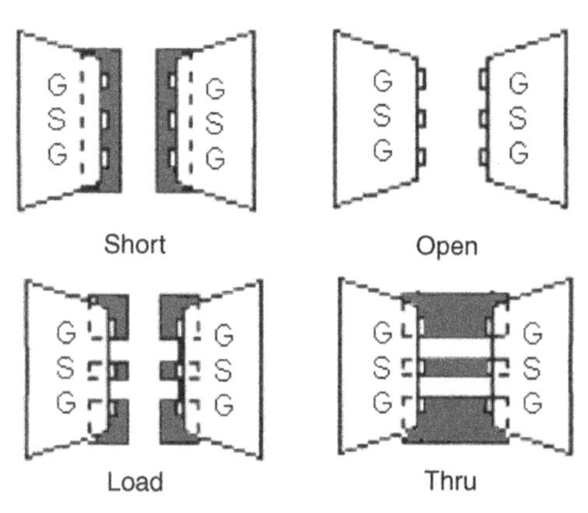

Figure 3.17. SOLT calibration

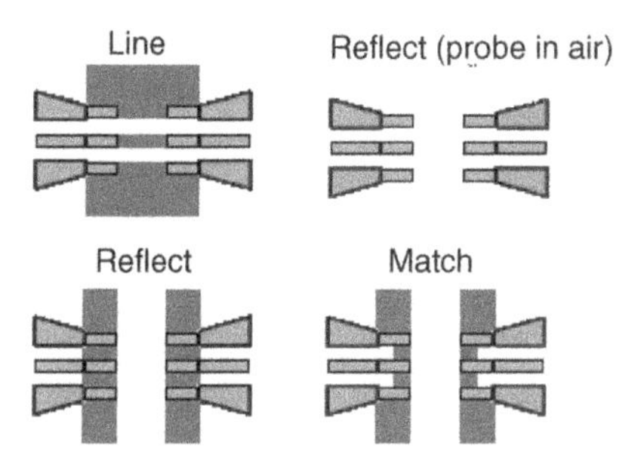

Figure 3.18. LRRM calibration

(or load) standard. Match acts as infinitely high loss line. The LRRM method has one key advantage over the LRM method: it avoids inconsistency between the load reactances seen by the port-1 and port-2 probes.

The TRL (thru-reflect-line) method uses thru, reflect, and one or more delay lines (see Figure 3.19). The TRL method uses the characteristic impedance of the delay lines to set the reference impedance (Z). TRL requires multiple probe spacing and is not suitable for fixed spacing probes. More details of calibration process can be found in [33].

As mentioned before, the measurement reference plane after calibration is at the probe tip. The measured result is the response of the device and the parasitics associated with the pads. Thus, de-embedding from OPEN and SHORT helps eliminate residual calibration errors.

If the measurement and calibration are made using connectors, users can employ 1.85 mm connectors (also called "V connectors") for 60 GHz measurement and 1 mm connectors for measurement up to 110 GHz. The calibration principle for connectors is the same as that for probes.

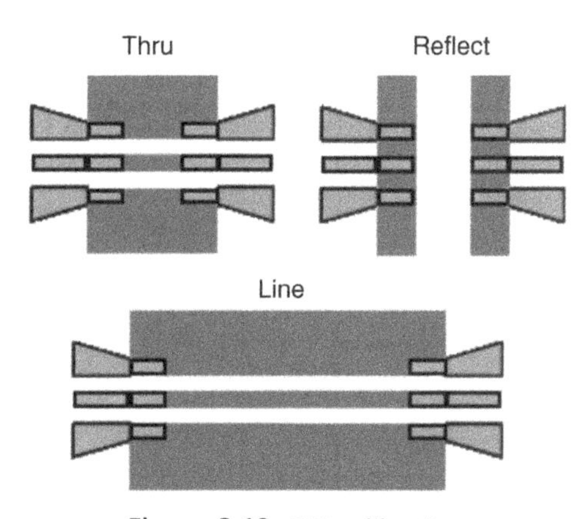

Figure 3.19. TRL calibration

3.6 RESEARCH TREND: TRANSCEIVER SILICONIZATION

Silicon-based transceivers provide a low-cost solution for millimeter wave communication systems. In 2006, the research team at IBM demonstrated two-chip transceiver front-ends in a SiGe BiCMOS (0.13-μm) technology [34]. A dual-conversion superheterodyne radio architecture was selected over a homodyne approach due to its lower carrier feed-through in the transmitter and better I/Q quadrature accuracy. The receiver power conversion gain is 38–40 dB and the noise figure is 5–6.7 dB. The image rejection is 30–40 dB, the third-order input-referred intercept point IIP3 is −30 dBm, and input P_{1dB} is −36 dBm. The transmitter conversion gain is 42–36 dB. P_{1dB} is 110–112 dBm while Psat is 116–117 dBm and the conversion gain is 34–37 dB. At P_{1dB}, the transmitter consumes 190 mA from 2.7 V and 72 mA from 4 V (power amplifier).

The millimeter wave transceivers can also be implemented in CMOS [35–37]. Since 2007, SiBEAM has demonstrated and delivered the first 60 GHz all-CMOS phased array system with 4 Gbps in 10-m non-LOS environments [38]. The fully integrated chip sets enable the first WirelessHD compliant systems in the market, supporting uncompressed and lossless wireless audio/video streaming. The twochip sets achieve complete antennas-to-bits integration, requiring minimal additional support hardware for integration into existing consumer electronics equipment (Blu-ray and HD DVD players and HD set-top boxes and displays). The radio chip's packaging includes integrated 60 GHz antennas, simplifying board and system design by containing all high-frequency routing within the CMOS die and chip package. Achieving full 10-m NLOS coverage at the required data rates requires many independent antennas and partial radio chains. Thus, this fully integrated radio chip, containing all radio chains and antennas, represents a significant advance in the degree of parallelism achieved at the RF level for high-volume consumer wireless communications products.

In 2009, a low-power 60 GHz integrated transceiver was presented. The single chip includes RF, LO, PLL, and baseband signal paths. The transceiver has been implemented in a standard 90-nm CMOS process and includes specially designed electrostatic-discharge protection on all millimeter wave pads. With a 1.2 V supply, the chip consumes 170 mW while transmitting 10 dBm and 138 mW while receiving. Data transmission has been measured up to 5 Gbps on each of *I* and *Q* channels. Data reception over a 1-m wireless link is at 4 Gbps QPSK with less than 10^{-11} BER [39, 40].

REFERENCES

[1] IEEE 802.15-05-0493-27-003c, "TG3c selection criteria." Jan. 2007.

[2] R. C. Hansen, *Phased Array Antennas*, Hoboken, NJ: Wiley Interscience, 1998.

[3] *SysCalc*, Arden Technologies, Inc., Goode, VA, USA.

[4] W. F. Egan, *Practical RF System Design*, Hoboken, NJ: Wiley-IEEE Press, 2003.

[5] A. Abidi, "Direct conversion radio transceivers for digital communications." *IEEE JSSSC*, Vol. 30, No. 12, pp. 1399–1410, 1995.

[6] F. Aschwanden, "Direct conversion – how to make it work in TV tuners." *IEEE Trans. Consum. Electron.* Vol. 42, No. 3, pp. 729–751, 1996.

[7] J. Wenin, "Ics for digital cellular communication." *European Solid State Circuits Conf.*, Ulm, Germany, pp. 1–10, 1994.

[8] M. Siddiqui, M. Quijije, A. Lawrence, B. Pitman, R. Katz, P. Tran, A. Chau, D. Davison, S. Din, R. Lai, and D. Streit, "GaAs components for 60GHz wireless communication applications." *2002 GaAsMANTECH Conference*, GaAsMAN- TECH, pp. 1–4, 2002.

[9] M. Micovic, A. Kurdoghlian, H. Moyer, P. Hashimoto, M. Hu, M. Antcliffe, P. Williadsen, W. Wong, R. Bowen, I. Milosavljevic, Y. Yoon, A. Schmotz, M. Werzel, C. McGuire, B. Hughes, and D. Chow, "GaN MMIC PA for e band radio." *Proc. IEEE Compound Semiconductor Integrated Circuit Symp. (CSISC)*, pp. 10–13, 2008.

[10] E. Skafidas, F. Zhang, B. Yang, B.N. Wicks, Z. Liu, C. M. Ta, Y. Mo, K. Wang, G. Felic, P. Nadagouda, T. Walsh, W. Shieh, I. Mareels, and R. J. Evans, "A 60GHz transceiver on CMOS." *International Topical Meeting on Microwave Photonics*, 2008, pp. 306–309. Sep. 2008.

[11] A. Tomkins, R. A. Aroca, T. Yamamoto, S. T. Nicolson, Y. Doi, and S. P. Voinigescu, "A zero-IF 60GHz transceiver in 65nm CMOS with > 3.5Gb/s links." *IEEE Custom Integrated Circuits Conference, 2008* CICC 2008, pp. 471–474, Sep. 2008.

[12] R. G. Vaughan, N. Scott, and D. White, "The theory of bandpass sampling." *IEEE Trans. Signal Process.*, Vol. 39, No. 9, pp. 1973–1984, 1991.

[13] Y. Zhao, C. Viereck, J. F. Frigon, R. G. Bosisio, and K. Wu, "Direct quadrature phase shift keying modulator using sixport technology." *IEE Electron. Lett.*, Vol. 41, No. 21, pp. 1180–1181, 2005.

[14] J. Li, R. G. Bosisio, and K. Wu, "A six-port direct digital millimeterwave receiver." *IEEE MTT-S Int. Microwave Symp. Dig.*, San Diego, CA, pp. 1659–1662, May 1994.

[15] J. Li, R. G. Bosisio and K. Wu, "Computer and measurement simulation of a new digital receiver operating directly at millimeterwave frequencies." *IEEE Trans. Microw. Theory Tech.*, Vol. 43, No. 12, pp. 2766–2772, 1995.

[16] A. Honda, K. Sakaguchi, J. I. Takada, and K. Araki, "A study of nonlinearity calibration for six-port direct conversion receivers." *IEEE Topical Conference on Wireless Communication Technology*, 2003. pp. 40–41, Oct. 2003.

[17] X. Y. Xu, R.G. Bosisio, and K. Wu, "Analysis and design implementation of six-port software defined radio." *IEEE Trans. Microw. Theory Tech.*, Vol. 54, No. 7, pp. 2937–2943, 2006.

[18] R. Shrum, "SDR regulatory issues." *Software Defined Radio Forum, IMS 2005 Workshop*, Long Beach, CA, Jun. 2005.

[19] J. F. Luy, "Software configurable receiver." *Proc. 32nd European Microwave Conf.*, Milan, Italy, pp. 1–8, Oct. 2002.

[20] C. H. Wang, H. Y. Chang, P. S. Wu, K. Y. Lin, T. W. Huang, H. Wang, and C. H. Chen, "A 60GHz low-power six-port transceiver for gigabit software-defined transceiver applications." *Proc. IEEE Int. Solid-State Circuits Conf.*, San Francisco, pp. 192–596, Feb. 2007.

[21] Y. Y. Zhao, J. F. Frigon, K. Wu, and R.G. Bosisio, "Multi six-port impulse radio for ultra-wideband." *IEEE Trans. Microw. Theory Tech.*, Vol. 54, No. 4, pp. 1707–1712, 2006.

[22] S. Y. Suh, W. Stutzman, W. Davis, A. Waltho, and J. Schiffer, "A novel CPW-fed disc antenna." *Proc. IEEE Antennas Propagation Soc. Int. Symp.*, pp. 2919–2922, Jun. 2004.

[23] Z. C. Wang, M. Ratni, D. Krupezevic, and J.-U. Juergensen,"Adaptive autocalibration method for the quadrature demodulator of a five-port receiver Document Type and Number." European patent EP1398872B1, Feb. 2008.

[24] N. K. Mallat and S. O. Tatu, "Six-port receiver in millimeter-wave systems." *IEEE International Conference on Systems, Man and Cybernetics, 2007* ISIC, pp. 2693–2697, Oct. 2007.

[25] R. G. Bosisio, Y. Y. Zhao, X. Y. Xu, S. Abielmona, E. Moldovan, Y. S. Xu, M. Bozzi, S. O. Tatu, C. Nerguizian, J. F. Frigon, C. Caloz, and K. Wu, "New-wave radio." *IEEE Microw. Mag.*, Vol. 9, No. 1, pp. 89–100, 2008.

[26] Z. Y. Song, T. Serioja, M. R. Soleymani, K. Wu, and R. G. Bosisio, "Initial coding results with a millimeter-wave six-port QPSK receiver." *Microw. Optical Technol. Lett.*, Vol. 36, No. 6, pp. 465–467, 2003.

[27] J. Li, R. G. Bosisio, and K. Wu, "A six-port direct digital millimeter wave receiver." *IEEE Microwave Symposium Digest*, Vol. 3, pp. 1659–1662, May 1994.

[28] S. O. Tatu, E. Moldovan, K. Wu, and R. G. Bosisio, "A new direct millimeter- wave six-port receiver." *IEEE Trans. Microw. Theory Tech.*, Vol. 49, No. 12, pp. 2517–2522, 2001.

[29] N. K. Mallat, E. Moldovan, and S. O. Tatu, "Comparative demodulation results for six-port and conventional 60GHz direct conversion receivers." *Progress In Electromagnetics Research, PIER 84*, 2008, pp. 437–449, 2008.

[30] Z. C. Wang and U. Masahiro,"Frequency-shift-keying demodulator and method of frequency-shift-keying." European patent EP1643704B1, Oct. 2007.

[31] H. Komurasaki, T. Sano, T. Heima, K. Yamamoto, H. Wakada, I. Yasui, M. Ono, T. Miwa, H. Sato, T. Miki, and N. Kato, "A 1.8-V operation RF CMOS transceiver for 2.4-GHz-band GFSK applications." *IEEE J. Solid-State Circuits*, Vol. 38, No. 5, pp. 817–825, 2003.

[32] S. Byun, C.-H. Park, Y. Song, S. Wang, C. S. G. Conroy, and B. Kim, "A low-power CMOS Bluetooth RF transceiver with a digital offset cancelling DLL-based GFSK demodulator." *IEEE J. Solid-State Circuits*. Vol. 38, No. 10, pp. 1609–1618, 2003.

[33] Agilent application note AN 1287-9 "In-fixture measurements using vector network analyzers." Agilent Technologies, Inc.

[34] S. K. Reynolds, B. A. Floyd, U. R. Pfeiffer, T. Beukema, J. Grzyb, C. Haymes, B. Gaucher, and M. Soyuer, "A silicon 60GHz receiver and transmitter chipset for broadband communications." *IEEE J. Solid-State Circuits.*, Vol. 41, no. 12, pp. 2820–2831, 2006.

[35] S. Pinel, S. Sarkar, P. Sen, B. Perumana, D. Yeh, D. Dawn, and J. Laskar,"60GHz single chip 90 CMOS radio." *ISSCC Dig. Tech. Papers*, Feb. 2008.

[36] J. Lee, Y. Huang, Y. Chen, H. Lu, and C. Chang,"A low-power fully integrated 60GHz transceiver system with OOK modulation and on-board antenna assembly." *ISSCC Dig. Tech. Papers*, pp. 316–317. Feb. 2009.

[37] C.-H. Wang, H.-Y. Chang, P.-S. Wu, K.-Y. Lin, T.-W. Huang, H. Wang, and C.-H. Chen," A 60GHz low power six-port transceiver for gigabit software-defined transceiver." *ISSCC Dig. Tech. Papers*, Feb. 2007.

[38] J. M. Gilbert, C. H. Doan, S. Emami, and C. B. Shung, "A 4-Gbps uncompressed wireless HD A/V transceiver chipset." *IEEE Micro,* Vol. 28, No. 2, pp. 56–64, 2008.

[39] C. Marcu, D. Chowdhury, C. Thakkar, et al. "A 90 nm CMOS low-power 60GHz transceiver with integrated baseband circuitry." *IEEE J. Solid-State Circuits.*, Vol. 44, No. 12, pp. 3434–3447, 2009.

[40] A. M. Niknejad, "Siliconization of 60GHz." *IEEE Microw. Mag.*, Vol. 11, No. 1, pp. 78–85, 2010.

4

MILLIMETER WAVE ANTENNAS

In this chapter, we present an overview of a millimeter wave antenna technology approach that offers significant promise of enabling gigabit/per second wireless communication rates. We will describe and discuss the requirements for the system components of a (nominally) 60 GHz free space point-to-point communications system. We also consider the fact that such a technology will have user-specific constraints (such as size and bulk) and that these will dictate or steer the direction of the design philosophy.

Several main constraints are considered for millimeter wave antenna design. The first constraint is that the 60 GHz channel is lossy (due to oxygen absorption), but is otherwise benign. The excess loss at 60 GHz is approximately 15 dB/km, and we need to provide a means to overcome oxygen absorption and ensure that sufficient margin exists to overcome other losses, such as rain-induced fading. Here we can compensate by increasing the transmitter or receiver antenna gain. We can, for example, use a directional antenna gain to substitute for the raw transmitter power and receiver noise figure. There is thus the prospect of system optimization by trading off the requirements in these different areas and this aspect will be considered in Section 4.1.

The second constraint is that there is a strong multipath effect in an indoor environment. In other words, the line-of-sight (LOS) signal and the reflected signal will arrive at the receiver via different paths. When the path difference is $n \times \lambda/2$ ($n = 1, 2, 3\ldots$), there is destructive interference between the signals and this causes a notch in the

Millimeter Wave Communication Systems, by Kao-Cheng Huang and Zhaocheng Wang

frequency spectrum. For example, if the path difference is 2.5 mm, there will be a notch at 60 GHz, and such a notch can cause an unstable wireless link with only a slight physical displacement of the terminals or scatterers, which will reduce the quality of the communication link. To minimize the multipath effect, a narrow beam antenna is therefore preferred. We will discuss beamwidth optimization in Section 4.2.

The third constraint is that there is a space limit for portable devices, such as handsets. It is essential to know how much gain we can achieve for an antenna in such a limited space. This limit will be discussed in Section 4.3.

For a predominantly LOS wireless link, circular polarization is useful to filter out the first reflected (multipath) signal. Additionally, the wireless communication data rate (the capacity of the link) can be increased by using multi-polarization in a multi-transmitter, multi-receiver system. Each polarization state can deliver different information channels and thus the data rate will be increased. Conversely, if the major concern is the robustness of the link, frequency reuse by polarization can be employed to support multiple copies of the information channel. The properties and performance of polarization will be discussed in Section 4.4.

Fourthly, we have the problem of noise and interference. The reliability of the communications link is defined by the signal-to-noise ratio (SNR). In a general sense, any undesired power appearing in the communications channel is noise and degrades the performance of the link. There are various sources of noise, which can be due to environmental radiators (either passive or active) or generated by equipment in the transceiver. This last component can only be controlled by good design practice and is not the main consideration of this work. The environmental component can, however, be mitigated by the antenna performance, which will be considered in the following chapter.

Finally, a wireless link can be interrupted by a blocking object (such as a human body), which introduces additional shadowing loss. To avoid this shadowing loss, a beam steering function (Section 4.5) or multibeam antennas can be considered to cope with the shadowing. The design consideration and production issues of these antennas are discussed in Section 4.6 and Section 4.7, respectively.

4.1 PATH LOSS AND ANTENNA DIRECTIVITY

For a wireless local area network (LAN), it is generally assumed that the signal arriving at the receiver consists of many copies of the information carrying signal, which have been generated by scattering and other environmental processes. Each path will have a specific delay, and arrival times will vary according to the dimensions of the environment.

For a specific path, as shown in Figure 4.1, the delay profile of the channel is determined by the delay time, path gain, and phase of each path. In order to reduce the multipath effect, it is usual to receive each path and its corresponding time shift (and/or phase shift) in order to maximize the received power and reduce the signal distortion. It is usually necessary to reduce the effects of the major paths, which have the most power (usually no more than the first four or five).

In a gigabit wireless system, the channel model is often assumed to be quasi-optical and line of sight. Therefore, in this case the major power is in the direct path. A typical

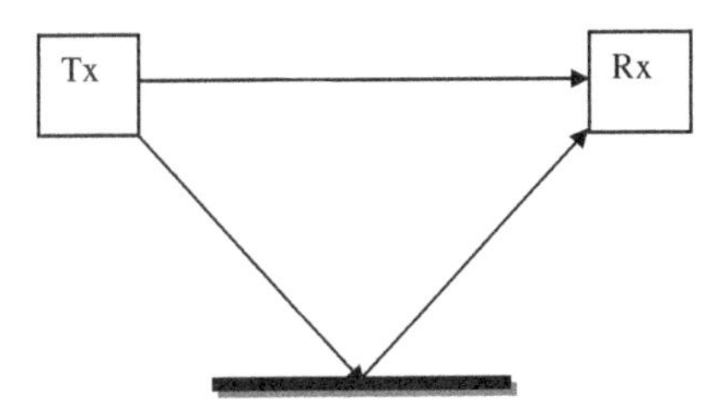

Figure 4.1. Path channel model

indoor measurement for 60 GHz propagation is shown in Figure 4.2. As can be seen in Figure 4.2, there are significant delayed components due to reflections, and if there is no direct line of sight, then comparable power may be distributed over numerous reflected paths.

To combat these effects, we need to focus or direct the radiated power from the antennas in a given direction. The power flux density in that direction will be greater than if an omnidirectional antenna were used to transmit the same power (the power presented at the antenna input terminals), and the ratio between these values (that is, the degree to which the antennas enhance the power flux density relative to an isotropic radiator) is called the antenna gain [1]. On the other hand, the degree to which the power is confined is called the directivity (or how directional the antenna is). These two quantities are closely related by the radiation efficiency of the antenna and can be

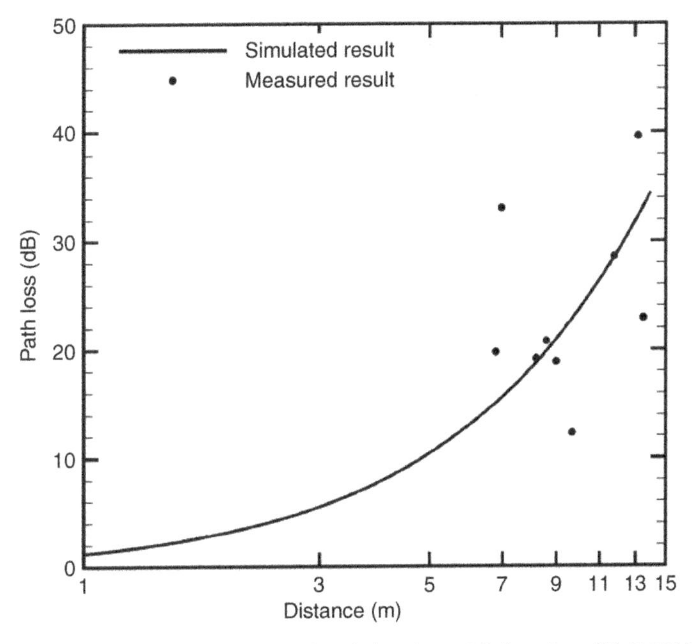

Figure 4.2. Simulated and measured path loss in NLOS situations [2] (©2006 IEEE)

expressed as follows:

$$\text{Gain} = \text{efficiency} \times \text{directivity} \quad (4.1)$$

A typical antenna is able to couple energy to and from free space with an efficiency of approximately 65%.

The usual approach to establishing the power received by an antenna is to consider an isotropic radiator transmission power P_T so that this power is distributed over the surface of an expanding sphere as the wave propagates. At the receiver, the power flux density (power per unit area) is then $P_T/4\pi R^2$, where R is the distance between the transmitting and receiving antennas. The received power is then determined by the effective capture area A_E of the receiving antenna, so that the power received is

$$P_R = A_E P_T / 4\pi R^2 \quad (4.2)$$

This effective capture area can in turn be related to the gain of the antenna [1], which can be written as

$$G = 4\pi A_E / \lambda^2 \quad (4.3)$$

where λ is the operating wavelength.

As we have seen, the directivity of an antenna is given by the ratio of the maximum radiation intensity (power per unit of solid angle) to the average radiation intensity (averaged over a sphere). The directivity of any source, other than an isotrope, is always greater than unity. The maximum gain of an antenna is simply defined as the product of the directivity and its radiation efficiency. If the efficiency is not 100%, the gain is less than the directivity. When the reference is a lossless isotropic antenna, the gain is expressed in dBi (decibels relative to an isotrope).

To aid in designing the appropriate antenna for our application, Table 4.1 lists the major technologies for millimeter wave antennas and provides a comparison of their features.

In gigabit wireless communications, a low-profile design is attractive because of the ease of fabrication, and such a design has the potential to be built at a low cost.

TABLE 4.1. Comparison of Different Types of Antenna

	Radiation Pattern	Power Gain	Polarization
Microstrip antenna	Endfire	Medium	Linear
Slotted Antenna	Broadside	Low/medium	Linear
Rod Antenna	Endfire	High	Linear/circular
Dipole	Broadside	Low	Linear
Multi-element Dipole	Broadside	Low	Linear
Flat Panel Antenna	Broadside	Medium	Linear/circular
Horn Antenna	Broadside	High	Linear/circular

Furthermore, the structures can be lighter than reflector antennas of similar performance, as well as easier to install.

In recent years, several configurations have been proposed for this type of application that produce high broadside directivity (the direction perpendicular to the antenna's length) [1, 3]. Studies have shown that resonant defects in an electromagnetic band gap material could be used to produce high directivity outside the crystal [4, 9]. In practice, in these studies the reflective superstrate was replaced by a single or multiple layers of electromagnetic band gap material, but was still placed over a resonating cavity.

The antennas described above could be excited by a single source located inside the cavity, such as a coaxial probe, a microstrip patch, a slot in the ground plane, or a waveguide horn. Some examples in the context of prototype fabrication illustrate the performance of such antennas in terms of bandwidth, aperture efficiency, and so on. The above optical concepts have been applied to the design of the Fabry-Perot cavity (FPC) in Figure 4.3, which hosts a dual polarized array with sparse elements [10, 11].

Research has shown that a material made of wires exhibits the electromagnetic behavior of a plasma, with a plasma cutoff frequency that depends on the radius of the wires and the period of the structure. A similar idea was reported in [12], which studied and realized a metamaterial made of wire grids that produces high directivity. A detailed explanation of the phenomena was presented in [13], which also provided design criteria for directivity and bandwidth. It was shown that a highly directive beam in this class of antennas is also produced by an excited leaky wave with a small attenuation constant and large phase velocity. In [14], the leaky wave model of [13] was compared with a ray-optic description, as illustrated in Figure 4.4 [14]. A larger class of metamaterials for directivity enhancement was analyzed in [15], where low and high permittivity and permeability materials, and the concept of low and high impedance materials, were analyzed to produce enhanced directivity in a given direction. We now briefly compare this class of antennas with those having a partially reflective surface, as it was reported in [16], where the figure of merit was taken to be the product of the directivity and bandwidth.

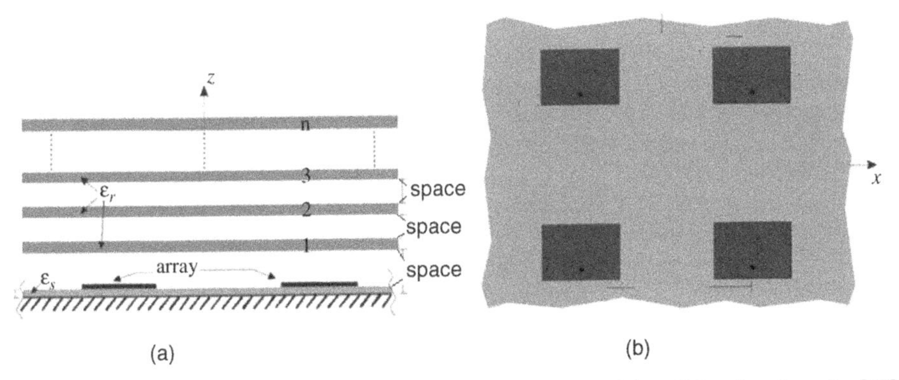

Figure 4.3. (a) Lateral and (b) top view of a sparse array with a Fabry-Perot cavity [10] (©2006 IEEE)

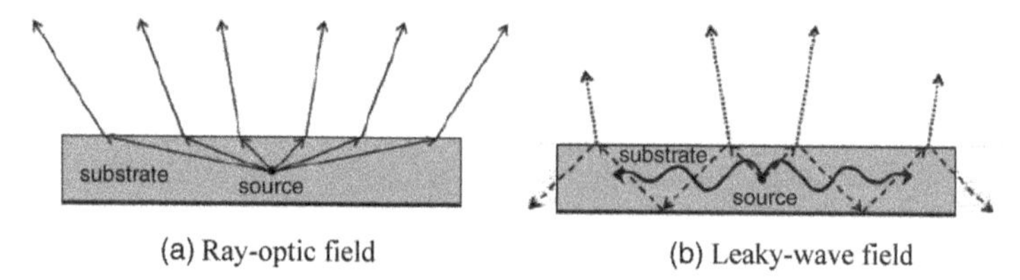

(a) Ray-optic field (b) Leaky-wave field

Figure 4.4. Radiation mechanisms for the attainable broadside radiation. (a) Ray-optic model showing the refractive lensing effect at the top interface. (b) Leaky-wave model showing a propagating leaky mode that is excited by the line source.

For an antenna with a 20° half-power beamwidth, the directivity can be approximately calculated as

$$D = \frac{41253}{\theta_{HPBW}\phi_{HPBW}} = \frac{41253}{20 \times 20} = 103.1325 \text{ or about } 20.13\text{ dBi} \tag{4.4}$$

where θ_{HPBW} is half-power beamwidth in one principal plane and ϕ_{HPBW} is half-power beamwidth in other principal plane,

One of the methods of achieving such high directivity and gain in a small size involves combining a planar antenna with a three-dimensional antenna. Table 4.2 summarizes the results of various studies.

We now discuss some examples from Table 4.2 in more detail.

A dipole antenna can be combined with a horn antenna, as illustrated in Figure 4.5. Similarly, a patch-fed horn antenna has been experimentally investigated at microwave and millimeter wave frequencies [20]. The results indicated that for a 70° flare-angle horn, horn apertures from 1.0 λ-square to 1.5 λ-square, with dipole positions between 0.36 and 0.55 λ, yield good radiation patterns with a gain of 10–13 dB at 60 GHz, and a cross-polarization level lower than −20 dB on boresight. It was also found that the impedance measurements can be safely used for two-dimensional horn arrays, but that the radiation patterns differ because of the Floquet modes [23] associated with the array environment. The integrated horn antenna is a high-efficiency antenna suitable for applications in millimeterwave imaging systems, remote-sensing, and radio astronomy.

The antenna gain can be increased by integrating slot antennas with a dielectric lens. One example is shown in Figure 4.6. The lens is made out of a low-cost, low-permittivity Rexolite material. The single-beam lens achieves a gain of 24 dBi at 30 GHz, a front-to-back ratio of 30 dB, and an axial ratio of 0.5 dB is maintained within the main lobe [25].

TABLE 4.2. Combination of 2D and 3D Antennas to Increase Antenna Gain

3D antenna 2D antenna	Rod	Reflector	Lens	Horn
Patch	Ref. [17]	Ref. [18]	Ref. [19]	Ref. [20]
Dipole/Slot		Ref. [21]	Ref. [22]	Figure 4.5
Yagi				Figure 4.7

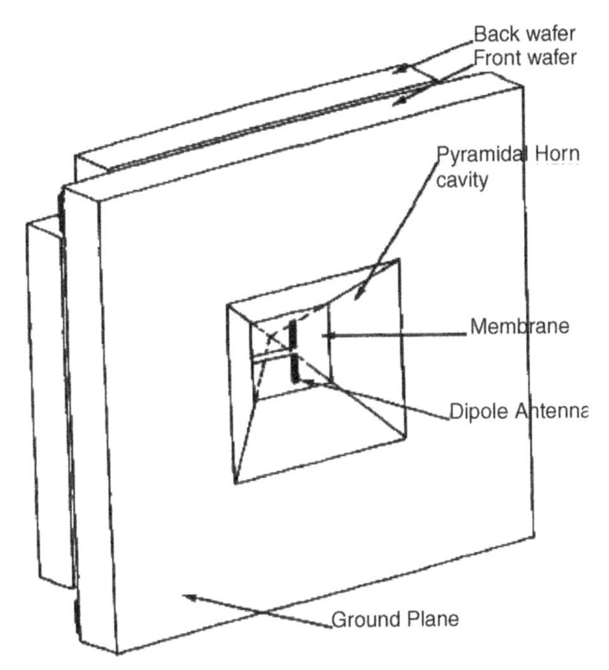

Figure 4.5. An integrated horn antenna in an infinite ground plane [24] (©1992 IEEE)

The measured impedance bandwidth is 12.5% within a standing wave ratio (SWR) of 1.8:1. The single-beam antenna is well suited for broadband wireless point-to-point links.

Figure 4.6 shows that a lens fed by multiple slots can radiate multiple beams with a minimum 3-dB level of overlap between adjacent beams. The coverage of the lens antenna system has been optimized through the utilization of a multiple slot arrangement, leading to a broad scan coverage. The multiple-beam lens antenna is suitable for an indoor wireless access point or for use as a switched-beam smart antenna in a portable device.

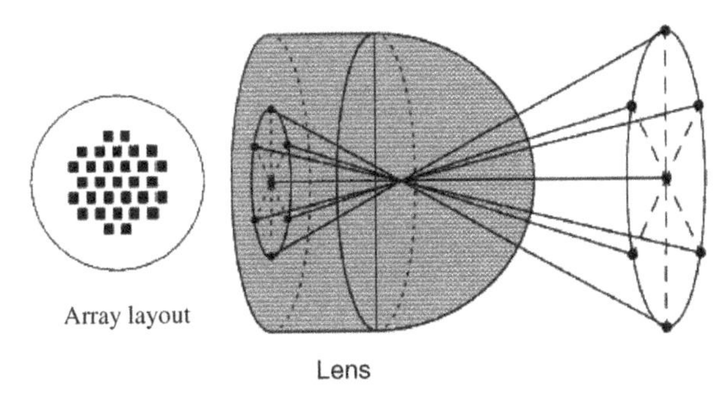

Figure 4.6. Multiple-beam launching through substrate lens antenna [25] (©2001 IEEE)

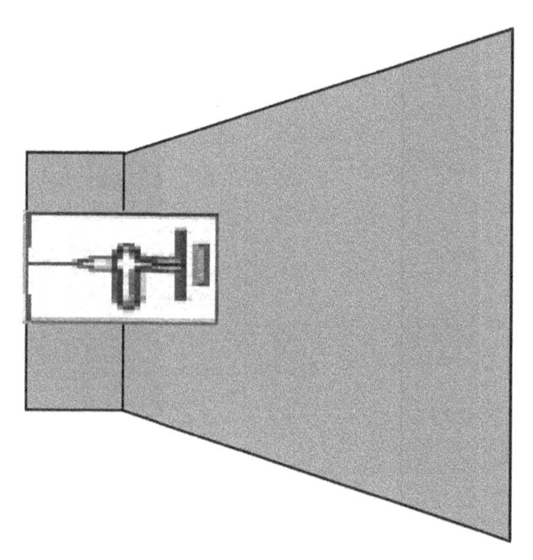

Figure 4.7. Cross-section of a circular horn antenna with a quasi-Yagi inside [26] (©2001 IEEE)

It is also possible to excite a circular horn antenna with a quasi-Yagi antenna, as illustrated in Figure 4.7 [26]. Single-mode operation is achieved by placing the circular waveguide transition in the horn, which suppresses the potential excitation of higher order modes. For a single-mode circular horn antenna, a typical aperture efficiency of 60% at 60 GHz can be achieved due to the high radiation efficiency of the quasi-Yagi antenna. The measured antenna gain and radiation patterns of the longer horn correspond to optimum horn characteristics with a waveguide input. Wider bandwidth can be achieved by doing the transition in the waveguide that feeds the horn.

The integration of a quasi-Yagi antenna with a horn makes this antenna a symmetric two-port device, regardless of the angle of reception, which can be realized in balanced receivers and transmitters. The edge diffraction from the incoming horn aperture is reduced, which can be of use in corrugated horns. The single-mode operation of the antenna allows the integration of a polarizer directly at the aperture.

4.2 ANTENNA BEAMWIDTH

The radiation pattern of an antenna is essentially the Fourier transform (linear space to angle) of its aperture illumination function. In a radiation pattern cut containing the direction of the maximum of a lobe, the angle between the two directions in which the radiation intensity (power) is one-half the maximum value is known as the half-power beamwidth (HPBW) (see Figure 4.8). This is also commonly referred to as the 3-dB beamwidth.

The beamwidth of an antenna is a measure of its directivity and is usually defined by the angles where the pattern drops to one-half of its peak value, which are known as the 3 dB points. For a circular aperture antenna of diameter D, if the antenna is uniformly

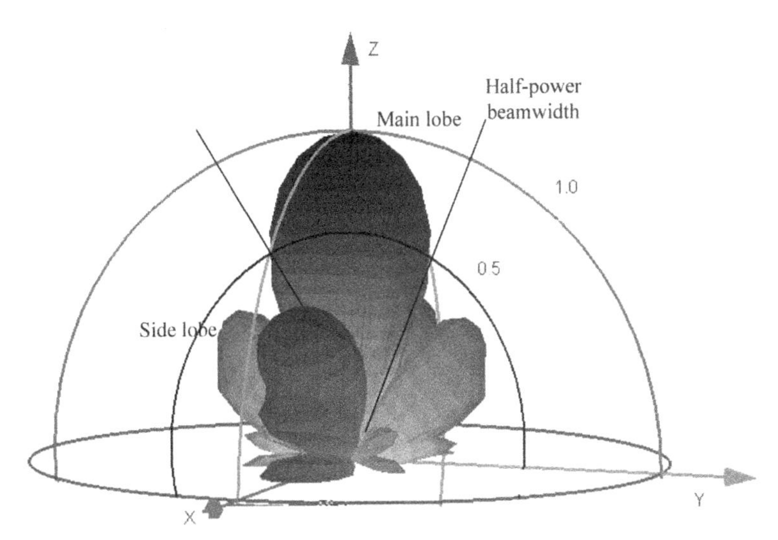

Figure 4.8. Beamwidth in a normalized power pattern (the radial scale is logarithmic)

excited, this beam width is about 70 × wavelength/D (the exact beamwidth actually depends on the aperture illumination function). The next lobe in the pattern, usually called the first sidelobe, will be about 1/20 (13 dB less) of the value of the main lobe and any sidelobes that are further away will be of even lesser value. The rate of decay of these sidelobes is an important parameter in many antenna applications and is used in many international standards as a defining parameter of antenna performance. In general, the maximum gain can be approximated by the following formula:

$$\mathrm{G} = \frac{27000}{BW_h \times BW_v} \tag{4.5}$$

where G is the power gain (linear) of the antenna, BW_h is the horizontal beamwidth of the antenna, and BW_v is the vertical beamwidth of the antenna [1].

As an example, considering an antenna that has a vertical beamwidth of 27° and a horizontal beamwidth of 10°, it will have a power gain of 100 (linear) or 20 dBi. This would also give a vertical dimension of about 2 wavelengths and a horizontal dimension of about 5 wavelengths if the antenna is uniformly excited.

The total received signal is normally expressed as a closed form expression, known as the Friis equation [1]

$$P_{RX} = P_{TX} \frac{G_{TX} G_{RX} \lambda^2}{16\pi^2 R^2 L} \tag{4.6}$$

where P_{Rx} is the received power, P_{Tx} is the transmitted power, G_{Tx} and G_{Rx} are the transmitting antenna gain and receiving antenna gain, respectively, and λ, R, and L are the wavelength, separation, and other losses, respectively. The allocations we give to each of these components constitute what we generally call the link budget.

There are four types of antenna configurations in a communication systems:

1. Tx: omnidirectional antenna vs. Rx: omnidirectional antenna
2. Tx: omnidirectional antenna vs. Rx: directional antenna
3. Tx: directional antenna vs. Rx: omnidirectional antenna
4. Tx: directional antenna vs. Rx: directional antenna

Omnidirectional antennas have signals that radiate in all directions and are useful when multi-path is needed for communication purposes. In case 1, a sophisticated equalizer is needed to recover data from multi-path signals. More details about adaptive equalization can be found in Chapter 8. A directional antenna has a narrow beam in the designed direction and weak signals in undesired directions. It is useful when multi-path is not required. Since a directional antenna has a small coverage area, it is necessary to incorporate a beam steering function to gain wider coverage. However, the narrower the antenna beamwidth is, the more complex the beam tracking function would need to be. In other cases, a sophisticated beam steering algorithm, such as multiple-input -multiple-output (MIMO), or adaptive multi-user detection is needed to improve communication quality. More details about these topics can be found in Chapter 4 and Chapter 5. Hence it is necessary to consider the balance between the complexities of a beam-forming antenna and a tracking function.

4.3 MAXIMUM POSSIBLE GAIN-TO-Q

The quality factor Q of an antenna is an important overall parameter specifying the antenna performance and the inherent physical limitations of antenna size. In particular, a high value of Q means that large amounts of reactive energy are stored in the near zone field. This in turn implies large currents, high ohmic losses, narrow bandwidth, and a large frequency sensitivity. Knowledge of the antenna Q leads to a reasonably definite assessment of antenna performance because of its clearer physical implications.

In millimeter wave applications, we need to maximize antenna gain and bandwidth (i.e., to minimize the quality factor Q for a lossless high Q antenna) simultaneously. Therefore, it is important to optimize the gain-to-Q ratio in antenna design. It is clear that the optimization of the gain-to-Q ratio will yield a greater minimized Q than the minimum possible Q, since it demands that the gain be maximized at the same time.

The first general study was published by Chu [27], who derived theoretical values of Q for an ideal antenna enclosed in an imaginary sphere. The Q of an electrical network at resonant frequency ω can be defined as

$$Q = \frac{\omega W_1}{P_1} \tag{4.7}$$

where W_l is the time-averaged energy stored in the network, and P_1 is the power dissipated in an electrical network. When the network is not resonant, the time-averaged magnetic energy stored in the network is not the same as the time-averaged electric energy. The input impedance is proportional to $P + 2j\,\omega\,(W_m - W_e)$, where W_m and W_e are the time-averaged magnetic and electric energy stored in the network, respectively.

To make the input impedance resistive, some additional energy storage must be added so that the net reactive energy vanishes. If now it is agreed that the network is always to be operated with an additional ideal lossless reactive element, so that the input impedance is purely real, then the Q of the resultant network is defined as

$$Q = \frac{2\omega W_2}{P} \tag{4.8}$$

where W_2 is the larger of W_m or W_e, and P is the power dissipation in the original untuned electrical network.

Therefore, the antenna is normally tuned to resonance by the addition of a reactive element. If the added reactive element does not dissipate any energy, the Q of the resultant system is then given by (4.8). Consequently, (4.8) can be considered to be the upper bound on the Q of an antenna system that is tuned to resonance by the addition of a single reactive element. Any loss in the tuning element would reduce the Q below the value given by (4.8). The parameter defined by (4.8) will be referred to as the antenna Q, even though in the usual network sense it is the Q that results only when the antenna is tuned to resonance by an ideal reactive element.

Extremely large field intensities could be found by looking at the field in the vicinity of a high-gain, small-size radiator. In a physical antenna, this would result in prohibitive heat loss. In an ideal lossless antenna, this would result in large energy densities, and thus a high Q in the antenna. There is a concern about whether the required current distribution could be obtained in practice, since it is determined by the solution to a boundary value problem.

Assuming that the antenna geometry and excitation are arbitrary, new upper limits can be derived for a directional antenna in this general situation.

Using spherical coordinate system (r, θ, ϕ) (see Figure 4.9), the field external to a sphere containing all sources can be expanded in terms of spherical wave functions [28]

$$\psi_n = h_n^{(2)}(kr)P_n^m(\cos\theta)e^{jm\phi} \tag{4.9}$$

with n and m being integers, where k is wave number, and $h_n^{(2)}$ and P_n^m are a spherical Hankel function of the second kind and a Legendre function, respectively [29].

With the miniaturization of electronic devices, there is a continual demand for reductions in antenna size and profile. A question that is frequently asked is how small an antenna can be made while at the same time maintaining good performance (i.e., the highest gain and bandwidth at the same time).

For a small antenna with $ka \ll 1$ (k as wave number and a is the radius of the imaginary sphere enclosing the antenna), the optimized gain-to-Q ratio can be approximated as [30]

$$\max\frac{G}{Q}\bigg|_{dir} \approx \frac{6(ka)^3}{2(ka)^2+1} \tag{4.10}$$

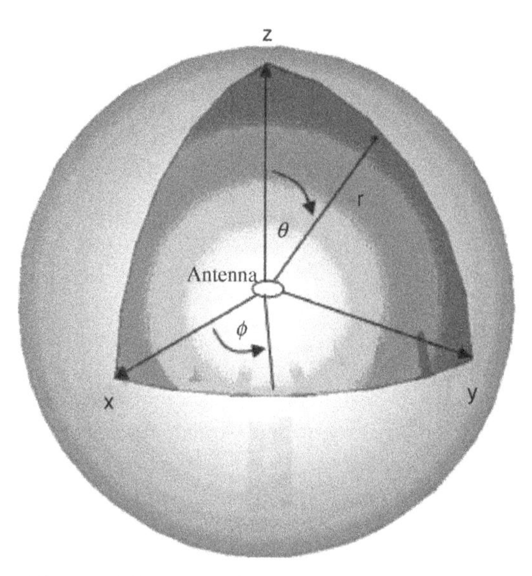

Figure 4.9. Spherical coordinate system (r, θ, ϕ)

It should be noted that an infinitesimally small dipole itself has an extremely narrow bandwidth since its real Q would be much higher than the minimum possible Q (mathematically it should be infinity).

The above relationships represent the best overall performance that a small antenna can achieve, and it can be used to determine the smallest possible antenna size once the required antenna bandwidth is given, and vice versa. To determine the best antenna performance for a large antenna, Figure 4.10 depicts the maximum possible gain-to-Q ratio. The plot of $\max G/Q|_{dir}$ shows that it is a monotonically increasing function of ka.

Figure 4.11 shows a plot of the maximum gain of an antenna. Again, it is a monotonically increasing function of ka.

In general, the curve in Figure 4.11 is applicable to all antennas. It can be used to determine the best overall performance once the maximum antenna size is given, or to determine the smallest possible required antenna size to get the best overall performance,

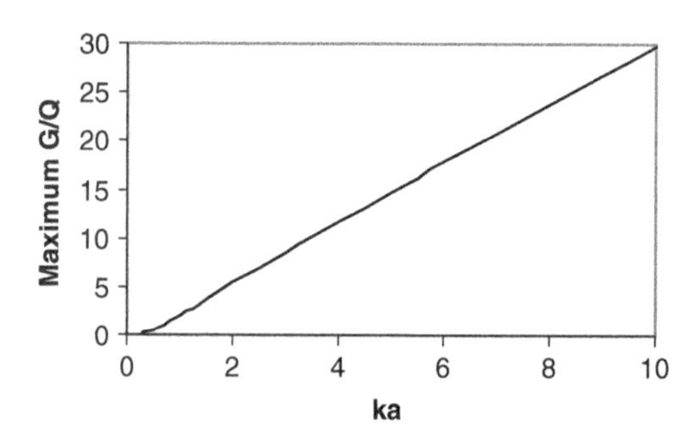

Figure 4.10. Maximum ratio of gain-to-Q for directional antenna

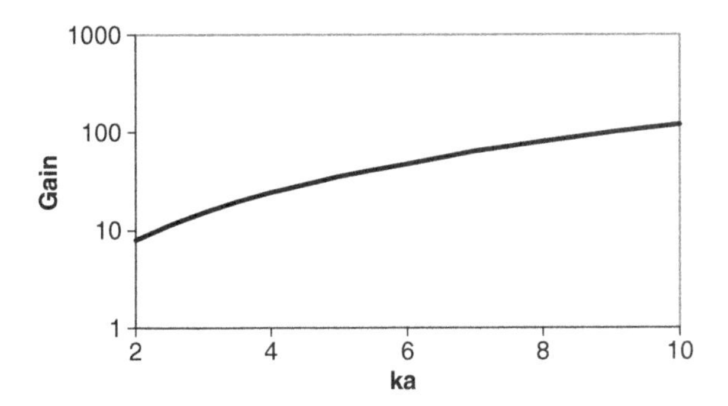

Figure 4.11. Maximized achievable gain [31] (©1969 IEEE)

as was discussed previously for small antennas. It is well known that there is no mathematical limit to the gain that can be obtained from currents confined to an arbitrary small volume. But a small antenna with extremely high gain will produce a high field intensity in the vicinity of the antenna, which results in high heat loss or high stored energy. By artificially truncating the spherical wave function expansions of the fields to the order N, the maximum obtainable gain is shown as [32]

$$G_{max} = N \times (N+2). \tag{4.11}$$

Although Harrington [32] obtained this result by considering a linearly polarized source, it can be easily proved that this result generally holds for an arbitrary current source. Hence, as N increases (equivalently the antenna complexity increases) the maximum gain also increases. Since the magnitude of the spherical Hankel function decreases very slowly for $n < ka$ and very rapidly for $n > ka$, the approximate transition point $ka = n$ can be considered to be the point of gradual cutoff [28].

The normal gain is often introduced and defined by letting $N = ka$ [27, 28, 31], so $G_{norm} = ka \times (ka + 2)$. Antennas with larger gains than normal have been called supergain antennas. It is believed that a supergain antenna will result in a high Q and therefore is not very practical. Therefore, the normal gain is the maximum gain achievable without incurring a high Q. The plot of G_{norm} is shown in Figure 4.11.

It is clear that the definitions of the normal gain and supergain are somewhat ambiguous, since the cutoff point $n = ka$ is an approximate transition point and, in addition, the Q is not clearly specified.

Assuming that there is an infinite ground plane and that the radius of the sphere that can enclose the antenna is a = 5 mm, we can calculate the free space wavelength at 60 GHz and the wavenumber, which allows us to evaluate ka.

$$\lambda = \frac{c}{f} = \frac{3 \times 10^8 \times 10^3}{60 \times 10^9} = 5 \quad mm \tag{4.12}$$

$$k = \frac{2\pi}{\lambda} = 1.26 \quad radians/mm \tag{4.13}$$

Therefore

$$ka = 1.26 \times 5 = 6.3 \tag{4.14}$$

Using Equations (4.10) and (4.11), we can calculate the maximum gain-to-Q factor and maximum gain as

$$\max \frac{G}{Q}\bigg|_{dir} \approx \frac{6(6.3)^3}{2(6.3)^2 + 1} = 18.7 \tag{4.15}$$

$$\text{maxGain} = 6.3 \times (6.3 + 2) = 52.3 = 17\ \text{dB} \tag{4.16}$$

4.4 POLARIZATION

Much has been published over the years on the subject of antenna polarization. The precise definition can be complex, and radiating and receiving structures respond varyingly, both in frequency and in the incident and transmitted wave angles. Here we shall confine ourselves to a simple treatment, and refer the reader to texts that deal with the topic in much greater depth [1]. We shall consider only "far-field" radiation (since the wavelength is small compared with the dimensions of the radiators), and for the cases discussed here, we shall also assume plane wave propagation.

In free space, the energy radiated by any antenna is conveyed by a transverse electromagnetic wave that is comprised of electric and magnetic fields. These fields are orthogonal to each another and also orthogonal to the direction of propagation. The electric field of the electromagnetic wave is used to define the polarization plane of the wave and, therefore, describe the polarization state of the antenna.

"Ludwig definition 3" is commonly used to describe antenna patterns [33]. In this definition, the reference and cross-polarizations are defined to be the parameters measured when antenna patterns are taken in the usual manner, as illustrated in Figure 4.12.

This wave is said to be linearly polarized, that is, the electric field vector is confined to a single plane. Two independent linearly polarized waves at the same frequency can therefore exist and propagate along the same path. This feature has been used for many decades in free space links, which utilize frequency re-use in order to double the capacity of a link for a given bandwidth. In this case, each polarization carries different information and is transmitted and received independently. In cases where the relative angular orientation of the transmitter and receiver is not defined, using two linear polarizations becomes a problem, with the alignment of the receiver and the transmitter being essential. Systems have been deployed in which dynamic control of the receiver is attained using the incident linear polarization, but these are beyond the scope of this book.

When the two polarizations carry the same information, and the two components possess a specific phase relationship with each other, we can construct a waveform in

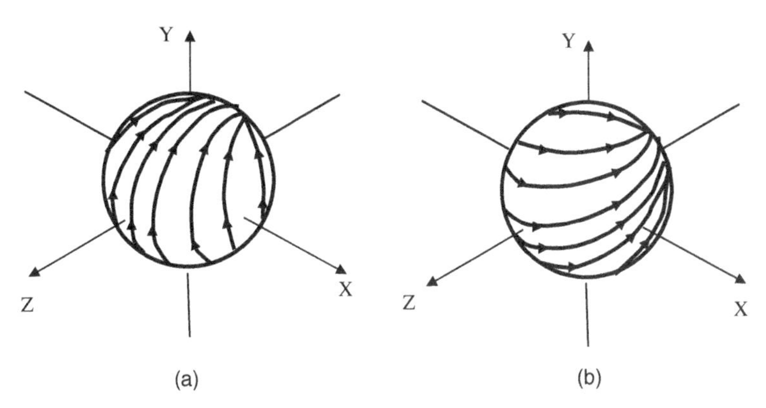

Figure 4.12. (a) Direction of the reference polarization. (b) Direction of the cross-polarization (Ludwig 3).

which the electric field vector rotates as the wave propagates. If the relative phase of the two components is fixed at $\pm 90°$, and the amplitudes of the components are equal, the electric vector describes a circle as the wave propagates. Such a wave is said to be circularly polarized. The sense or handedness of the circular polarization depends on the sense of the phase shift. In general, the two linear components of the propagating wave can have an arbitrary (though constant) phase relationship and also different amplitudes. Such waves are said to be elliptically polarized.

The majority of electromagnetic waves in real systems are elliptically polarized. In this case, the total electric field of the wave can be decomposed into two linear components that are orthogonal to each other, and each of these components has a different magnitude and phase. At any fixed point along the direction of propagation, the electric field vector would trace out an ellipse as a function of time. This concept is shown in Figure 4.13, where, at any instant in time, E_x is the component of the electric field in the x-direction and E_y is the component of the electric field in the y-direction. The total electric field E is the vector sum of E_x plus E_y, the projection along the line of propagation.

Therefore, from the above discussion we may consider two special cases of elliptical polarization, linear polarization and circular polarization. The term used to describe the

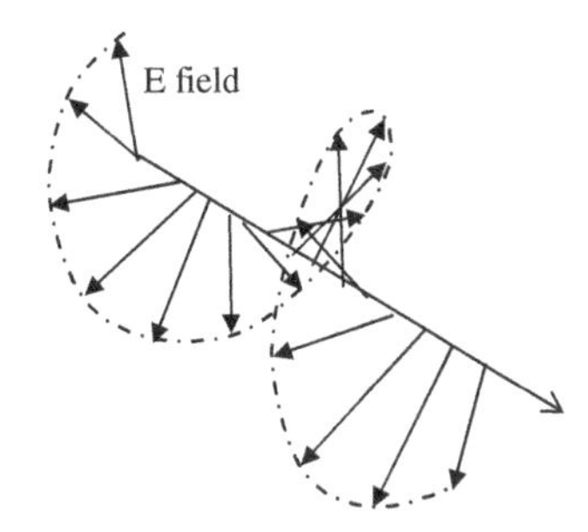

Figure 4.13. Propagation of elliptical polarization

relationship between the magnitudes of the two linearly polarized electric field components in a circularly polarized wave is axial ratio. In a pure circularly polarized wave, both electric field components have equal magnitude and the axial ratio, AR, is 0 dB or 1. Thus, in a pure linearly polarized wave the axial ratio is ∞. In this case, the polarization ellipse traced by the wave is a circle.

In order to transfer maximum energy or power between a transmitting antenna and a receiving antenna, both antennas must have the same angular orientation, the same polarization sense, and the same axial ratio. When the antennas are not aligned, or do not have the same polarization, there will be a reduction in the energy or power transfer between them. This reduction in power transfer will reduce the overall signal level, system efficiency, and performance.

It has been reported that circular polarization can significantly reduce the power of a reflected path in a millimeter wave LOS link [34]. Owing to the boundary conditions on the electric field, the in-plane and normal components of the electric field suffer a differential phase shift of 180° on reflection. This causes the sense of the circular polarization to be changed at each surface reflection. Thus, for an odd number of reflections the reflected wave attains a polarization state orthogonal to the incident wave. When there are an odd number of reflections, a left-hand circularly polarized wave would become a right-hand circularly polarized wave, and vice versa. On the other hand, the direction of circular polarization remains the same when there is an even number of reflections and there is only a reduction in power due to reflection loss.

Circular polarization has an advantage in some user scenarios. For example, if a user holds the terminal at an arbitrary tilt angle to the transmitted signal, there would be a signal strength degradation with a linearly polarized signal. However, such degradation is not present in the case of circular polarization for a direct LOS path and an arbitrary terminal tilt angle. In addition, the terminal will receive less multipath (with single polarization) and circular polarization offers the possibility of frequency reuse, albeit with the complication of crosspolar interference due to multipath reflections. Clearly the magnitude of the multipaths depends on the reflection coefficients of the materials in the environment, and on the material properties of the reflecting objects.

Conventional short-range systems normally use linearly polarized antennas to reduce cost. When the transmitting and receiving antennas are both linearly polarized, any physical antenna misalignment will result in a polarization mismatch loss, which is given by $\mathrm{PLF} = \cos^2(\theta)$, where θ is angular misalignment or tilt angle between the two antennas. The polarization mismatch loss for any angular alignment θ between major axes can be calculated from

$$\mathrm{PLF}(dB) = 10\log\left[\frac{1+\rho_T^2\rho_R^2+2\rho_T\rho_R\cos 2\theta}{(1+\rho_T^2)(1+\rho_R^2)}\right] \tag{4.17}$$

Where ρ_T: the circular polarization ratio of the transmitted wave

ρ_R: the circular polarization ratio of the receiving antenna

Figure 4.14 illustrates some typical mismatch loss values for various misalignment angles.

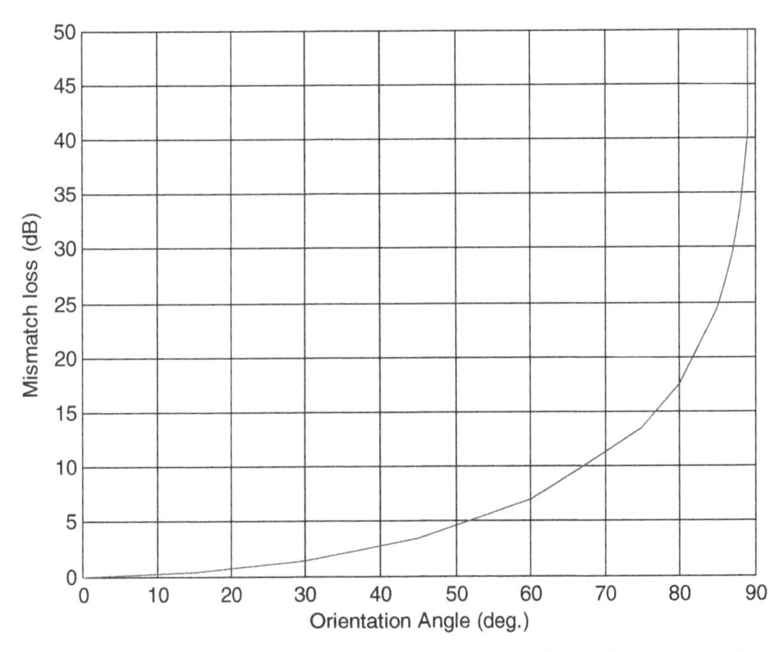

Figure 4.14. Polarization mismatch between two linearly polarized waves as a function of angular orientation

In a situation where the transmitting antenna in a wireless link is circularly polarized and the receiving antenna is linearly polarized, it is generally assumed that a 3-dB system loss will result because of the polarization difference between the two antennas. In reality, the polarization mismatch loss between these two antennas will only be 3 dB when the circularly polarized antenna has an axial ratio of 0 dB. The actual mismatch loss between a circularly polarized antenna and a linearly polarized antenna will vary depending upon the axial ratio of the (nominally) circularly polarized antenna.

When the axial ratio of the circularly polarized antenna is greater than 0 dB (i.e., it is in fact elliptically polarized), one of the two linearly polarized axes will generate a linearly polarized signal more effectively than the other component. When a linearly polarized receiver is aligned with the polarization ellipse's major axis, the polarization mismatch loss will be less than 3 dB. When a linearly polarized wave is aligned with the polarization ellipse's minor axis, the polarization mismatch loss will be greater than 3 dB. Figure 4.15 illustrates the minimum and maximum polarization mismatch loss potential between an elliptically polarized antenna and a linearly polarized antenna as a function of the axial ratio. Minimum polarization loss occurs when the major axis of the polarization ellipse of the transmitter (receiver) is aligned with the plane of the linearly polarized wave of the receiver (transmitter). Maximum polarization loss occurs when the weakest linear field component of the circularly polarized wave is aligned with the linearly polarized wave.

An additional issue to consider with circularly polarized antennas is that their axial ratio will vary with observation angle [33, 34]. Most manufacturers specify the axial ratio

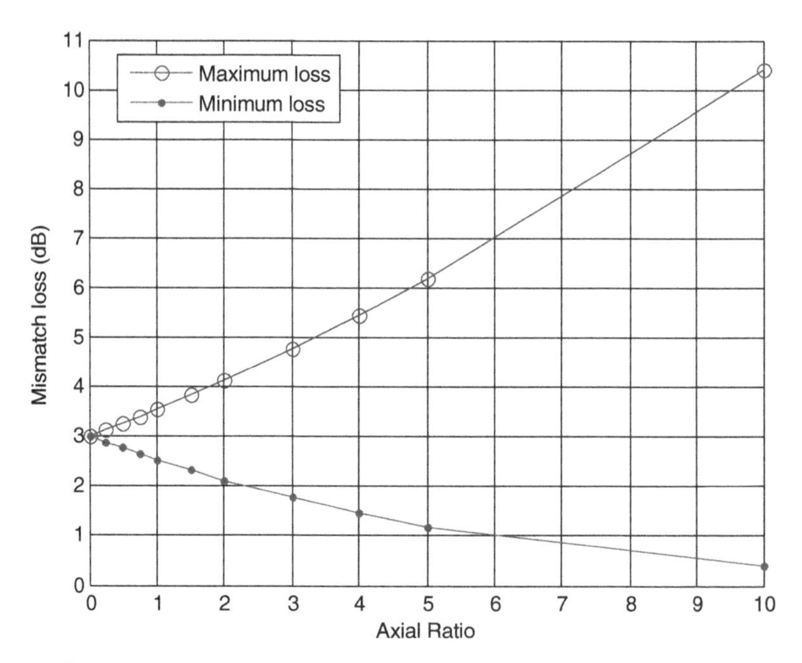

Figure 4.15. Polarization mismatch between a linearly and circularly polarized wave as a function of the circularly polarized wave's axial ratio

at the antenna boresight or as a maximum value over a range of angles. This range of angles is generally chosen to represent the main beam of the antenna. In order to measure the axial ratio, antenna manufacturers measure the antenna radiation pattern with a spinning linearly polarized source. As the source antenna spins, the difference in amplitude between the two linearly polarized wave components radiated or received by the antenna is evident. The resulting radiation pattern will describe the antenna's axial ratio characteristics for all observation angles.

Figure 4.16 presents a typical axial ratio pattern for a circularly polarized antenna. From the antenna radiation pattern, it can be seen that the axial ratio at boresight is about 0.9, while at an angle of $+60^\circ$ off boresight, it is about 0.14. As the axial ratio varies with the observation angle, the polarization mismatch loss between a circularly polarized antenna and a linearly polarized antenna will also vary with the observation angle.

In most cases, the polarization mismatch loss issue is much more complex. Based on the discussion of polarization mismatch loss, it would be conceivable that communication in a wireless system would be nearly impossible when, for instance, the antenna of a mobile device was orthogonal to the antenna of an access point.

Obviously, this is unlikely to happen in the real world. In any mobile handset communications link, the signal between the handset antenna and the base station antenna is generally comprised of a direct LOS signal and a number of multipath signals. In many instances, the LOS signal is not present and the entire communications link is established with multipath signals.

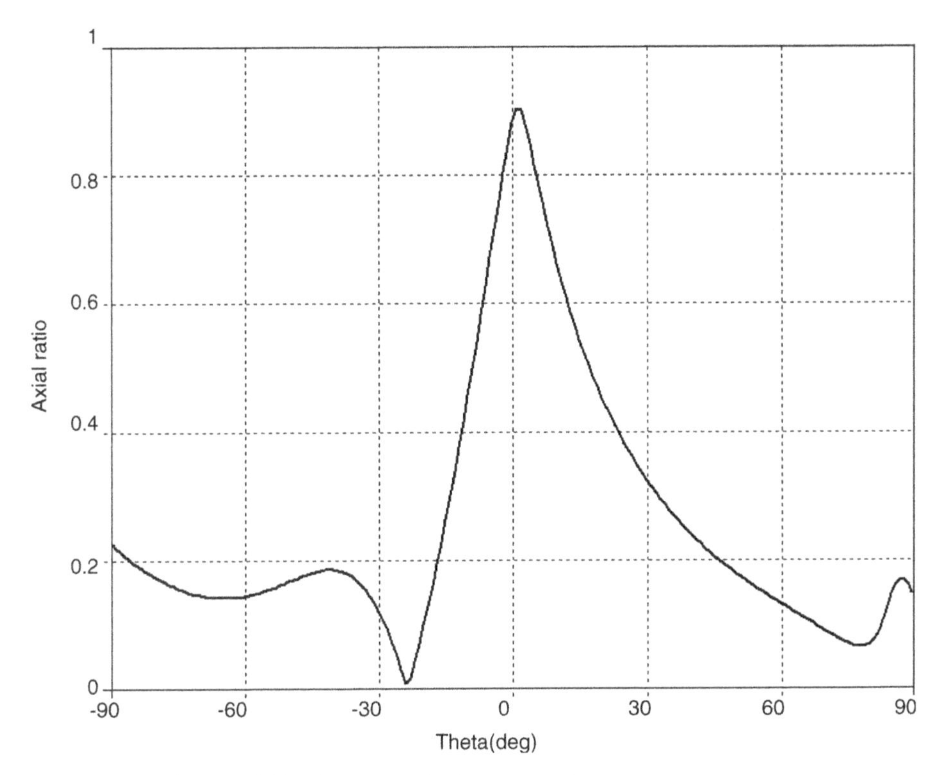

Figure 4.16. Typical axial ratio pattern for a helix antenna

Multipath signals arrive at the antennas of mobile devices via reflections of the direct signal off nearby and distant objects or surfaces. If the reflecting objects are oriented in such a way that they are not aligned with the polarization of the incident wave, the reflected wave will experience a polarization state change. The resultant or total signal available to the receiver at the end of the communications link will be the vector sum of the direct signal and all of the multipath signals. In many instances, there will be a number of signals arriving at the reception site that are not aligned with the assumed standard polarization of the system antenna. As the receiving antenna rotates from vertical to horizontal, it simply intercepts or receives power from these multiple signals, and in fact will receive different multipath signals as the angle of orientation varies.

4.4.1 Polarization Diversity

In order to improve or extend system performance, some system designers use polarization diversity techniques at the receiver in an effort to enhance signal reception. In these systems, a circularly polarized or dual linearly polarized antenna is used at the reception site to take advantage of the fact that many linearly polarized multipath signals, with different orientations, exist at the reception site. A dual polarized antenna can accept the orthogonal signals and combine them in the receiver. They thus have a greater probability of receiving more total power than a single linearly polarized antenna.

Typically, in polarization diversity systems that use a dual linear polarized antenna, the receiver samples and tracks the polarization output that provides the strongest signal level (selection combining). Each output will provide a total signal that is the combination of all the incident signals arriving in that polarization. This combined signal will be a function of the amplitude and phase of each signal, as well as the polarization mismatch of each signal, as described by Equation (4.17).

In polarization diversity systems that use a circularly polarized antenna, the receiver only samples a single output. The total signal developed at the output will be a combination of all the signals arriving at the antenna. Again, this is a function of the individual amplitude and phase of each signal, as well as the polarization mismatch loss between the circularly polarized and linearly polarized signals, as described in Figure 4.15.

It is difficult to determine whether it is better to use a circularly polarized antenna or a dual linearly polarized antenna. This choice is really a function of the make-up of the total signal arriving at the reception site. If the total signal arriving at the reception site is predominantly contained in a linearly polarized wave, then a dual linearly polarized antenna may be the preferred choice. However, antenna alignment is critical in determining the total received signal.

With a dual-polarized antenna, the signal loss, due to polarization mismatch, will therefore be between 0 and 3 dB. If a circularly polarized antenna is used (assuming a 1- dB maximum axial ratio over the main beam), the signal loss due to polarization mismatch will be between 2.5 and 3.5 dB. If the total signal arriving at the antenna is comprised of a random sample of multiple linearly polarized signals, a circularly polarized antenna will be able to detect the waves, and may be the correct choice for a receiver in a dense scattering environment. With a dual linearly polarized antenna, the polarization mismatch loss will generally be greater than 3 dB.

A number of researchers have shown that dual polarization diversity, using vertical and horizontal polarizations, can improve the received signal to noise ratio [35, 36]. In addition, the performance of a three-branch orthogonal polarization diversity system in a scattering environment has been investigated and compared to that of a dual channel polarization diversity system. The results show that the former system has a 2-dB advantage over the latter [37]. The results also suggest that the use of horizontal polarization at the transmitter results in a 2-dB improvement over a vertically polarized transmitter. This observation could be explained by the fact that in an indoor environment (where the receiving antennas were located), the majority of reflectors are horizontal (floors and ceilings). Thus, there are clear benefits to using a three-branch polarization diversity scheme to improve the link budget.

Diversity schemes have been shown to be highly efficient in mitigating the effects of multipath fading. Three-branch polarization diversity schemes [34, 35] can also be applied to millimeter wave antenna systems. These particular schemes use three orthogonal antennas at the receiver to increase the link budget by more than 6 dB and 2 dB for the Rayleigh and dual-channel cases, respectively. The approach has an added advantage of being relatively small and compact, since the antenna elements can be co-located, making it suitable for applications where space is limited. The ability of the system to provide three uncorrelated copies of the transmitted signal also implies that it can potentially be deployed at both the transmitter and receiver in a conventional MIMO arrangement, to enhance both the capacity and the link budget of the channel. However, in this section we emphasize the operation of the scheme as a diversity system (for a robust channel), while the spatial multiplexing MIMO analysis will be reported in a future publication. The performance of the triple polar scheme was analyzed for an indoor environment in [37]. It was envisaged that users requiring high-speed data services will typically be located in this environment and will be relatively stationary, with data terminals larger than current mobile phones and capable of accommodating the diversity antennas [38]. The polarization diversity scheme can then be implemented at the access points and/or at the mobile devices to enhance the link budget.

The three-branch polarization diversity system employs three orthogonal antennas, which may be implemented as either electric or magnetic elements (a total energy antenna can be constructed by using both electric and magnetic elements). In this configuration, three orthogonal electric field detectors are used. One of the antennas is in a vertical position (V) and the other two are in orthogonal horizontal positions (H1 and H2), as shown in Figure 4.17. As mentioned above, one of the major benefits of this configuration is that the antennas are co-located and can be designed to occupy a minimal space. This is particularly important in access points where space is limited, and also in handsets where the device size cannot accommodate spatially separated antennas. One of the problems of having closely spaced antennas is the mutual coupling between the elements, which can adversely affect the application of the array in a diversity system (the communications channels then become correlated). However, by careful design, sufficient isolation can be achieved. In a reported work, a measurement was conducted in an anechoic chamber to quantify the isolation between the elements [37, 39].

A measurement is made to compare the single polarization 4-by-4 subsystems with the multiple polarization 4-by-4 subsystem [40]. The results are shown in Figure 4.18. The horizontal-vertical (HV) subsystem outperforms the single-polarization (H_1, H_2) subsystems in most locations.

When multiple omni-directional antennas are used to form an array in an indoor environment, multiple polarizations do not help the communication link as there are too many 60 GHz reflections (i.e., multipath), which confuse the original polarization. The

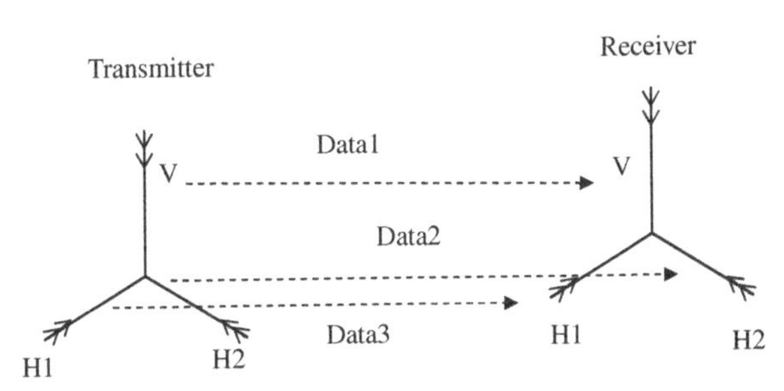

Figure 4.17. Configuration of the three-branch polarization diversity scheme at a transmitter and receiver (V, vertical polarization; H1, horizontal polarization 1; H2, horizontal polarization 2)

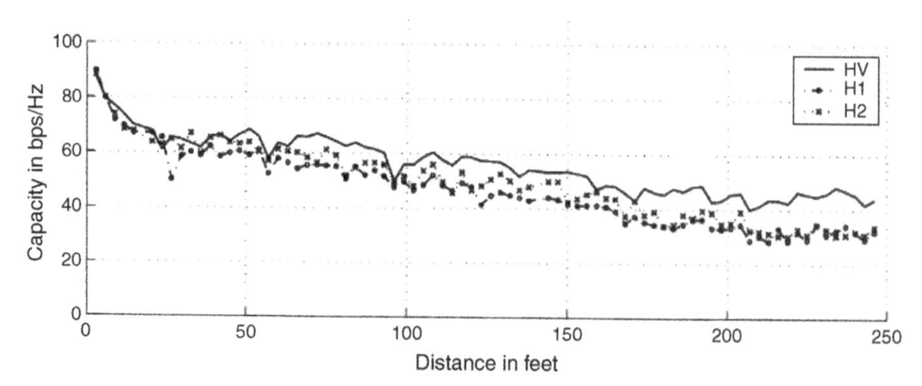

Figure 4.18. Comparison of single polarization versus multiple polarization in a 4-by-4 subsystems for 0° oriented receivers [40] (©2002 IEEE)

receiving result is similar to multi-omni-directional beam with single polarization. However, if multiple directional antennas are used in an indoor environment, multiple polarizations can improve the communication link with an antenna array or with multiple-input-multiple-output system (see Chapter 5).

4.5 BEAM STEERING ANTENNA

Multipath fading presents as delays that differ by $\lambda/2$, and frequency selective fading is a natural phenomenon. Thus, the multipath signals give rise to intersymbol interference (ISI). When the data rate increases, the symbol length decreases. If the multipath delay spread is greater than tens of the symbol length, the equalization complexity increases significantly. To overcome this issue, a beam-steered antenna reduces the spatial extent of the signal, resulting in an improved multipath profile. Antenna directivity thus reduces the multipath fading effect to constrain the ISI problem.

For a single antenna element with an antenna gain of more than 30 dBi with a half-power beamwidth of approximately 6.5°, a reliable communication link is difficult to establish even for LOS at 60 GHz. This, as we have seen, is due to the movement of people, who can easily block and attenuate such a narrowbeam signal. To overcome this problem, a switched-beam antenna array or adaptive antenna array is required to perform search or beamforming, in order to capture the available signal path. The array is subsequently required to track the signal path either continuously or periodically, depending on the stationarity of the link. One major parameter used to describe the performance of the link is how many antenna elements are required to achieve the intended antenna gain. We can separate this consideration from the array gain, which refers to the performance improvement in terms of the SNR for a single antenna. On the other hand, the gain of the antenna array can be described as the product of the array's directivity and efficiency. We are also interested in the angular resolution or beamwidth of such antennas, since this defines the number of multipaths that the antenna sees in a scattering environment. The directivity of a linear array is given by [41]

$$D = \frac{4\pi}{\iint |F_n(\phi,\theta)|^2 \sin\theta d\theta d\phi} \tag{4.18}$$

where $F_n(\theta, \phi)$ is the normalized field pattern, which can be expressed as a product of the normalized element pattern and normalized array factor. The variables ϕ and θ represent the azimuth and elevation angle, respectively. For a uniform linear array, the normalized array factor can be expressed as

$$f_n(\phi,\theta) = \frac{\sin((N/2)(kd\cos\theta + \beta))}{N\sin((1/2)(kd\cos\theta + \beta))} \tag{4.19}$$

where N, d, and β are the number of antenna elements, the antenna spacing between adjacent elements, and the phase shift between adjacent elements, respectively. For an omnidirectional antenna, it can be shown that up to 100 elements are required to achieve a gain of only 23 dBi, which is far from the required specification shown previously. Hence a more directive element is required to improve the overall gain of the array.

Many types of antenna structures are considered to be unsuitable for 60 GHz WPAN/WLAN applications due to the requirements for low cost, small size, light weight, and high gain. In addition, it is also necessary to operate 60 GHz antennas with an approximately constant gain and high efficiency over a broad frequency range (57–66 GHz). Beamsteering can be achieved by the use of either switched beam arrays or phase arrays. Switched beam arrays have multiple fixed beams that can be selected to cover a given service area. Their implementation is much easier than phase arrays, which require the capability of continuously varying the progressive phase shift between the elements.

The complexity of phase arrays at 60 GHz typically limits the number of elements. In [42], a 2×2 beam steering antenna with circular polarization at 61 GHz was developed. Its gain was approximately 14 dBi with 20° HPBW. Similarly in [43], another 60 GHz integrated 4-element planar array was developed. Each antenna was integrated with a subharmonic I/Q mixer to make high-speed signal processing, such as adaptive beamforming, more convenient. However, the implementation of a larger phase array presents some technical challenges, such as the requirement for a higher feed network loss, more complex phase control network, stronger coupling between antennas, as well as feedlines, and so forth. These challenges make the design and fabrication of larger phase arrays more complex and expensive. Hence, more research is required to develop a low-cost, small-size, light-weight, and high-gain steerable antenna array that can be integrated into the RF front-end.

To achieve the desired high speeds, the design approach should focus on either:

1. accepting the presence of multipath (with delays on the order of the room size) and mitigating it with multitone or equalization techniques, or
2. using line of sight links with narrow beam antennas to eliminate virtually all multipaths, allowing the use of simple unequalized modulation schemes, such as frequency shift keying (FSK) or phase shift keying (PSK).

In the first case, the design effort will emphasize sharp beam antenna design techniques, whereas in the second case, the work will concentrate on antenna/beam steering techniques. These techniques must be used because multipath delays in the typical indoor environment are on the order of the target bit duration (tens of nanoseconds) and would cause intersymbol interference. The delays are related to the size of the room and the density of the scatters within the illuminated space.

We assume that at high speed, a simple two-or-four level FSK or PSK system will be used, because complex modulation methods that require equalization, diversity, or multicarrier techniques are deemed to be impractical or too expensive for a 60 GHz system. For such a simple system to work reliably, the channel impulse response should not contain significant multipath components, ensuring that the data rate is not limited by multipath effects. We also make the initial assumption that high-speed and high-capacity WLANs can use a femtocellular architecture, with a single cell for each small room, and multiple cells for a large open office area. The consequences of using non-LOS links are also considered. Indoor propagation measurement show that multipath can be reduced with sharp beam antennas.

For this "LOS with sharp beam antennas" approach, the amount of multipath will depend on the number of beam paths between the transmitter and receiver, which in turn will depend on the directivity of the antennas at the transmitter and receiver, as well as on environmental details. It will also depend on the ability of the antenna to resolve the angles of the multipaths. If omnidirectional antennas are used at both the transmitter and receiver, there will be many possible beam paths, whereas if highly directional antennas are used, there may only be a single LOS path. Once the beamwidth is sufficiently narrow, there is no significant multipath in most practical circumstances. (However, if the transmitter to receiver LOS is perpendicular to a pair of parallel reflectors, one will in fact get an infinite number of multipaths.)

To explore the consequences of this approach, we investigate three different antenna designs.

4.5.1 Phased Array

In an ideal phased array, beams can be formed with most versatility to almost arbitrary directions. But even an 8×8 phased array antenna with beam steering at 60 GHz requires complex phase shifters and therefore it is subject to high losses at 60 GHz. These losses reduce the effective gain of the antenna array. In addition, currently no phase shifter MMIC is available in the market for 60 GHz, so we would need a hybrid tee and real weights (antennuators) to build phase-shifting functions. In addition, the beam shape becomes asymmetric when the beam direction moves away from the z-axis (this is generally called aberration). This means that the sidelobes of the radiation pattern will grow when the beam is away from broadside. Also, with a phase-shifted array, circular polarization at wide angles is almost impossible to achieve. Circular polarization is unlikely to be achieved as the phased array becomes complex. The main challenge of this design is how to incorporate a complex phase shifter and to achieve low loss. Finally, it is also very tricky to achieve good circular polarization in all directions.

4.5.2 Beam-Switching Array

For beam switching, beams can be formed toward only a finite number of predefined directions. This design uses a minimum number of elements (4×4) to achieve $\pm 100°$ coverage [44, 45]. No phase shifter is required. Each element generates an independent beam. The configuration operates in a different manner to that of the phased array, and there is no size limit for each element. Each element can be optimized individually to meet the specifications for the individual links. Sidelobe levels can therefore be controlled by the design of a single horn. The gain of each element can be improved by adding a superstrate to a horn or using stacked patches. More details about gain enhancement can be found in Chapter 2. The feed network needs to have the correct amplitude excitation for each element, but the phase is less critical. The circular polarization can be improved by a tilted waveguide or helical element.

4.5.3 2×2 Horn Array Plus Beam Switching

This design combines the features of array and beam switching. The gain of this design is limited by the size and separation distance of the horns. In Figure 4.19, each unit, consisting of a 2×2 element array, acts as an independent source [46]. This design requires a multi-bit phase shifter and generates good circular polarization. By adding several tilted horns, this design can attain $\pm 100°$ coverage. The feeding network needs to have the correct amplitude and phase in the two orthogonal linear polarizations in order to generate good circular polarization. The main challenge in this design is reducing the sidelobe level caused by the 2×2 elements.

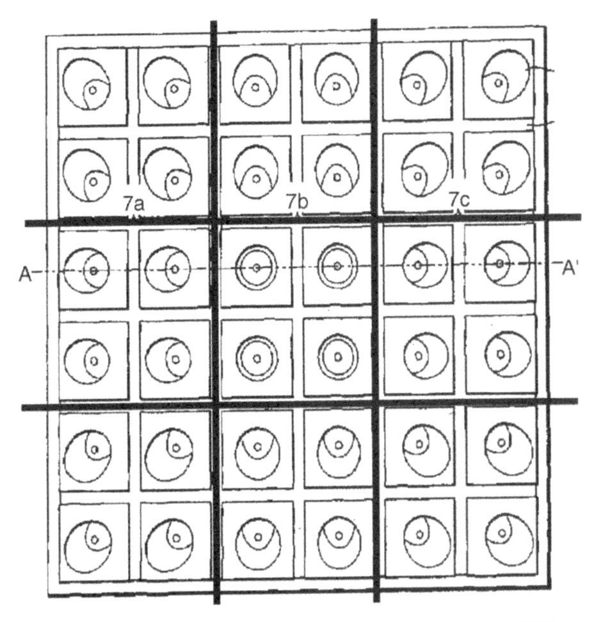

Figure 4.19. Nine units of 2×2 horn antennas [46]

TABLE 4.3. Comparison of Three 60 GHz Antenna Designs

	8 × 8 Phased Array	4 × 4 Beam Switching Array	2 × 2 Array Plus Beam Switching
High Gain	Yes (but the loss of the phase shifters is also high)	Yes	Yes
HPBW 20°	Yes	Yes	Yes
Sidelobe < −10 to −20 dB	Difficult when beam is tilted toward endfire direction.	Yes	Not easy
Circular polarization	Medium	Possible	Possible
Beam Steering Range	Beam direction is controlled by phase shifting. Sidelobe level increases when the beam is away from broadside	Beam direction is controlled by switches	Beam direction is controlled by the height of the horns and by phase shifters
Feeding Point Design	Amplitude, phase	Amplitude	Amplitude, phase
Phase Shifters	Complex	No	2-bit
Challenge	Complex phase shifter, low loss phase shifter	High gain with small size	Side lobe reduction

In all of these cases, number of antenna elements in the array can be adaptive for millimeter wave communications. That is, the actual number of antenna elements required is determined by the training process during operation.

The main defining parameters for these designs are compared in Table 4.3.

4.6 MILLIMETER WAVE DESIGN CONSIDERATION

When a low dielectric substrate (e.g., $\varepsilon_r = 3$) is used for millimeter wave antenna design, one should note that the width of 50 ohm feeding line is close to the width of a 60 GHz patch (see Figure 4.20). In other words, the radiation of feeding line should be considered in antenna measurement. This gives challenges of impedance mismatch. Although it is possible to add an external impedance matching lines between the patch edge and the 50 ohm feeding line, the impedance matching lines could give rise to spurious radiation.

If a high dielectric substrate (e.g., $\varepsilon = 9$) is used, the width of 50 ohm feeding line is reduced and the coupling effect between feeding line and patch is not so

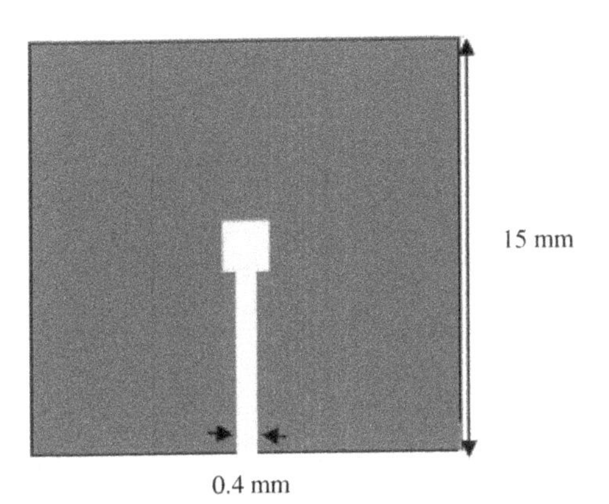

Figure 4.20. A 60 GHz patch antenna on Duroid substrate (size: 15 mm × 15 mm)

strong. Alternatively, one can use probe-feeding or proximity-coupled method to avoid coupling issue.

When designing for millimeter wave applications, the thickness of the dielectric between the patch or patch stack and the ground plane has a strong influence on the bandwidth of the antenna [47] and, consistently, simulations and measurements have shown a wider impedance bandwidth for a circular single-element (Figure 4.21(a)) than for a stacked two-element patch antenna (Figure 4.21(b))[48].

The patch can be designed with a diameter of 890 μm, with the spreading of the silver paste adding an extra 60 μm to this diameter. This explains why the observed center frequency spots are sometimes detuned below 60 GHz. The 25-μm-shifted feed point position corrects, to some degree, the effect of this increase in the realized patch size. Measured radiation patterns of two 60 GHz patches in E-plane are shown in Figure 4.22.

The spreading of the silver paste during the firing stage of the LTCC (Low-temperature co-fired ceramics) [49] processing has occurred quite often with Ferro A6-S materials [50]. It is difficult to predict how much compensation for this spreading should be included in the layout design in each case. On the other hand, research has shown that when gold paste is used in a Ferro A6-S system, it has practically no spread at all, and hence it is the preferred choice for these types of applications.

The transition can be designed by using a coupling slot with a coplanar-waveguide line, as shown in Figure 4.23. In general, a slot transition cannot be used if the wideband

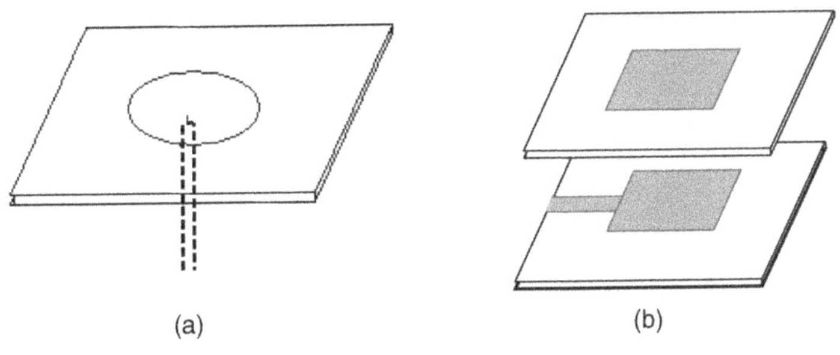

Figure 4.21. Patch antenna: (a) single patch, (b) stacked patches

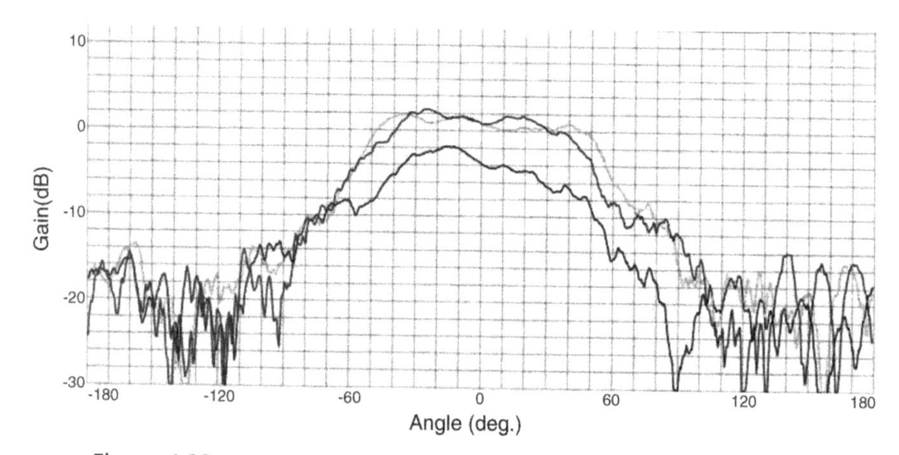

Figure 4.22. Measured radiation patterns of two 60 GHz patches in E-plane

function is desired, but in this case that is not regarded as a problem. The design of the slot transition is quite simple. The width of the slot is selected so that its realization in the LTCC processing is feasible. For this implementation, 150 μm is considered reasonable. The length of the slot from end to end is close to the corresponding electrical half wavelength in the dielectric medium used, which is about 1.0 mm in this case. Both the coplanar-waveguide and strip line continue slightly over the slot region and, hence, form two stubs in the transition. The exact dimensions of the slot and stubs can be defined with a simulator to achieve adequate return loss and bandwidth, which are trade-offs. For this transition, the simulated return loss is about −39 dB and the insertion loss is close to −0.62 dB at 60 GHz. The impedance matching bandwidth for a return loss of at least −15 dB is in the range of 54–66 GHz.

The minimum insertion loss for a single transition seems to be 1.1 dB, and there might be different frequency responses in terms of the S_{11} and S_{22} parameters if there are variations in the physical dimensions of the realized conductor patterns, which cause some asymmetry [48]. Despite the deteriorated performance caused by the dimensional tolerances, the functioning of the transition can be regarded as adequate for the aimed antenna radiation gain measurements.

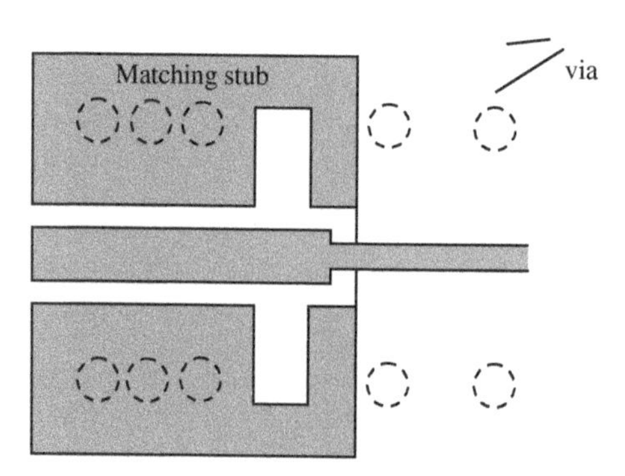

Figure 4.23. Coplanar waveguide-to-stripline transition

4.7 PRODUCTION AND MANUFACTURE

Millimeter wave production allows small tolerance and low loss. This section introduces several manufacturing technologies that are suitable for millimeter wave antenna production in terms of reducing the loss and increasing the printing accuracy.

4.7.1 Fine Line Printing

The fine line technique is based on print-and-etch techniques. It uses a standard dry film etch resist [51]. This allows large areas of very fine printed circuit boards (PCBs) to be manufactured on a range of laminates with higher reliability and at a lower cost than alternative techniques. Tracks and gaps can be fabricated on various laminates, including standard FR4, FR5, and Flex [52], as well as soft boards (e.g., Teflon™) and ceramic substrates.

The fine line technique has particular application to higher-density microelectronic packaging for mobile portable electronics, particularly for telecom applications, where functionality is at a premium. Among the key areas where this interconnect technology will help is "flip chip" assembly, which is essential for many applications where high-speed signal performance and packaging are demanded. This is particularly important in the mobile telecom field.

The new technology allows 25-μm tracks and gaps to be reliably produced at a lower cost than current techniques. It extends this printing circuit technology into the millimeter wave range and addresses the needs of a wide range of microelectronic applications, from laptops to medical instruments.

The move toward very small devices, such as chip scale packages and flip chip assemblies, is restrained by the density limitations of conventional PCB technology. In most of the imaging processes used in PCB technology, it is difficult to achieve a pitch below 100 μm. It is necessary to progress from the old technologies to the new ones. One alternative to conventional imagining is the new technology of laser structuring. It creates a structure in resists (tin or organic coating) by direct laser ablation. After ablating the resist, the structure is etched and the resist is stripped. The direct structuring of tin and organic coating has been studied to create masks for the etching process. The target is to achieve lines/spaces below 125 μm [53].

4.7.2 Thick Film

Millimeter wave circuitry requires high-resolution etching technology and its functional density is increasing. It is common to use ceramic materials, including thick film conductors on aluminates and circuits using Green Tape™ LTCC [49]. Conductor patterning techniques include conventional screen-printing, Fodel® photoimaging, and etching processes. Thin film deposition has also been considered for the fabrication of millimeter wave circuits, since it is easy to control and monitor the tolerance in the production process [51].

It is important to note that the fabrication of millimeter wave circuits requires the close involvement of the designer, from circuit layout to final production, because of the special geometries required for the transmission and reception lines, for waveguides, to limit reflection and propagation losses, as well as for the construction of specific components, such as antennas, couplers, and dividers.

Since the mid-1990s, more opportunities have arisen for the use of ceramic circuitries for high frequency modules, enhanced by the availability of more information on their performance in the millimeter wave range and by new material offerings for such applications, including LTCC and FODEL® photodefinable conductors [54].

4.7.3 Thin Film

Thin film technology has been the traditional method used to manufacture microwave circuitry for many years. An outstanding line resolution and excellent conductor edge definition, combined with superb ceramic substrate properties in regard to high frequency and thermal behavior, are ideal prerequisites for these applications. On the other hand, this approach can be critical in terms of cost. Complex modules are often assembled in a special hermetic housing using a patchwork arrangement of different substrates. The major reason for dividing the circuit into subcircuits is related to the yield figures obtained for large substrates. The positioning accuracy of these substrates is crucial to avoid gaps and related impedance changes. The housing itself needs to have expensive hermetic RF-interconnections. Multilayer substrates based on LTCC [49] offer a variety of design options for microwave designs. DC-connections and digital control functions can be implemented in separate layers. Chip-tailored cavities can improve the return loss of the signal interconnections. Various transmission line types, as well as waveguides, are available. And the hermetic substrate itself can be used as a part of the package, with integrated feedthroughs. Embedded resistors and capacitors are additional features that can be used to further shrink the designs. However, the fine line printing resolution and associated tolerances might be the bottleneck. Although lines and spaces down to 50 microns are achievable, this is not sufficient for certain elements, like edge coupled filters, couplers, etc.

A recent approach, FINEBRID, which combines the advantages of both the thick film and thin file technologies, was developed and evaluated within a funded program [51]. Thin film structures are applied on fired LTCC substrates without special surface treatment. This process allows a combination of printed thick film and thin film structures on the surface. Hence, the combined technology offers improved features, such as smaller lines and spaces. Thin film features can be reduced to necessary areas and special thick film materials for hermetic sealing can be applied, thus providing options to reduce costs, size, and weight.

4.7.4 System-on-Chip

Increasing demand for low-cost, broadband, high-speed, and small wireless communication devices, especially in the millimeter wave frequency range, has turned the SoC (system on-chip) solution into an important technique to satisfy these demands [55]. One of the most important performance problems for on-chip antennas in the millimeter wave range is substrate loss. By using micromachining techniques [56], it is possible to remove unwanted regions of the substrate and thus reduce substrate loss.

Considering the silicon implementation process, (100) silicon substrates with a thickness of 550 μm are typically used for the realization of on-chip antennas [56]. The fabrication of these devices requires two steps, involving back- and front-side processing. The back-side etching is performed in a KOH (potassium hydroxide) solution with a concentration of 8-moles and at a temperature 52–58°C. During this step, silicon is removed through the openings in the masking layer over a period of 25–30 hours. Since this etching step is rather extended, it is important to use a masking layer that is capable of withstanding long exposures to etching chemicals.

REFERENCES

[1] R. S. Elliott, Antenna Theory and Design, IEEE Press Series on Electromagnetic Wave Theory, Wiley-IEEE Press; New Jersey, U.S. Rev Sub edition, 2003.

[2] R. Janaswamy, "An indoor pathloss model at 60GHz based on transport theory." *IEEE Antennas and Wireless Propagation Letters*, Vol. 5, No. 1, pp. 58–60, 2006.

[3] G. V. Trentini, "Partially reflecting sheet arrays." *IEEE Trans. Antennas Propag.* Vol. 4, No. 4, pp. 666–671, 1956.

[4] M. Thévenot, C. Cheype, A. Reineix, and B. Jecko, "Directive photonic bandgap antennas." *IEEE Trans. Microw. Theory Tech.*, Vol. 47, No. 11, pp. 2115–2122, 1999.

[5] B. Temelkuaran, M. Bayindir, E. Ozbay, R. Biswas, M. M. Sigalas, G. Tuttle, and K. M. Ho, "Photonic crystal-based resonant antenna with a very high directivity." *J. Appl. Phys.*, Vol. 87, No. 1, pp. 603–605, 2000.

[6] R. Biswas, E. Ozbay, B. Temelkuran, M. Bayindir, M. M. Sigalas, and K.-M. Ho, "Exceptionally directional sources with photonic-bandgap crystals." *JOSA B*, Vol. 18, No. 11, pp. 1684–1689, 2001.

[7] A. Fehrembach, S. Enoch, and A. Sentenac, "Highly directive light sources using two-dimensional photonic crystal slabs." *Appl. Phys. Lett.*, Vol. 79, No. 26, pp. 4280–4282, 2001.

[8] C. Cheype, C. Serier, M. Thevenot, T. Monediere, A. Reinex, and B. Jecko, "An electromagnetic bandgap resonator antenna." *IEEE Trans. Antennas Propag.*, Vol. 50, No. 9, pp. 1285–1290, 2002.

[9] Y. J. Lee, J. Yeo, R. Mittra, and W. S. Park, "Application of electromagnetic bandgap (EBG) superstrates with controllable defects for a class of patch antennas as spatial angular filters." *IEEE Trans. Antennas Propa.*, Vol. 53, No. 1, pp. 224–235, 2005.

[10] R. Gardelli, M. Albani, and F. Capolino, "Array thinning by using antennas in a Fabry-Perot cavity for gain enhancement." *IEEE Trans. Antennas Propag.*, Vol. 54, No. 7, pp. 1979–1990, 2006.

[11] A. R. Weily, L. Horvath, K. P. Esselle, B. C. Sanders, and T. S. Bird, "A planar resonator antenna based on a woodpile EBG material." *IEEE Trans. Antennas Propag.*, Vol. 53, No. 1, pp. 216–223, 2005.

[12] S. Enoch, G. Tayeb, P. Sabouroux, N. Guérin, and P. Vincent, "A metamaterial for directive emission." *Phys. Rev. Lett.*, Vol. 89, pp. 213902-1–213902-4, Nov. 2002.

[13] G. Lovat, P. Burghignoli, F. Capolino, D. R. Jackson, and D. R. Wilton, "Analysis of directive radiation from a line source in a metamaterial slab with low permittivity." *IEEE Trans. Antennas Propag.*, Vol. 54, No. 3, pp. 1017–1030, 2006.

[14] G. Lovat, P. Burghignoli, F. Capolino, and D. R. Jackson, "High directivity in low-permittivity metamaterial slabs: rayoptic vs. leaky-wave models." *Microwave Opt. Techn. Letters*, Vol. 48, No. 12, pp. 2542–2548, 2006.

[15] G. Lovat, P. Burghignoli, F. Capolino, and D. R. Jackson, "Combinations of low/high permittivity and/or permeability substrates or highly directive planar metamaterial antennas." *IEEE Trans. Microw. Antennas Propag.*, Vol. 1, No. 1, pp. 177–183, 2007.

[16] G. Lovat, P. Burghignoli, F. Capolino, and D. R. Jackson, "Highly-directive planar leaky-wave antennas: a comparison between metamaterial-based and conventional designs." *EuMA (European Microwave Association) Proceedings*, Vol. 2, pp. 12–21, 2006.

[17] K. Huang and Z. Wang, "V-band patch-fed rod antennas for high data-rate wireless communications." *IEEE Trans. Antennas Propag.*, Vol. 54, No. 1, pp. 297–300, 2006.

[18] W. Menzel, D. Pilz, M. Al-Tikriti, "Millimeter-wave folded reflector antennas with high gain, low loss, and low profile." *IEEE Antennas Propag. Mag.*, Vol. 44, No. 3, pp. 24–29, 2002.

[19] F. Colomb, K. Hur, W. Stacey, and M. Grigas, "Annular slot antennas on extended hemispherical dielectric lenses." *Antennas and Propagation Society International Symposium 1996 (AP-S)*, Vol. 3, pp. 2192–2195, Jul. 1996.

[20] W. Y. Ali-Ahmad, G. V. Eleftheriades, L. P. B. Katehi, and G. M. Rebeiz, "Millimeter-wave integrated-horn antenna. II. Experiment." *IEEE Trans. Antennas Propag.*, Vol. 39, No. 11, pp. 1582–1586, 1991.

[21] S. Sierra-Garcia and J.-J. Laurin, "Study of a CPW inductively coupled slot antenna." *IEEE Trans. Antennas and Propag.* Vol. 47, No. 1, pp. 58–64, 1999.

[22] L. Malland and R. B. Waterhouse, "Millimeter-wave proximity-coupled microstrip antenna on an extended hemispherical dielectric lens." *IEEE Trans. Antennas Propag.*, Vol. 49, No. 12, pp. 1769–1772, 2001.

[23] J. A. Besley, N. N. Akhmediev, and P. D. Miller, "Methods of periodic waveguides." *Optics Lett.*, Vol. 22, No. 15, pp. 1162–1164, 1997.

[24] G. M. Rebeiz, L. P. B. Katehi, W. Y. Ali-Ahmad, G. V. Eleftheriades, and C. C. Ling "Integrated horn antennas for millimeter-wave applications." *IEEE Antennas Propag. Mag.*, Vol. 34, No. 1, pp. 7–16, 1992.

[25] X. Wu, G. V. Eleftheriades, and T. E. van Deventer-Perkins, "Design and characterization of single- and multiple-beam mm-wavecircularly polarized substrate lens antennas for wireless communications." *IEEE Trans. Microw. Theory Tech.*, Vol. 49, No. 3, pp. 431–441, 2001.

[26] M. Sironen, Y. Qian, and T. Itoh, "A 60GHz conical horn antenna excited with quasi-Yagi antenna." *Microwave Symposium Digest, IEEE MTT-S International*, Vol. 1, pp. 547–550, 2001.

[27] L. J. Chu, "Physical limitations of omni-directional antennas." *J. Appl. Phys.*, Vol. 19, pp. 1163–1175, Dec. 1948.

[28] R. F. Harrington, "On the gain and beamwidth of directional antennas." *IRE Trans. Antennas Propag.*, Vol. 6, pp. 219–225, 1958.

[29] E. Kreyszig, *Advanced Engineering Mathematics*, 9th edition, Hoboken, NJ: Wiley, 2005.

[30] W. Geyi, "Physical limitations of antenna." *IEEE Trans. Antennas Propag.*, Vol. 51, No. 8, pp. 2116–2123, 2003.

[31] R. L. Fante, "Quality factor of general ideal antennas." *IEEE Trans. Antennas Propag.*, Vol. 17, No. 2, pp. 151–155, 1969.

[32] R. F. Harrington, "Effect of antenna size on gain, bandwidth, and efficiency." *J. Research National Bureau of Standards—D. Radio Propagation*, Vol. 64D, No. 1, 1960.

[33] A. C. Ludwig, "The definition of cross polarization." *IEEE Trans. Antennas Propag.* Vol. 21, No. 1, pp. 116–119, 1973.

[34] T. Manabe, K. Sato, H. Masuzawa, K. Taira, T. Ihara, Y. Kasashima, and K. Yamaki, "Polarization dependence of multipath propagation and high-speed transmission characteristics of indoor millimeter-wave channel at 60GHz." *IEEE Trans. Veh. Technol.*, Vol. 44, No. 2, pp. 268–274, 1995.

[35] L. Lukama, K. Konstantinou, and D.J. Edwards, "Polarization diversity performance for UMTS." *Proceedings of the Int. Conf. on Antennas and Propagation (ICAP2001)*, Manchester, England, Apr. 2001.

[36] R. G. Vaughan, "Polarization diversity in mobile communications." *IEEE Trans. Veh. Technol.*, Vol. 39, No. 3, pp. 177–186, 1990.

[37] L. C. Lukama, D. J. Edwards, and A. Wain, "Application of three-branch polarisation diversity in the indoor environment." *IEE Proceedings Communications*, Vol. 150, No. 5, pp. 399–403, 2003.

[38] L. C. Lukama, K. Konstantinou, and D. J. Edwards, "Performance of a three-branch orthogonal polarization diversity scheme." *IEEE Vehicular Technology Conference Proceedings,* pp. 2033–2037, 2001.

[39] L. Lukama, K. Konstantinou, and D. J. Edwards, "Three-branch orthogonal polarization diversity scheme." *Electron. Lett.*, Vol. 37, No. 20, pp. 1258–1259, 2001.

[40] P. Kyritsi, D. C. Cox, R. A. Valenzuela, and P. W. Wolniansky, "Effect of antenna polarization on the capacity of a multiple element system in an indoor environment." *IEEE J. Sel. Areas Commun.*, Vol. 20, No. 6, pp. 1227–1239, 2002.

[41] C. A. Balanis, *Antenna Theory: Analysis and Design*, 2nd edition, John Wiley & Sons, New York, 1997.

[42] K.-C. Huang and Z. Wang, "Millimeter-wave circular polarized beam-steering antenna array for gigabit wireless communications." *IEEE Trans. Antennas Propag.*, Vol. 54, No. 2, part 2, pp. 743–746, 2006.

[43] J.-Y. Park, Y. Wang, and T. Itoh, "A 60GHz integrated antenna array for high-speed digital beamforming applications." Proceedings of IEEE MTT-S International Microwave Symposium Digest, Vol. 3, pp. 1677–1680, Philadelphia, PA, Jun. 2003.

[44] K. Huang and Z. Wang,"Antenna array comprising at least two groups of at least one rod antenna." European patent EP1703590B1, Sep. 2008.

[45] K. Huang and D. Edwards "60GHz multi-beam antenna array for gigabit wireless communication networks." *IEEE Trans. Antennas Propag.*, Vol. 54, No. 12, pp. 3912–3914, 2006.

[46] K. Huang and S. Koch,"Circular polarization antenna." US patent, US7212163B2 May 2007.

[47] K. R. Carverand J. W. Mink "Microstrip antenna technology." *IEEE Trans. Antennas Propag.*, Vol. 29, No. 1, pp. 18–20, 1981.

[48] A. Vimpari, A. Lamminen, and J. Säily "Design and measurements of 60GHz probe-fed patch antennas on low-temperature co-fired ceramic substrates." *Proceedings of the 36th European Microwave Conference 2006,* pp. 854–857, 2006.

[49] Wikipedia "Low temperature co-fired ceramic." http://www.wikipedia.org/.

[50] Ferro "LTCC A6 system for wireless solutions." Ferro® Electronic Material.

[51] G. Vanrietvelde, E. Polzer, S. Nicotra, J. Mueller, and A. Brokmeier, "Microwave and mm-meter wave applications: a new challenge for ceramic thick film technology." *IEE Seminar Microwave Thick Film Materials and Circuits*, (2002/097), pp. 6–10, 2002.

[52] R.S. Khandpur, *Printed Circuit Boards,* McGraw-Hill Professional, New York, 2005.

[53] W. Falinski, G. Koziol, and J. Borecki, "Laser structuring of fine line printed circuit boards." *Electronics Technology: Meeting the Challenges of Electronics Technology Progress, 2005. 28th International Spring Seminar*, pp. 196–201, May 2005.

[54] DuPont electronic material data sheet "Fodel® 6778 conductor." http://www. dupont.com/

[55] Y. P. Zhang, "Recent advances in integration of antenna on silicon chip and ceramic package." *IEEE International Workshop on Antenna Technology*, pp. 151–154, 2005.

[56] G. P. Gauthier, J.-P. Raskin, L. P. B. Katehi, and G. M. Rebeiz, "A 94 GHz aperture-coupled micromachined microstrip antenna." *IEEE Trans. Antennas Propag.*, Vol. 47, No. 12, pp. 1761–1766, 1999.

5

MILLIMETER WAVE MIMO

Multiple-input-multiple-output (MIMO) systems apply multiple antennas at both the transmitters and receivers to improve communication performance. MIMO technology offers significant increases in data throughput and link range without additional bandwidth or transmission power. It achieves this by higher spectral efficiency (more bits per second per hertz of bandwidth) and link reliability or diversity (reduced fading).

Combining millimeter wave and MIMO technologies has become important. The use of MIMO technology improves the reception and allows for a better reach and diversity. The implementation of MIMO also gives millimeter wave applications a significant increase in spectral efficiency. A MIMO configuration can be negotiated dynamically between individual base and mobile stations, such as with mobile WiMAX. Thus it could support a mix of mobile stations with different MIMO capabilities. This helps to maximize the sector throughput by leveraging the different capabilities of a diverse set of mobile stations.

In the previous chapters, we have discussed path loss and multipath issues. Solutions are provided in Chapter 4. This chapter will present solutions for the shadowing issue. When a person walks across a wireless link, or any other obstacle blocks the wireless link, millimeter wave communication will stop immediately. Neither a sharp beam antenna with fixed direction nor a wide beam antenna can support non-line-of-sight (NLOS) Gbps transmission over the 60 GHz band for the following two reasons:

Millimeter Wave Communication Systems, by Kao-Cheng Huang and Zhaocheng Wang

1. A loss of more than 30 dB is caused by an obstruction from a human body or other obstacles (Figure 1.11). This makes it difficult to have a sufficiently large Eb/N_0 in the link budget.
2. Severe channel distortion is caused by an obstruction from a human body or other obstacles. To solve the issue, one solution is to use a single-carrier solution with a complex receiver with two FFTs (fast fourier transform) and a frequency domain equalizer. The other is to use an OFDM solution with the challenge of high peak-to-average ratio.

A sharp beam antenna with a beam steering function is required for NLOS Gbps wireless data transmission at 60 GHz. A sufficiently strong signal with little distortion can be acquired from ceiling and wall reflections.

Table 5.1 summarizes millimeter wave antenna design issues and solutions.

The issue of human shadowing in the wireless environment can be overcome by a MIMO system. MIMO builds multiple channels and then selects the channel with best signal quality for communications. In addition, it has the following features:

- No complex baseband circuit
- Simple, noncoherent receiver chain
- Simple antenna feeding network
- Easy-to-optimize beam shape

5.1 SPATIAL DIVERSITY OF ANTENNA ARRAYS

The antenna capture area for a 60 GHz radio is small compared to that for a lower-frequency wireless system. Thus, it becomes possible to implement an antenna array in portable devices and to improve the antenna directivities. While it is possible to increase the antenna gain for a single antenna (e.g., using mechanical structures such as a horn antenna), it is more desirable to increase the directivity by employing an antenna array or MIMO system, as shown in Figure 5.1. For a fixed antenna aperture size a, the directivity

TABLE 5.1. Summary of Millimeter Wave Antenna Issues and Solutions

Issues	Solution	Benefit	Challenge	Note
Path Loss	High gain	Compensates for path loss	Antenna size	Section 4.1
Multipath	Beam steering	Reduces multipath effect, smaller notch depth, wider notch period	Steering speed	Section 4.2
				Section 4.5
Shadowing	NLOS link with multibeam	Reduces loss from obstacles, shadows.	Precise tracking for Tx and Rx antennas.	Section 5.2

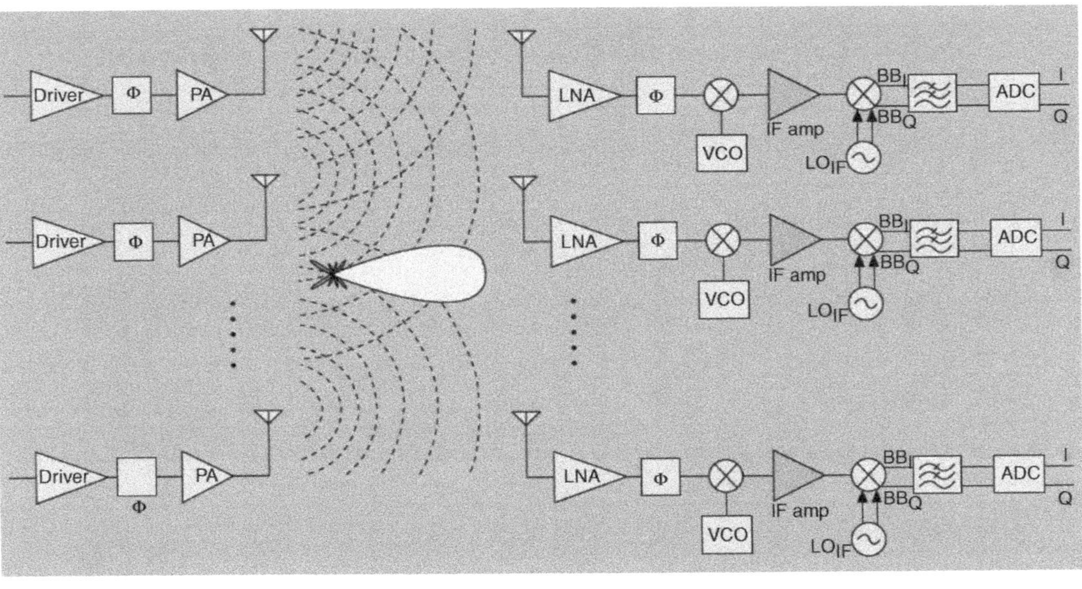

Figure 5.1. A generic multiple transceiver architecture with beam-steering antennas [1] (©2004 IEEE)

is $D = 4\pi a/\lambda^2$ and, from Equation (1.2), it can be seen that there is actually an improvement in the received power by moving to higher frequencies for a fixed antenna form factor. For example, a 60 GHz system with a 16-element antenna array has a 3-dB gain over a 5-GHz omnidirectional system, while occupying only 10% of the antenna area.

For land-mobile radio, the diversity method has been proven to be very effective. In fact, spatial diversity with receiving antennas spaced typically 10 or more wavelengths apart on cellular telephone towers is in widespread use. Base station spatial diversity is discussed in [7]. Polarization diversity has been found to be nearly as effective as spatial diversity for base stations [8–10], and provides great space and cost savings. Polarization diversity also compensates for any polarization mismatch due to random handset orientation [11]. Pattern diversity using multiple directional beams has also been investigated for use at base stations that have switched-beam smart antennas [12].

5.1.1 Transceiver Architecture

A generic adaptive beamforming multiple antenna radio system is shown in Figure 5.1. We now assume that antenna elements are small enough to be directly integrated into the package, or even potentially on-chip. The main benefit of the multi-antenna architecture used here is the increased gain that the directional antenna pattern can provide, which, as we have seen, is needed in order to support multi-gigabit-per-second data rates at typical indoor distances. In addition to the antenna gain, the use of antenna arrays also provides spatial (or angular) diversity, automatic spatial power combining, and an electronic beam steering function. The transceiver architecture in Figure 5.1 depicts N independent transmission and reception chains. Such an approach would enable a flexible MIMO system that could fully exploit a multipath-rich environment for increased capacity and/or robustness [1].

The main disadvantages of this arrangement are the high transceiver complexity and power consumption, since there is little sharing of the hardware components. Measurements of the 60 GHz channel properties indicate that most of the received energy is contained in the specular path [2], so a full MIMO solution targeting capacity may not be able to fully benefit from this channel. A more efficient implementation would be to use a phased array that takes the identical RF signal and shifts the phase for each antenna to achieve beam steering. Essentially, communication systems can select one strong path and apply an angular or spatial filter to form a narrow beam in the direction of the chosen signal [3]. This approach significantly reduces hardware costs, as most of the transceiver can be shared, with the addition of controllable phase shifters between the transceiver and antenna array.

The large spectral efficiencies associated with MIMO channels are based on the premise that a rich scattering environment provides independent transmission paths from each transmitting antenna to each receiving antenna. Therefore, for single-user systems, a transmission and reception strategy that exploits this structure achieves capacity on approximately $min\ (N,M)$ separate channels, where M is the number of transmitting antennas and N is the number of receiving antennas. Thus, capacity scales linearly with min (M,N), relative to a system with just one transmitting and one receiving antenna. This capacity increase requires a scattering environment such that the matrix of the channel gains between the transmitting and receiving antenna pairs has full rank and independent entries, and perfect estimates of these gains are available at the receiver.

5.1.2 Frequency Selective Fading Channels

While flat fading is a realistic assumption for narrowband systems where the signal bandwidth is smaller than the channel coherence bandwidth, broadband communications involve channels that experience frequency selective fading. Research on the capacity of MIMO systems with frequency selective fading typically takes the approach of dividing the channel bandwidth into parallel flat fading channels and constructing an overall block diagonal channel matrix, with the diagonal blocks given by the channel matrices corresponding to each of these subchannels. Under perfect channel state information at the receiver (CSIR) and channel state information at the transmitter (CSIT), the total power constraint then leads to the usual closed-form waterfilling solution. Note that the waterfill is done simultaneously over both space and frequency. Even single-input-single-output (SISO) frequency selective fading channels can be represented by the MIMO system model in this manner [4]. For MIMO systems, the matrix channel model was derived by Bolcskei, Gesbert, and Paulraj in [5], based on an analysis of the capacity behavior of OFDM-based MIMO channels in broadband fading environments. Under the assumption of perfect CSIR and CDIT for the zero-mean spatially white (ZMSW) model, their results show that, in the MIMO case, unlike the SISO case, frequency selective fading channels may provide advantages over flat fading channels, not only in terms of ergodic capacity but also in terms of capacity versus outage. In other words, MIMO frequency selective fading channels are shown to provide both higher diversity gain and higher multiplexing gain than MIMO flat-fading channels. The measurements in [6] show that frequency selectivity makes the cumulative distribution function (CDF) of the capacity steeper and, thus, increases the capacity for a given outage as compared with the flat fading case, but the influence on the ergodic capacity is small.

Spatial diversity for reception uses two or more antennas separated in space. Diversity for a transmitter is also possible, but systems are often uplink-limited and link improvements are primarily needed for the receiver side of the base station. In multipath propagation conditions, as encountered with a blocked or shadowed direct line-of-sight (LOS) path, each receiving antenna experiences a different and independent fading environment. Thus, it is always the case that if one antenna is in a deep fade, then the other one without fading can provide a sufficient signal level. Base stations require wide antenna spacing for proper diversity operation because the multipath arrival is over a narrow angle spread [1]. However, as shown in [13], under wide multipath angle spread conditions, diversity spacing can be small. This occurs in outdoor urban environments and for the indoor operation of mobile/personal terminals.

It was found that antenna spacings as small as 1/10 th of a wavelength provide a diversity gain of up to 8 dB at a 1% outage level. Similar large gains for polarization and pattern diversity with small antenna spacings were also observed. Measurements, including operator effects, showed a diversity gain of over 8 dB using antennas spaced

0.25 wavelengths apart, with the operator's head next to the antennas. The results of this investigation can be used to design effective diversity antennas for handheld radios.

This discussion is confined to reception diversity, although transmission diversity can have similar effects. Diversity antennas provide two major benefits. First, reliability is improved in multipath channels, where interference from reflected signals causes fading of the received signal. The fade level experienced on average for a given outage probability (percentage of down time) is decreased through diversity. Systems that use diversity combining can provide a diversity gain (to be defined below) of 10 dB or greater for the worst 1% of cases. Second, the overall average received signal power is increased. Systems that use polarization or angle diversity automatically match the antenna characteristics to the received signal and increase the efficiency of the radio link. These gains can be dramatic. A radio without polarization diversity can easily experience a 10 to 20 dB decrease in mean received signal power due to polarization mismatch. A simple polarization diversity system can provide at least half the best-case received signal power for even the worst polarization mismatch.

The use of diversity antennas results in improvements in system performance. Diversity gains from antennas permit the use of lower transmission power for a given level of reliability. Thus, diversity reduces interference to and from other users, and reduces the probability that a hostile party will intercept the signals. Battery life is increased in peer-to-peer handheld systems. The outputs from diversity antennas can be selected or combined in several ways to optimize the received signal power or signal-to-noise ratio (SNR). These techniques include maximal ratio combining, equal gain combining, selection diversity, and switched diversity. These methods are described in Chapter 6, and more details can be found in [14]. In this investigation, maximal ratio combining and selection diversity were considered. The two techniques do not differ greatly in performance, with maximal ratio combining providing about a 1.5 dB higher diversity gain at the 1% cumulative probability (99% reliability) level. Adaptive beamforming algorithms can also provide diversity gain, in addition to rejecting interfering signals [15].

In general, multiple diversity dimensions can be exploited dynamically in receivers. The diversity dimensions that are available are (see also Figure 5.2):

- Spatial—multiple antenna elements occupy separate locations in the radio
- Polarization—the antenna(s) provide dual orthogonal polarizations
- Pattern or angle—directional antennas discriminate over angle space

5.2 MULTIPLE ANTENNAS

In a multipath environment, antennas with a narrow beam can be used to reduce the number of multipath signals, and therefore to minimize the root-mean-square delay spreads. Moreover, a narrow-beam antenna has a high directivity, which directs or confines the power or reception to a given direction and thus extends the communication range. Furthermore, the gain of the antennas can partly reduce the required gain of

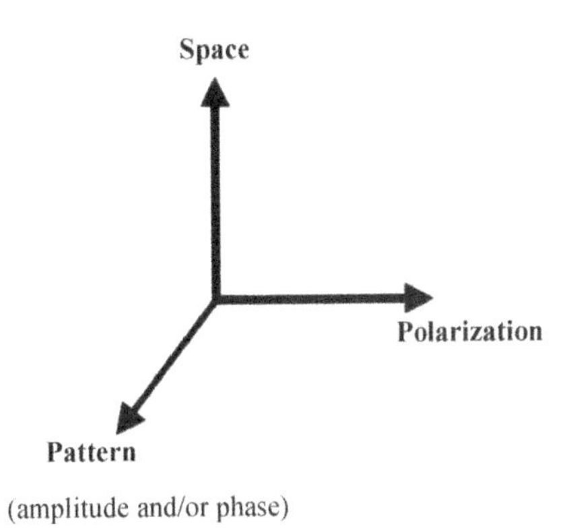

Figure 5.2. Three "dimensions" of antenna diversity

millimeter wave power amplifiers by supplying more captured power to the output terminals of the antenna, potentially reducing the power consumption in the millimeter wave circuitry.

Conventional millimeter wave links can be classified based on whether or not an uninterrupted LOS is established between the transmitter and receiver. For indoor applications, NLOS scenarios, also called diffuse links, are very common [17]. Conventional millimeter wave communication systems, whether LOS or NLOS, mostly employ a single antenna. This section discusses a method of improving the performance by using a multi-beam directional array, which utilizes multiple elements that are pointed in different directions [17]. Overall, such an angle-diversity antenna array can have a high directivity and larger information capacity.

The concept also offers the possibility of reducing the effects of co-channel interference and multipath distortion, because the unwanted signals are angularly filtered out (by the individual narrow beams). The multibeam antenna array can be implemented using multiple dielectric rods that are oriented in different directions. A conventional rod antenna is fed by waveguides [18], an arrangement that is too bulky for the particular array structure under consideration here. A more attractive approach for the dielectric rod configuration is to use a patch-fed method [19], which has the advantage of creating a rod array that can be integrated with a variety of planar circuits.

The geometry of an antenna array is shown in Figure 5.3. The central rod, antenna No. 1, is in an upright position, which is perpendicular to the plane of the patches, while the other rods (Nos. 2, 3, 4, 5, 6, and 7) are tilted at a polar angle (θ) of 40° relative to the surface normal, toward the plane of the patches. The tilted antennas are rotated with respect to the central rod and have an azimuthal angular spacing ($\Delta\Phi$) of 60°. Seven rod antennas are fitted into a metal plate with corresponding feeds. Each rod has different angular radiation coverage. When the antennas cover a given spatial area, each rod covers a nominally nonoverlapping cell in an arrangement that is similar to a cellular system, as shown in Figure 5.3(c).

The center frequency of the antenna described here is designed to be 60 GHz. As shown in Figure 5.4, the rods are made of Teflon® with a 3-mm diameter cylindrical base. The upper part of each rod is tapered linearly to a terminal aperture with 0.6-mm diameter

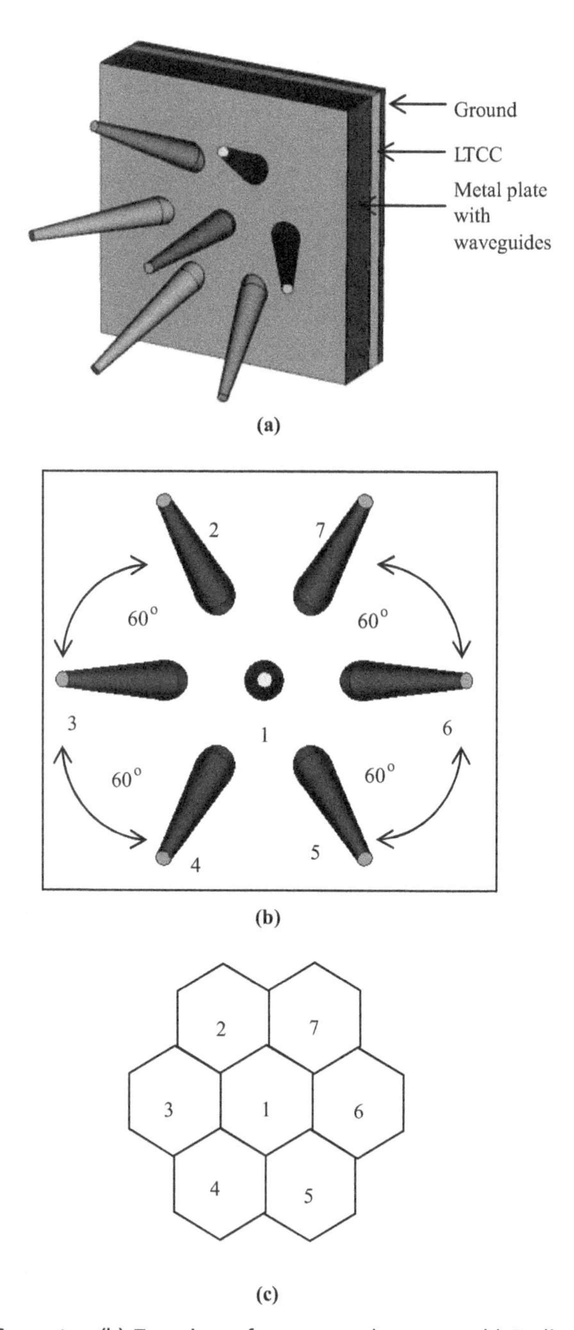

Figure 5.3. (a) Geometry. (b) Top view of a seven-rod antenna. (c) Radiation coverage in hexagonal configuration [16] (©2006 IEEE)

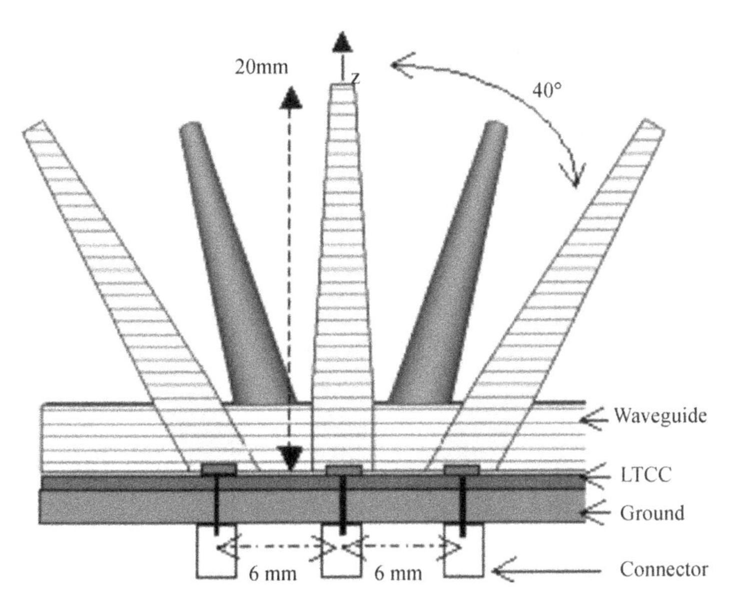

Figure 5.4. Side view of the rod antennas [16] (©2006 IEEE)

to reduce minor lobes in the radiation pattern. The total height of the central rod antenna is 20 mm to ensure a high antenna gain. Each rod antenna is designed to have a 40° half-power beamwidth. Therefore, the total radiation of the seven rod antennas covers a polar angle of approximately 60° with respect to the z-axis, which is the axis perpendicular to the plane of the patches.

The fields at the rod surfaces are derived using equivalent electric and magnetic current sheets, and the radiation field is simulated from these currents. The relative electric field pattern E as a function of the polar angle θ from the axis is derived by the following formula [19],

$$E(\theta) = (\sin \Phi)/\Phi \tag{5.1}$$

where $\Phi = H_\lambda \, \pi \, (\cos \theta - 1) - 0.5 \, \pi$, and H_λ is the height of a rod in free space wavelengths.

Each tapered rod can be treated as an impedance transformer, and reduces the reflection that would be caused by an abrupt discontinuity [20]. The rods are fed by patches on low temperature cofired ceramics substrates (LTCC) [21, 22]. The patch-fed method can be used to adapt rod antennas to most planar circuits. It saves feeding space and brings more design flexibility to the array structure. Additionally, it increases the directivity and bandwidth of conventional patch antennas alone.

The patches are either circular in shape or elliptical in shape to match the base of the rods. The patches are individually energized by probes connected to coaxial connectors. The probe feed can provide smaller sidelobes in comparison with microstrip-line feeds, because the coupling between the antennas and feeding lines is limited by the ground plane.

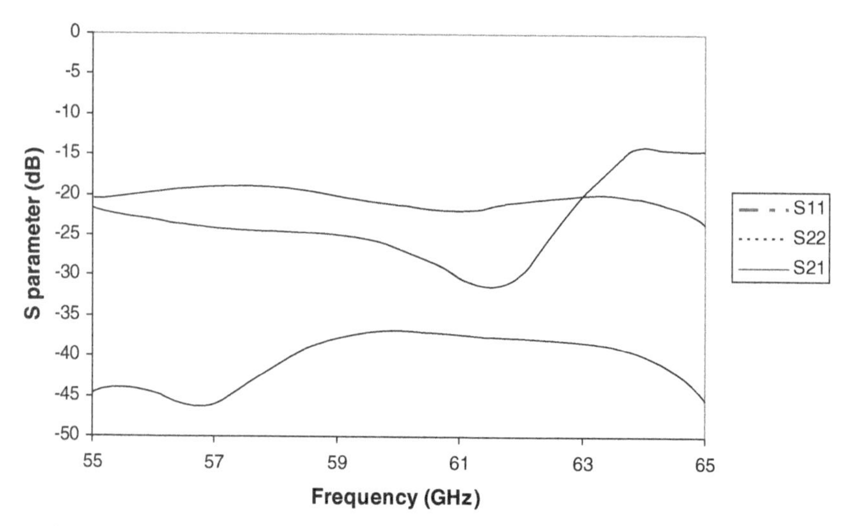

Figure 5.5. Measured S parameters for the upright rod, No. 1 (S11), one of the tilted rods, No. 2 (S22), and their mutual coupling (S21) [16] (© 2006 IEEE)

The performance of the above antennas was simulated using CST Microwave Studio, which is a simulator based on the finite integration time domain method. The spacing between adjacent rods was set to 6 mm to allow enough space between the test connectors. As can be seen in Figure 5.5, the antenna can be operated from 55 GHz to 65 GHz with a return loss of about −23 dB. Rod antennas show a broad impedance bandwidth and are suitable for wireless personal area networks, such as IEEE 802.15.3c-related applications [23]. The coupling between adjacent rods (e.g., No. 1 and No. 2, No. 3 and No. 2) was measured to be approximately −40 dB. Because of the symmetric configuration, any two pairs of rods with the same relative spacing have similar coupling coefficients. Appropriate grounding of the conducting plate is important to reduce the coupling from surface-wave propagation.

The 60 GHz radiation pattern is shown in a Cartesian plot in Figure 5.6. It depicts the 60 GHz radiation pattern of the upright rod (No. 1) and a tilted rod (No. 3) at a $\theta = 0°$ plane. The main beam is in the direction of $\theta = 0°$ and –40° for the upright rod and the tilted rod, respectively. It shows that the maximum gain of a rod (No. 3) radiates in the same direction as the rod's physical axis. The upright rod has a half-power beamwidth at $\theta = -20°$, while the tilted rod has a half-power beamwidth between $\theta = -14°$ and $-53°$. Because of the asymmetric shape of the tilted rod, one side has a longer height than the other. The side with the longer height can radiate and receive more energy than the short side. Thus, its radiation pattern is asymmetric, as shown in Figure 5.6. The curves are normalized to the pattern of the upright rod to show the relative power. For the tilted rod, there is a sidelobe at 13° with a level of about −8.5 dB. Measurement results for the proposed antennas were obtained to confirm the theoretical predictions.

The frequency response was measured by comparing it with a V-band standard horn antenna. As can be seen in Figure 5.7, an average antenna gain of about 11.5 dBi and 9.8 dBi was measured between 57 and 65 GHz for the upright and tilted rods,

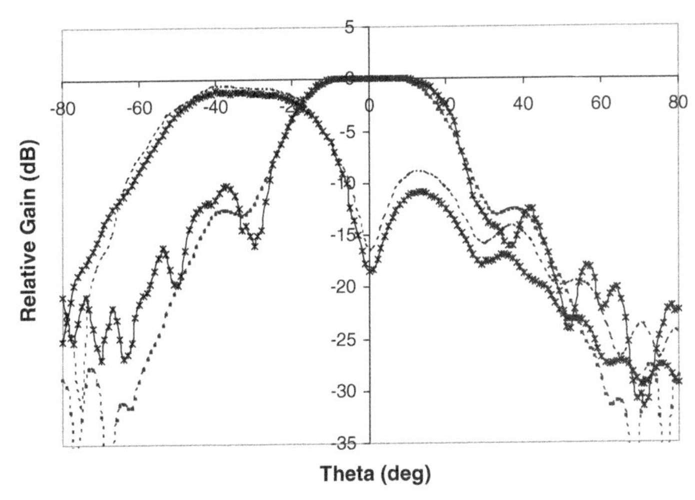

Figure 5.6. Simulated (--) and measured (-*-) radiation pattern at a phi = 0° plane as a function of the elevation angle at 60 GHz when the main beam is in the direction of theta = 0° and −40°, respectively. These curves are normalized to that of the central rod, No. 1 [16] (©2006 IEEE)

respectively. The 3-dB bandwidth of both the upright antenna and the tilted antenna are approximately 19% of the center frequency, which is higher than the 11% bandwidth reported in [24]. By comparing the measured antenna gain with the directivity, the radiation efficiency of the implemented prototype was estimated to be 80% and 73% for the upright and tilted rods, respectively, while the aperture efficiency was 74% and 57%, respectively. Note that the main beam is fairly circularly symmetric in its half-power beamwidth region and, therefore, the interelement angle spacing ensures a 3-dB radiation pattern overlapping in any principal cut in the azimuthal plane.

This antenna configuration is useful for transmitting antenna switch diversity. The procedure can be established as follows. Transmitter informs receiver of the number of antennas. At first transmission, transmitter subsequently selects one out of seven antennas for transmitting. At each stage, receiver will compare the received SNR with

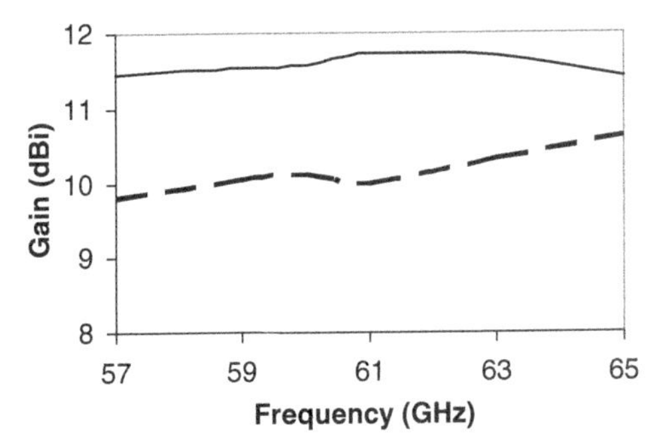

Figure 5.7. Maximum gain of upright rod (solid line) and tilted rod (dashed line) between 57 GHz and 65 GHz [16] (©2006 IEEE)

a predetermined threshold. If the SNR is larger than the threshold, the receiver will feedback a "no switch" signal. If the SNR is smaller than the threshold, the receiver will feedback a "switch" signal. If the SNR for all seven antennas is less than the threshold, the receiver will then feedback the antenna number with largest SNR. At last, the transmitter will use this antenna for transmission. A closed-loop iterative training algorithm can be used to determine the weights for each antenna element. A training procedure will be run periodically to compensate for channel change.

Interestingly, having a larger number of antennas does not need to result in a larger number of RF chains. By using antenna selection techniques [25] it is possible to retain the benefits of a large MIMO array with just a subset of antennas being active at the same time [26].

It is important to note that, as the number of antennas increases, the diversity effect will give diminishing returns. The data rate gain of spatial multiplexing remains linear with the number of antennas. Therefore, for a larger number of antennas it is expected that more weight has to be put on spatial multiplexing and less on diversity.

5.3 MULTIPLE TRANSCEIVERS

Point-to-point wireless links exist that can reach speeds on the order of Gbps. For example, a 1.25-Gbps point-to-point link using the 60 GHz band was reported in [27], and similar products are available in the marketplace [28]. MIMO technology can be used to increase such data rates by more than an order of magnitude, to 10–40 Gbps. In effect, MIMO technology provides the ability for an array to support many independent communication channels, as long as the elements in the array can "see" a separate link to a specific element in the transmitting array. The elements in the channel transfer function matrix **H** are (essentially) independent [30]. In addition to the natural application for communication infrastructure recovery after disasters, such wireless links offer tremendous commercial potential, as they can be used interchangeably with optical transmission equipments. For commercial applications, perhaps the greatest advantage of 10–40 Gbps wireless links is their lower cost, as they can be used to provide bridge connections between optical links, where difficult terrain, such as mountains and rivers, need to be crossed, or where installation costs are prohibitive, as in city centers.

In a system with M transmitters and N receivers, we can assume that the same signal is transmitted by each antenna. It is possible to get approximately an MN-fold increase in the SNR, yielding a channel capacity equal to

$$C \approx B \cdot \log_2(1 + MN \cdot SNR_0) \tag{5.2}$$

where B is bandwidth, SNR_0 is signal-to-noise ratio.

Thus, we can see that the channel capacity for a MIMO system is higher than that of a multiple-input-single-output (MISO) or single-input-multiple-output (SIMO) system. However, we should note here that in all four cases (i.e., SISO, MISO, SIMO, and MIMO) the relationship between the channel capacity and the SNR is logarithmic. This means that trying to increase the data rate by simply transmitting more power is extremely costly [29].

When different signals are transmitted by each antenna, it is assumed that $N \geq M$, so that all the transmitted signals can be decoded at the receiver. One interesting idea in MIMO is that we can send different signals using the same bandwidth and still be able to decode correctly at the receiver. Thus, it is like we are creating a separate channel for each of the transmitters. The capacity of each of these channels is roughly equal to [30].

$$C_{single} \approx B \times \log_2\left(1 + \frac{N}{M} SNR_0\right) \tag{5.3}$$

But, since we have M of these channels (M transmitting antennas), the total capacity of the system is

$$C \approx M \times B \times \log_2\left(1 + \frac{N}{M} SNR_0\right) \tag{5.4}$$

As we can see from Equation (5.4), a linear increase in capacity is obtained with respect to the number of transmitting antennas. Thus, the key principle here is that it is more beneficial to transmit data using many different low-powered channels than by using one single, high-powered channel [31]. Figure 5.8 shows the information capacity for one-input-one-output, two-input-two-output, and three-input-three-output systems,

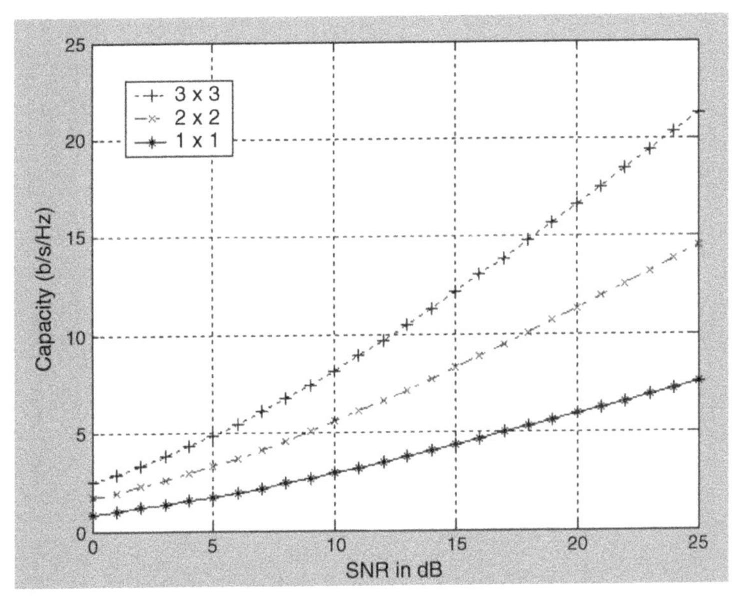

Figure 5.8. Information capacity for one-input-one-output, two-input-two-output, and three-input-three-output

respectively. As the number of inputs and outputs in a MIMO system increases, the information capacity increases accordingly.

In order to reduce the manufacturing cost for future 60 GHz MIMO, a high integration level is necessary [32]. Single-chip 60 GHz transmitter (Tx) and receiver (Rx) MMICs have been designed and characterized in a 0.15 m (120 GHz/max 200 GHz) GaAs metamorphic High Electron Mobility Transistor (mHEMT) and Monolithic Microwave Integrated Circuit (MMIC) process. The size of the receiver chip is 5.5 × 4.6 mm^2 and the size of the transmitter chip is 4.0 × 3.0 mm^2 [33]. It occupies small chip area and has high Rx gain (12.9 dB) and high Tx output power (5.6 dBm) at only half the DC power consumption (Tx: 420mW, Rx: 450mW). For a silicon approach, a 0.13 μm SiGe BiCMOS technology was used to make a double-conversion superheterodyne receiver and transmitter chipset [34]. The size of receiver chip is 3.4 × 1.7 mm^2. It achieves a 6-dB noise figure, 30 dBm IIP3 (third-order intermodulation intercept point), and consumes 500 mW. The size of transmitter chip is 4 × 1.6 mm^2. It achieves output P_{1dB} (1 dB compression point) of 10–12 dBm and consumes 800 mW. All these single-chip technologies can be applied to MIMO systems. Research has successfully shown that transmitter and receiver front-ends for the 60 GHz band are implemented in 90 nm CMOS [35]. Output power of the transmitter can go up to 6 dBm. The same CMOS technology can apply to the integration of a low-power beam former with signal processing [36]. With 64 antenna elements, the beam-former achieves a gain of 20 dBi. Such technology is very suitable for MIMO applications.

Alternatively, a MIMO system can be made using a six-port solution, as shown in Figure 5.9 [37]. The details of six-port technology have been introduced in Section 3.3.

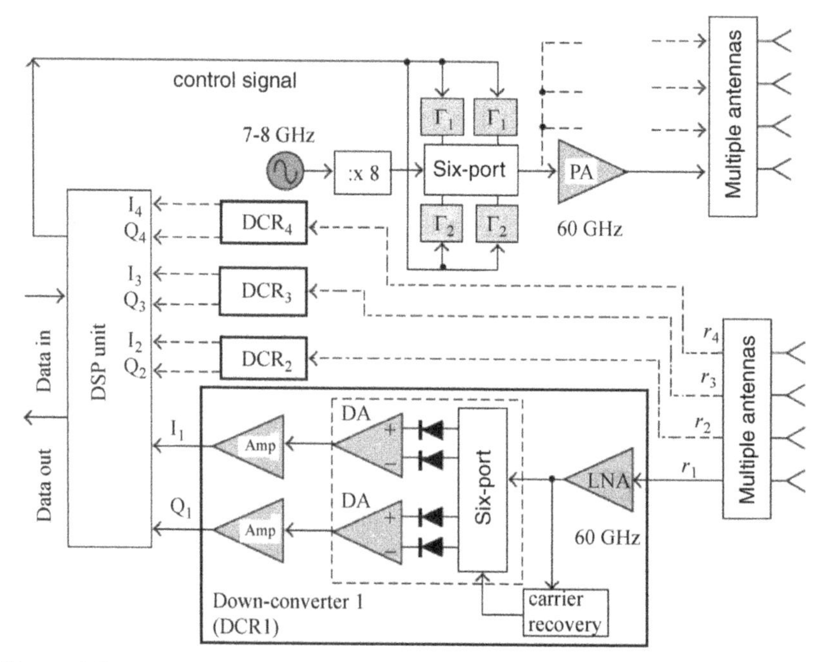

Figure 5.9. The simplified block diagram of a multiport direct conversion transceiver

Two-phased arrays based on Butler matrices are used. This solution appears optimal because a few discrete beam directions are generally sufficient in indoor wireless local area networks (WLANs). A 20-GHz microwave oscillator and a frequency multiplier generate the 60 GHz signal. The DSP unit modulates the carrier using a six-port direct modulator. According to the proposed 60 GHz standard, the required amplifier output power is + 10 dBm. In order to obtain control of the transmitted beams, the DSP will activate one of the millimeter wave amplifiers to feed the transmitter six-port antenna array. The corresponding direct conversion receiver is composed by a low-noise amplifier, a carrier recovery circuit, a six-port down-converter, power detectors, and differential amplifiers (DA), and two baseband amplifiers. The access to the inphase (I) and quadrature (Q) signals will enable significant additional capabilities, increasing the phase measurement accuracy and offering a straightforward correspondence between the baseband phasor rotation frequency and the Doppler shift if the same oscillator is used in the receiver part. The carrier recovery circuit is used as reference signal and will compensate the Doppler shift in a hardware approach.

The gain of MIMO systems can be analyzed in two categories:

1. Spatial multiplexing methods yield capacity gain. The capacity gain increases linearly with the number of transmitting antennas [38].
2. Diversity methods yield link quality gain. In MIMO diversity, the gains actually grow with the number of receivers.

Most MIMO algorithms focus on diversity or data rate maximization rather than just increasing the average SNR at the receiver or reducing interference. Measurement shows that the relevant SNR metric is 6–12 dB higher for the diversity-based schemes than for spatial-multiplexing based schemes, subject to the numbers of transmitting and receiving antennas [38]. These differences can be translated into smaller required transmit power, more coverage, better quality, or trade-off of all three.

Figure 5.10 is an example of a millimeter wave MIMO configuration. At these small wavelengths at 60 GHz, it is possible to synthesize highly directive beams with moderately sized antennas, permitting significant spatial reuse and drastically limiting multipath. The key concepts behind the proposed system are as follows:

1. Adaptive beamforming: Beamforming systems thrive in near Los environments because the beams can be more easily optimized to match one or two multipaths than a hundred of them. By forming a highly directive beam, steerable over 10 times the half power beamwidth, we can simplify the task of installation. Directivity gains are obtained at both the transmitter and receiver by the use of adaptive antenna arrays, which are termed subarrays.
2. Spatial multiplexing: The transmitting and receiving nodes each consist of an array of subarrays, as shown in Figure 5.10. After transmission and reception beamforming using the subarrays, each subarray can be interpreted as a single virtual element in a MIMO system. As a consequence of the small wavelength at 60 GHz, the moderate separation between the subarrays ensures that each virtual

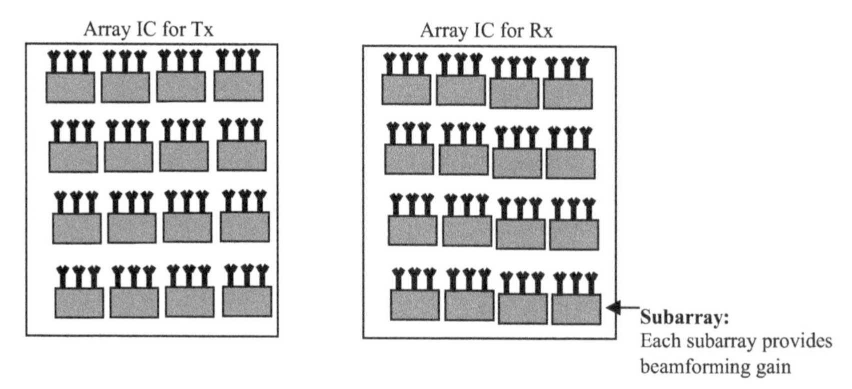

Figure 5.10. Configuration of a millimeter wave MIMO system. Each integrated circuit (IC) consists of an array of subarrays. Each subarray in a node steers a beam toward the node it is communicating with, providing beamforming gain and ISI reduction

transmitting element sees a sufficiently different response at the virtual receiving array. This enables spatial multiplexing: different virtual transmitting elements can send different data streams, with a spatial equalizer at the virtual receiving array used to separate the streams.

Figure 5.10 shows a 4×4 matrix of subarrays at each end with the following configurations:

- Each parallel spatial link employs quadrature amplitude modulation (QAM) with the full bandwidth, transmitting at 3 Gbps.
- Eight out of the 4×4 subarrays are selected to transmit parallel streams at 3 Gbps, resulting in an aggregate link speed of 24 Gbps.
- All 16 subarrays at the receiver are used in the spatial equalizer in order to separate out the 8 parallel data streams.

The signal processing underlying a millimeter wave MIMO system operates from a beamforming layer to a spatial multiplexing layer. At the beamforming layer, each subarray at the transmitter synthesizes a beam to point toward the receiver, and each subarray at the receiver synthesizes a beam to point toward the transmitter. Once these beams have been formed, the spatial multiplexing layer signal processing can continue for the resulting virtual MIMO system.

Possible low-cost implementations of millimeter wave MIMO systems depend on ongoing advances in modern CMOS technology for implementing millimeter wave RF circuits, as well as on cost-effective packaging techniques.

Beamforming and diversity using receiving antenna arrays are classical concepts in communication theory, but the important role played by transmitting antenna arrays, when used in conjunction with receiving arrays, was pointed out by the pioneering work of Telatar [39]. Since then, three major concepts for utilizing transmitting antenna arrays have emerged: spatial diversity (see Section 5.1), spatial multiplexing, and transmission beamforming.

Millimeter wave MIMO is different from other MIMO systems at lower frequencies in two aspects:

- Beam-forming layer: A beam-forming function is preferred for LOS or quasi-LOS channels, whereas diversity is preferred for NLOS channels so that the channel is robust.
- Spatial multiplexing layer: Spatial multiplexing is obtained by focusing the receiving antenna array on the different transmitting antenna elements, instead of relying on rich scattering.

Beam-forming layer signal processing for beamforming is discussed below. At millimeter wave frequencies, it is very challenging with current technology to perform analog-to-digital conversion of a signal with several GHz bandwidth at sufficient precision for beamforming on the complex envelope. The first step is therefore to consider an architecture for the beamforming layer that combines up/down conversion with antenna phase selection.

5.3.1 Beamforming Layer

The important device of a millimeter wave MIMO system is a beamsteering integrated subarray as shown in Figure 5.10. Each beamforming integrated circuit electronically steers an $M \times M$ antenna array. The required $M \times M$ is estimated to be between 4×4 and 10×10.

The antenna directivity is proportional to its effective aperture. The effective aperture of the subarray can be increased using multiple dielectric rods or a patch-fed horn implementation (see Figure 5.11), while maintaining the steerability of the antenna. This provides the necessary beamforming gains to offset the higher attenuation in millimeter waves and can be used to suppress multipath to the extent possible.

The gain of each subarray can be enhanced by:

1. Increasing the size of horns.
2. Increasing the height of rods.

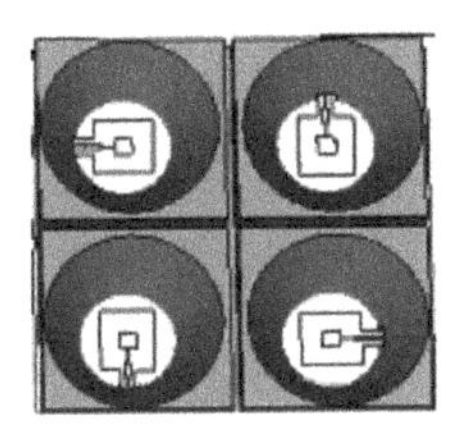

Figure 5.11. Steerable subarray configurations. (a) Patch-horn configuration. (b) rod antenna [21] (©2006 IEEE)

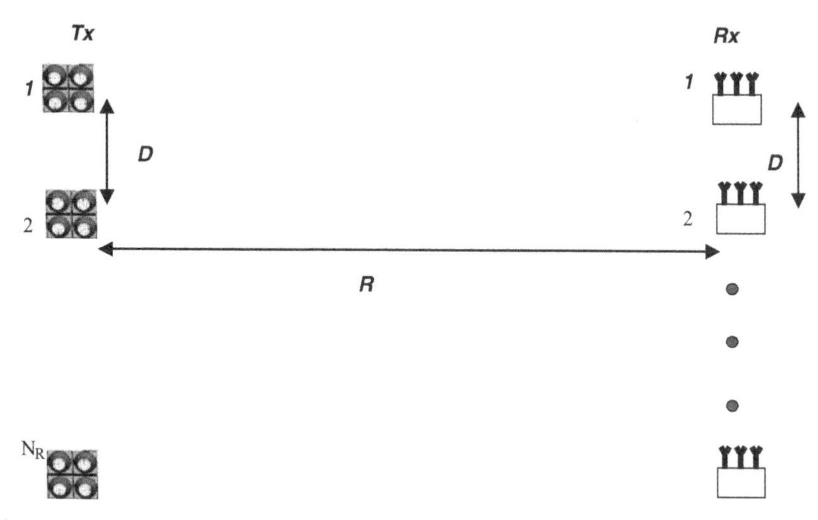

Figure 5.12. Geometry of a linear array MIMO system at the spatial multiplexing layer

While phased arrays at lower frequency can employ complex valued beamforming weights at baseband, such approaches do not scale to the symbol rates and carrier frequencies of interest in this book. Therefore, a matrix-type beamsteering IC, as depicted in Figure 5.13, is used, in which two multiphase local oscillators are mixed to synthesize the millimeter wave carrier for each antenna element. Thus, the phase of the (i, j)-*th* element of the array is given by

$$\phi(i,j) = \phi_h(i) + \phi_v(j), \qquad 1 \leq i,j \leq M \tag{5.5}$$

where $\phi_h(i)$ is the phase of the i-th row, and $\phi_v(j)$ is the phase of the j-th column, both chosen from a discrete set of values distributed uniformly around the unit circle. For the far-field regime, the transmitting subarrays beamform toward the receiver subarrays and vice-versa at the receiver, which can be accomplished efficiently using a two parameter search over the discrete set.

A special case of the matrix-type beamsteerer occurs when both the horizontal and vertical phases obey a linear profile, corresponding to steering a linear array in a specific direction. That is, $\phi_h(i) = i\delta_h$ and $\phi_v(j) = j\delta_v$, where $\delta_h = \frac{2\pi \sin\theta_h}{\lambda}$ and $\delta_v = \frac{2\pi \sin\theta_v}{\lambda}$ are the phase shifts for adjacent horizontal and vertical elements, respectively, corresponding to a horizontal steering angle of θ_h and a vertical steering angle of θ_v. Here the phase increments θ_h and θ_v must also be chosen from the discrete set allowed by the phase increments of $\pi/4$ or $\pi/8$, corresponding to the use of 8- and 16-phase oscillators, respectively. The minimum phase increment corresponds to the desired resolution in steering angle.

5.3.2 Spatial Multiplexing Layer

A millimeter wave MIMO antenna can be designed as an array of monolithic subarrays (see Figure 5.10). Spatial multiplexing is obtained by focusing the receiving subarrays on the different transmitting subarrays. Once the subarrays beamsteer in the desired direction, they can be considered antenna elements of a virtual MIMO system. An $N \times N$ array of subarrays with lateral spacing D has dimensions $(N-1)D \times (N-1)D$. To realize the desired spatial multiplexing, each of the N^2 virtual transmitting elements must see a sufficiently different N^2-ary (virtual) receiving array response in order to be able to separate out the different transmitted streams. The Rayleigh criterion in imaging [40] determines the minimum spacing between transmitting elements, so that they can be resolved by the receiving array with no coupling effect. In the case of sub-Rayleigh spacing, we could derive the correlation between the responses at the receiver to two different (virtual) transmitting elements. For uniform linear arrays (ULA) aligned to each other's broadside, as displayed in Figure 5.12, the spatial angular separation of the two transmitters is

$$\delta\theta = D/R \tag{5.6}$$

where R is the distance from transmitter to receiver.

Then, the signal phase separation at the receivers is

$$\delta\phi_e = \delta\theta \cdot 2\pi D/\lambda \tag{5.7}$$

If the phase difference at the receiver is $\delta\phi_e = \pi$ (e.g., $D = \sqrt{\lambda R/2}$), then the simple in-phase combining of receiver signals to aim the receiver array at the desired transmitter will result in 100% suppression of a signal from an undesired transmitter. This corresponds to the Rayleigh criterion in diffraction-limited imaging.

The above circuit functions are integrated in an array format as shown in Figure 5.13 to support an $M \times M$ antenna matrix. Consequently, the highly regular floor plan of the top-layer layout of the beamsteering IC eases the chip-to-board interface design, and thus accommodates the matching networks for the antenna matrix. This high layer of parallelism and complexity can be implemented using modern CMOS technology. The

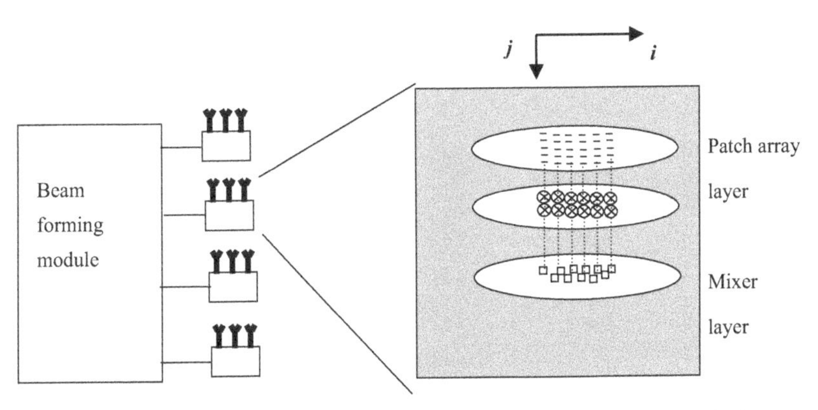

Figure 5.13. Configuration of a matrix-type beamsteering array [41]

main advantage of using CMOS for millimeter wave systems up to 100 GHz is its capability to integrate massively parallel transceiver arrays for directivity gain and adaptive beamforming. Such beamsteering ICs can be made using a 90-nm CMOS technology [42]. System-level simulations have shown a 90-dB gain with a 32×32 overall array (4×4 beamsteering ICs with each one supporting an 8×8 antenna matrix) [43]. Advanced research outcome has proved that a single-chip 60 GHz radio solution is possible using 45-nm digital CMOS technology [44]. This system is used to distribute uncompressed high-definition video with 16 antenna paths over a 10-m range. Only 1.6 W power is consumed for the complete receiver. This breakthrough paves the way to area-, cost- and power- efficient implementation of 60 GHz MIMO systems.

Figure 5.14 shows the median value of the capacity of the measured channel based on the number of elements per array for the cases of $[TX \cdot RX] = [2 \cdot 2], [3 \cdot 3], [4 \cdot 4]$, and $[5 \cdot 5]$, with three different SNR values of 10, 20, and 30 dB, respectively. It might be said that the capacity of the measured channel is slightly lower than that of the MIMO Rayleigh channel. The difference between the experimental results and the theoretical values increases as the SNR increases. This difference is mainly due to the fact, that when the distance between the elements is not infinite but is equal to one wavelength, the arrival and departure angles of the signals are not uniformly distributed, and the channel is not ideal (with a finite number of multipaths).

The antenna array configuration for the cases of $[5 \cdot 5]$ can be seen in Figure 5.15(a). A case of $[16 \cdot 16]$ is depicted in Figure 5.15(b). In this case, the antenna arrays at the transmitter and receiver are similar: they are squares with five "virtual antennas" per side. The space between any two elements is $\lambda/2$ or $\lambda/4$, depending on the measurement. Therefore, we get a 16×16 channel matrix (16 positions for the transmitter and 16 positions for the receiver). The scanner of the transmitter has a resolution of 2.5 μm. The scanner of the receiver has a resolution of 6.25 μm.

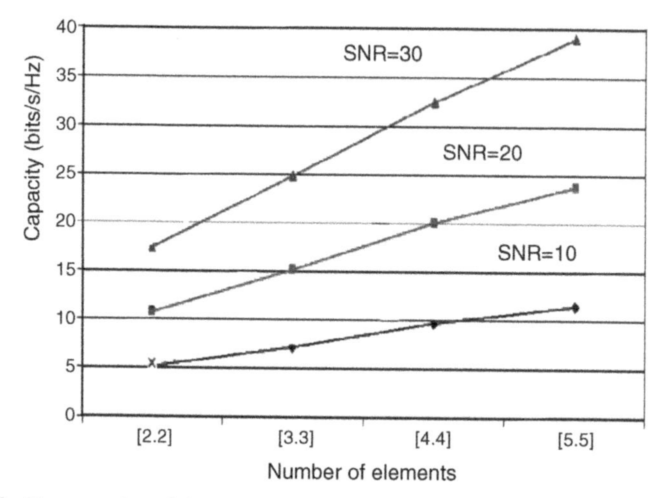

Figure 5.14. The capacity of the single point–point link for different modulation schemes, selecting two, three, four, and five elements at the Tx and Rx [45] (©2007 IEEE)

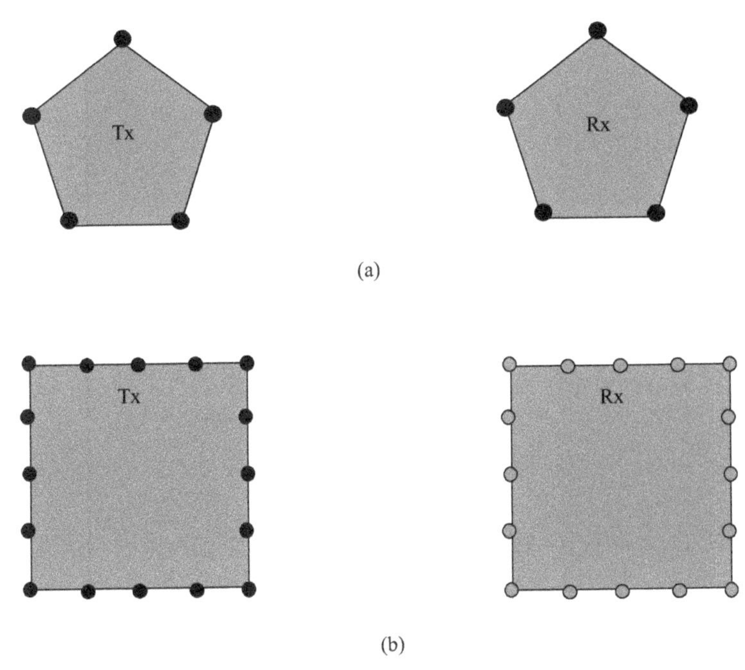

Figure 5.15. Antenna array configurations. (a) 5 to 5, (b) 16 to 16 [47] (©2005 IEEE)

NLOS measurements were made in order to demonstrate the wide measurement possibilities allowed by our system. Two identical antennas were used at the transmitter and receiver. In the measurement, the antennas were positioned 1.5m above the floor level. In the NLOS measurements, the receiver was kept in a room, whereas the transmitter was moved to different positions in a corridor. The distance between elements was $\lambda/2$. These NLOS measurements were used to calculate the capacity of the MIMO channel, as shown in Figure 5.16 [48]. Further measurements using a 5×5 antenna array at the Tx and a 5×5 antenna array at the Rx can be found in [49].

A millimeter wave ad hoc MIMO system utilizes distributed antennas that belong to several users, while conventional MIMO, or single-user MIMO, only utilizes antennas belonging to a self-terminal. This improves the performance of a wireless network by introducing the advantages of using multiple antennas, such as diversity, multiplexing, and beamforming. If the main interest lies in the diversity gain, this is known as cooperative diversity.

Ad hoc MIMO is a useful technique for future cellular networks, for which wireless mesh networking or wireless ad hoc networking are being considered. In wireless ad hoc networks, multiple transmitting nodes communicate with multiple receiving nodes. To optimize the capacity of ad hoc channels, the MIMO concept and techniques can be applied to the multiple links between the transmitting and receiving node clusters. Unlike the multiple antennas at a single-user MIMO transceiver, multiple nodes are located in a distributed manner. So, to achieve the capacity of this network, techniques to manage

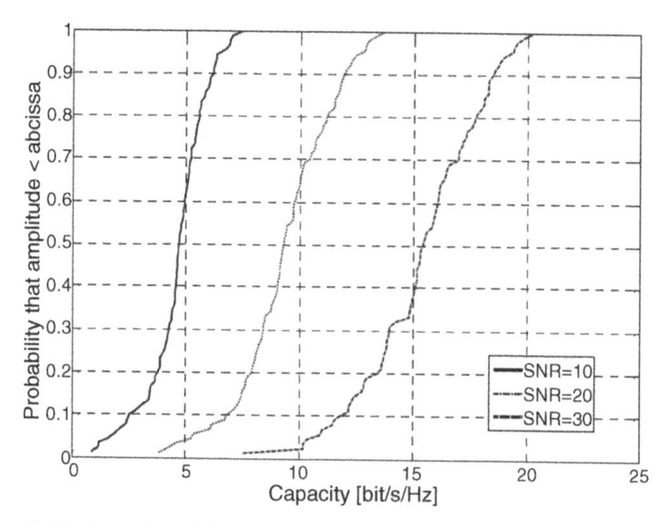

Figure 5.16. Capacity of the channel in a NLOS configuration [47] (©2005 IEEE)

distributed radio resources are essential, such as autonomous interference cognition, node cooperation, and network coding with the dirty paper coding concept.

An analogy to this trend of ad hoc MIMO is the evolution of computing cores. A single high-performance core system represented the first generation of CPU core evolution. This was then progressed to the recent computational environment, where a system consists of a few to many cores, in a centralized fashion. Many computing initiatives anticipate that cooperative work from multiple cores owned by different users will be made available to the individual user, in return for help with others' information processing, such as ambient intelligence, high-speed wireless ubiquitous computing, and the semantic web.

5.4 NOISE COUPLING IN A MIMO SYSTEM

Up until now, we have recongnized the signal coupling between antennas, but we should also note that the noise between antennas could be coupled. Because the size of millimeter wave circuits is small, the noise coupling effect can become an issue. Since SNR is affected by mutual coupling, noise behavior has to be considered in a multi-antenna system. This section only concentrates on the thermal noise effect. Other noise effects on the antenna array in a communication system can be found in [50, 51].

Thermal noise is produced by the thermal random motion of the atoms and electrons in all material. The performance of a communication system can be defined in terms of a system temperature T and a suitable noise temperature value T_N, chosen so that SNR can be written as

$$\frac{S}{N} = \frac{T}{T_N} \tag{5.8}$$

Thermal noise plays a key role in MIMO communication systems that use antenna arrays to increase the communication capacity [46]. High capacity could be achieved in these systems by providing independent channel matrix coefficients, a condition normally achieved with wide antenna element spacing. But the persistent miniaturization of consumer devices makes such large separations impossible, and the resulting antenna mutual coupling [52] significantly impacts the system performance. In [53], the noise that originated from the amplifier at the receiver end of a MIMO system was included in the consideration, but the thermal noise on the coupled antennas was not considered. The possibility that the thermal noise from a radiating body could be induced in an antenna was discussed in [54]. The topic of partially correlated noise sources that might be introduced into receivers from two closely spaced antennas was discussed in [55].

The impact of antenna mutual coupling on a MIMO system has been evaluated by examining how the coupled antennas change the signal correlation [53]. The modifications in the channel matrix coefficients are then used to assess the mutual coupling effects on the system capacity [56]. In addition, the radiated power at the transmitter and the power collection capability due to the effect of this mutual coupling in multi-antenna systems are presented in [57]. The effect of mutual coupling on the MIMO channel capacity in the context of the SNR was then presented in [58, 59].

The thermal noise that originates from the antenna material itself is called *self-noise* or *self-radiation*. Besides this self-radiated noise, induced thermal noise appears in the antenna from radiating bodies in the antenna's vicinity [60].

Thermal noise correlation due to mutual coupling effects in closely spaced antennas was a missing piece of the puzzle in the assessment of MIMO system performance with small antenna element separation, which is especially critical for customer units.

A multi-antenna system can be represented as a general linear network using a generalized form of Thevenin's theorem. The generalization of the theorem holds not only for coherent sources but also for thermal noise sources [61]. It is valid even for a general linear network that may contain a number of inaccessible (hidden) nodes, together with internal voltage and current sources, the locations of which may be unknown. However, as long as there are only N independent accessible nodes, such a system is indistinguishable from a noise-source-free network, with the same impedance or admittance matrix, together with a set of N nodal current generators of infinite internal impedance. The current from the generator of the *i-th* node, in such an equivalent network, is equal to the current flowing into the *r-th* node of the original network when all the nodes of the latter are short-circuited to ground. The internal sources may be alternatively represented by a set of N nodal voltage generators of infinite internal admittance, such that the voltage across the generator in the *r-th* node is equal to the voltage across the *r-th* node of the original network, when all the nodes of the latter are open-circuit. The nodal noise sources are not generally independent.

A multi-antenna system with $N = n_R$ antenna elements can be represented as a linear n_R-terminal-pair network containing internal prescribed signals or noise generators. It is specified completely with respect to its terminal pairs by its admittance matrix $\mathbf{Y}$ and a set of n_R nodal current generators $i_1, i_2, \ldots i_{nR}$. In matrix form, $\mathbf{Y}$ denotes a squared matrix of order n_R

$$\mathbf{Y} = \begin{pmatrix} y_{11} & y_{12} & & y_{1n_R} \\ y_{21} & y_{22} & & y_{2n_R} \\ & & & \\ y_{n_R1} & y_{n_R2} & & y_{n_Rn_R} \end{pmatrix} \tag{5.9}$$

The complex amplitudes of the thermal current generators are represented conveniently by a column vector $\boldsymbol{i}$

$$\mathbf{i} = \begin{pmatrix} i_1 \\ i_2 \\ \\ i_{n_R} \end{pmatrix} \tag{5.10}$$

The nodal noise sources are not, in general, independent. The spectral density of the squared current can then be written in the matrix form

$$\overline{\mathbf{ii}^+} = \begin{pmatrix} \overline{i_1 i_1{}^+} & \overline{i_1 i_2{}^+} & & \overline{i_1 i_{n_R}{}^+} \\ \overline{i_2 i_1{}^+} & \overline{i_2 i_2{}^+} & & \overline{i_2 i_{n_R}{}^+} \\ & & & \\ \overline{i_{n_R} i_1{}^+} & \overline{i_{n_R} i_2{}^+} & & \overline{i_{n_R} i_{n_R}{}^+} \end{pmatrix} \tag{5.11}$$

where the subscript $^+$ indicates the Hermitian transpose (complex conjugate transpose). The internal sources may be alternatively represented by a set of n_R nodal voltage generators of infinite internal admittance, such that the *r-th* voltage across the generator in the *r-th* node is equal to the voltage across the *r-th* node of the original network, when all the nodes of the latter are open-circuit.

The isolated receivers of two closely spaced antennas will receive partially correlated noise [62]. The magnitude correlation was calculated using a generalized form of Nyquist's thermal noise theorem, given in [63]. It was shown that, in general, a nonreciprocal network with a system of internal thermal generators all at temperature T is equivalent to a source-free network, together with a system of noise current generators I_r and I_s with infinite internal impedance [64]. The noise currents were correlated and their cross-correlation was given by

$$\overline{I_s I_T} df = 2kT(Y_{ST} + Y_{ST}^*)df \tag{5.12}$$

where k is the Boltzmann's constant $1.3806503 \times 10^{-23}\,\mathrm{m^2\,kg\,s^{-2}\,K^{-1}}$ and Y_{ST} is the mutual admittance.

Alternatively, the internal noise sources can be represented by a system of nodal voltage generators V_T and V_s, with zero internal impedance. The correlation of the nodal voltage generators is then given by

$$\overline{V_s V_T} df = 2kT(Z_{ST} + Z_{ST}^*)df \tag{5.13}$$

where Z_{ST} is the mutual impedance. The correlation is zero when the mutual coupling is purely reactive.

The application of the generalized thermal noise theorem of Nyquist allows us to determine the thermal noise power of coupled antennas in a multi-antenna system. The theorem states that for a passive network at thermal equilibrium it would appear to be possible to represent the complete thermal-noise behavior by applying Nyquist's theorem independently to each component element of the network. The coupling in a multi-antenna system is represented by an antenna's self-impedances and mutual impedances. In order to determine the thermal noise behavior, these self-impedances and mutual impedances should be taken into account.

The nodal network representation of a multi-antenna system with n_R antenna elements is shown in Figure 5.17. Define

$$\mathbf{I} = \begin{pmatrix} i_{L_1} + i_1 \\ i_{L_2} + i_2 \\ \\ i_{Ln_R} + i_{n_R} \end{pmatrix} \tag{5.14}$$

$$\mathbf{Y} + Y_L\mathbf{U} = \begin{pmatrix} y_{11} + Y_L & y_{12} & & y_{1n_R} \\ y_{21} & y_{22} + Y_L & & y_{2n_R} \\ \\ y_{n_R1} & y_{n_R2} & & y_{n_Rn_R} + Y_L \end{pmatrix} \tag{5.15}$$

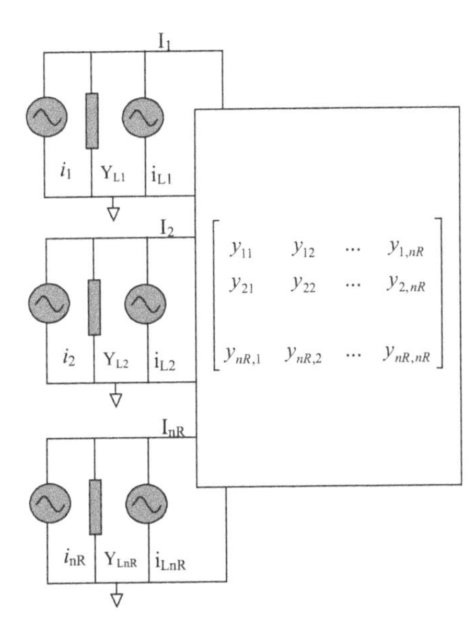

Figure 5.17. Nodal network representation of multi-antenna system [62]

where $\mathbf{U}$ denotes the $n_R \times n_R$ identity matrix, and

$$\mathbf{V} = \begin{pmatrix} V_1 \\ V_2 \\ \\ \\ V_{n_R} \end{pmatrix} \tag{5.16}$$

The system in Figure 5.17 can be expressed as follows

$$\mathbf{I} = \mathbf{i} + \mathbf{i}_L = (\mathbf{Y} + Y_L\mathbf{U})\mathbf{V} \tag{5.17}$$

Equation (5.17) can be rewritten as

$$\mathbf{V} = (\mathbf{Y} + Y_L\mathbf{U})^{-1}\mathbf{I} = (\mathbf{Y} + Y_L\mathbf{U})^{-1}(\mathbf{i} + \mathbf{i}_L) \tag{5.18}$$

The total noise power N for multiple coupled antenna elements can be written as (5.19)

$$\mathbf{N} = \frac{1}{2}(Y_L + Y_L{}^*)\overline{\mathbf{V}\mathbf{V}^+} \tag{5.19}$$

where $+$ and * represent the Hermitian transpose and the complex conjugate.

From (5.18), we have

$$\mathbf{V}\mathbf{V}^+ = (\mathbf{Y} + Y_L\mathbf{U})^{-1}\mathbf{I}\mathbf{I}^+\left((\mathbf{Y} + Y_L\mathbf{U})^{-1}\right)^+ \tag{5.20}$$

where the symbol $^+$ represents the Hermitian transpose.

Then, the square of the current can be expressed as

$$\mathbf{I}\mathbf{I}^+ = (\mathbf{i} + \mathbf{i}_L)(\mathbf{i} + \mathbf{i}_L)^+ \tag{5.21}$$

Based on Equation (5.12) the following relations are valid and can be put into (5.21)

$$\begin{aligned} &1. \quad \overline{i_j i_k{}^*}\, df = 2kT\left(y_{jk} + y_{jk}{}^*\right); \\ &2. \quad \overline{i_{L_j} i_k{}^*} = 0; \\ &3. \quad \overline{i_{L_j} i_{L_k}{}^*} = 0 \qquad j \neq k; \end{aligned} \tag{5.22}$$

Finally, the square of the currents (5.21) is expressed as

$$\mathbf{I}\mathbf{I}^+ = 2kT\left(\mathbf{Y} + \mathbf{Y}^* + (Y_L + Y_L{}^*)^*\mathbf{U}\right) \tag{5.23}$$

Substituting $Y_L + Y_L{}^* = 2G_L$ and $\mathbf{Y}_a = \mathbf{Y} + Y_L\mathbf{U}$, we can rewrite (5.19) as

$$\mathbf{N} = 2kTG_L\left[\mathbf{Y}_a^{-1}(\mathbf{Y}_a + \mathbf{Y}_a^*)(\mathbf{Y}_a^{-1})^+\right] \tag{5.24}$$

The thermal noise received in each antenna element includes two parts, self-thermal noise and induced thermal noise from the adjacent antenna elements. The total thermal noise power received from the antenna array in the receiver load is given in (5.24). The

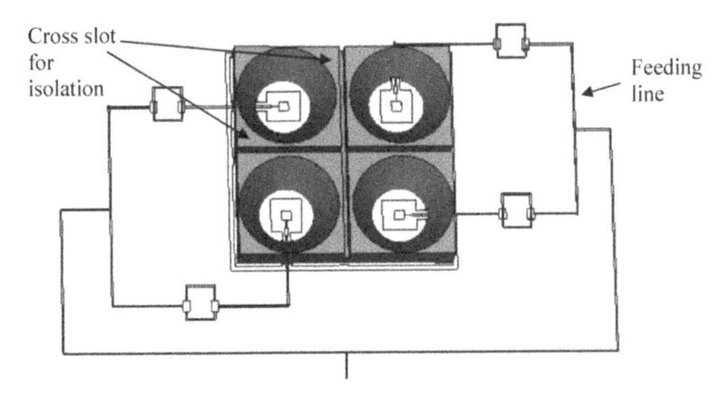

Figure 5.18. Top view of a 2 × 2 beam-steering array antenna and the size of a horn (DC power lines are not shown in the figure) [66] (©2006 IEEE)

total noise for two coupled antenna elements, in frequency bandwidth B, can be simplified to a sum of these noise powers

$$N_{total} = \int P_{L1} df + \int P_{L2} df \tag{5.25}$$

where P_{L1} is the thermal noise power absorbed in the receiver load of the first antenna and P_{L2} is thermal noise power absorbed in the receiver load of the second antenna.

The coupling between antenna elements is not negligible. To minimize the mutual coupling between the elements in a printed substrate, it is possible to create additional isolation between antenna elements using a cross-slot, as shown in Figure 5.18, or by using absorbing materials. It has been found that the isolation of elements can prevent the distortion of the radiation pattern and thus increase the output gain [65]. The same principles can be used to reduce the coupling between transmitters/receivers.

REFERENCES

[1] C. H. Doan, S. Emami, D. A. Sobel, A. M. Niknejad, and R. W. Brodersen, "Design considerations for 60GHz CMOS radios." *IEEE Commun. Mag.* pp. 132–140, Dec. 2004.

[2] M. R. Williamson, G. E. Athanasiadou, and A. R. Nix, "Investigating the effects of antenna directivity on wireless indoor communication at 60GHz." *8th IEEE Int'l. Symp. PIMRC,* pp. 635–39, Sep. 1997.

[3] R. C. Hansen, *Phased Array Antennas*, Wiley-Interscience, New York, U.S., 1998.

[4] G. Raleigh and J. M. Cioffi, "Spatio-temporal coding for wireless communication." *IEEE Trans. Commun.*, Vol. 46, pp. 357–366, Mar. 1998.

[5] H. Bolcskei, D. Gesbert, and A. J. Paulraj, "On the capacity of OFDM based spatial multiplexing systems." *IEEE Trans. Commun.*, Vol. 50, pp. 225–234, Feb. 2002.

[6] A. F. Molisch, M. Stienbauer, M. Toeltsch, E. Bonek, and R. S. Thoma, "Capacity of MIMO systems based on measured wireless channels." *IEEE J. Select. Areas Commun.*, Vol. 20, pp. 561–569, Apr. 2002.

[7] W. C. Y. Lee, "Antenna spacing requirement for a mobile radio base-station diversity." *Bell System Tech. Journal*, Vol. 50, pp. 1859–1876, Jul./Aug. 1971.

[8] W. C. Y. Lee and Y. S. Yeh, "Polarization diversity for mobile radio." *IEEE Trans. Commun.*, Vol. 20, No. 5, pp. 912–923, 1972.

[9] A. M. D. Turkmani, A. A. Arowojolu, P. A. Jefford, and C. J. Kellet, "An experimental evaluation of the performance of two-branch space and polarization diversity schemes at 1800 MHz." *IEEE Trans. Veh. Technol.*, Vol. 44, No. 2, pp. 318–326, 1995.

[10] F. Lotse, J.-E. Berg, U. Forseen, and P. Idahl, "Base station polarization diversity reception in macrocellular systems at 1800 MHz." *IEEE 46th Veh. Tech. Conf.*, Vol. 3, pp. 1643–1646, 1996.

[11] D. C. Cox, "Antenna diversity performance in mitigating the effects of portable radio telephone orientation and multipath propagation." *IEEE Trans. Commun.*, Vol. 31, No. 5, pp. 620–628, 1983.

[12] P. L. Perini and C. L. Holloway, "Angle and space diversity comparisons in different mobile radio environments." *IEEE Trans. Antennas Propag.*, Vol. 46, No. 6, pp. 764–775, 1998.

[13] R. H. Clarke, "A statistical theory of mobile-radio reception." *Bell System Tech. Journal*, pp. 957–1000, Jul.–Aug. 1968.

[14] W. C. Jakes, Ed., *Microwave Mobile Communications*, New York: Wiley, 1974. (Reprint IEEE Press, New York, 1993.).

[15] J. H. Winters, J. Salz, and R. D. Gitlin, "Adaptive antennas for digital mobile radio." *Proc. IEEE Long Island Section Adaptive Antenna Systems Symposium,* Long Island, NY, pp. 81–86, Nov. 1992.

[16] K. Huang and D. J. Edwards "60GHz multi-beam antenna array for gigabit wireless communication networks." *IEEE Trans. Antennas Proppag.*, Vol. 54, No. 12, pp. 3912–3914, 2006.

[17] M. Uno, Z. C. Wang, V. Wullich, and K.-C. Huang,"Communication system and method." Europe patent EP1659813B1, Apr. 2009.

[18] S. Kobayashi, R. Mittra, and R. Lampe, "Dielectric tapered rod antenna for millimetre wave applications." *IEEE Trans. Antennas Propag.*, Vol. 30, No. 1, pp. 54–58, 1982.

[19] J. Kraus and R. Marhefka, *Antennas for All Applications*, 3rd ed. New York: McGraw Hill, 2002.

[20] T. Ando, J. Yamauchi, and H. Nakano, "Demonstration of the discontinuity-radiation concept for a dielectric rod antenna." *Proc. IEEE Antennas and Propagation Society Int. Symp.*, Vol. 2, pp. 856–859, Jul. 2000.

[21] K. Huang and Z. Wang, "V-band patch-fed rod antennas for high datarate wireless communications." *IEEE Trans. Antennas Propag.*, Vol. 54, No. 1, pp. 297–300, 2006.

[22] K. Huang and Z. Wang,"Dielectric rod antenna and method for operating the antenna." Europe patent EP1703590B1, Sep. 2008.

[23] IEEE 802.15.3c standard http://www.ieee.org/

[24] T. Ando, J. Yamauchi, and H. Nakano, "Rectangular dielectric-rod by metallic waveguide." *IEE Proc. Microwave, Antennas Propag.*, Vol. 149, No. 2, pp. 92–97, Apr. 2002.

[25] A. Molisch, M. Z. W. J. Winters, and A. Paulraj, "Capacity of MIMO systems with antenna selection." *IEEE Intern. Conf. on Communications*, pp. 570–574, 2001.

[26] D. Gesbert and J. Akhtar, "Breaking the barriers of Shannon's capacity: An overview of MIMO wireless systems." *Telenor's Journal: Telektronik*, pp. 1–9, Jan. 2002.

[27] K. Ohata, K. Maruhashi, M. Ito, S. Kishimoto, K. Ikuina, T. Hashiguchi, K. Ikeda, and N. Takahashi, "1.25 Gbps wireless gigabit ethernet link at 60GHz-band." *2003 IEEE MTT-S International Microwave Symposium Digest*, Vol. 13, pp. 373–376, Jun. 2003.

[28] Proxim-Wireless™, "Gigalink Series–Alternative to Fiber up to Gigabit Speeds." DS 0806 GIGALINK USHR.pdf, 2006.

[29] A. Lozano, F. R. Farrokhi, and R. A. Valenzuela, "Lifting the limits on high-speed wireless data access using antenna arrays." *IEEE Commun. Mag.* pp. 156–162, Sep. 2001.

[30] J. G. Proakis. *Digital Communications*, 4th edition, New York: McGraw Hill, 2000.

[31] G. D. Durgin, *Space-Time Wireless Channels*,. Englewood Cliffs, NJ: Prentice Hall, 2003.

[32] H. Zirath, S. E. Gunnarsson, M. Ferndahl, R. Kozhuharov, C. Karnfelt, and C. C. Stoij, "Newly developed chip sets for 60GHz radio communication systems." *17th International Conference on Microwaves, Radar and Wireless Communications,* 2008. MIKON 2008, pp. 1–4. May 2008.

[33] S. E. Gunnarsson, C. Kärnfelt, H. Zirath, R. Kozhuharov, D. Kuylenstierna, and C. Fager, "60GHz single-chip front-end MMICs and systems for multi-Gb/s wireless communication." *IEEE J. Solid-State Circuits*, Vol. 42, No. 5, pp. 1143–1157, May 2007.

[34] S. K. Reynolds, B. A. Floyd, U. R. Pfeiffer, T. Beukema, J. Grzyb, C. Haymes, B. Gaucher, and M. Soyuer, "A silicon 60GHz receiver and transmitter chipset for broadband communications." *IEEE J. Solid-State Circuits*, Vol. 41, No. 12, pp. 2820–2831, 2006.

[35] M. Tanomura, Y. Hamada, and S. Kishimoto, "Tx and Rx front-ends for 60GHz band in 90 nm standard bulk CMOS." *2008 IEEE International Solid-State Circuits Conference*, Feb. 2008.

[36] S. Pinel, S. Sarkar, et al. "60GHz CMOS 90 nm radio." *2008 IEEE International Solid-State Circuits Conference*, Feb. 2008.

[37] S. O. Tatu, E. Moldovan, and S. Affes, "Low-cost transceiver architectures for 60GHz ultra wideband WLANs." *International Journal of Digital Multimedia Broadcasting, Article ID 382695*, pp. 1–6, 2009.

[38] S. Catreux, L. J. Greenstein, and V. Erceg, "Some results and insights on the performance gains of MIMO Systems." *IEEE J. Select Areas Commun.*, Vol. 21, No. 5, pp. 839–847, 2003.

[39] E. Telatar, "Capacity of multi-antenna Gaussian channels." Tech. Report, AT&T Bell Labs, 1995.

[40] J. D. Kraus, *Radio Astronomy*, 2nd ed. Cygnus-Quasar, pp. 6–19, Powell, Ohio, Cygnus-Quasar, 1986.

[41] E. Torkildson, B. Ananthasubramaniam, U. Madhow, and Mark Rodwell, "Millimeter-wave MIMO: wireless links at optical speeds." *Proceeding of the 44th Annual Allerton Conference* Sep. 2006.

[42] C. Carta, M. Seo, and M. Rodwell, "A mixed-signal row/column architecture for very large monolithic mm-wave phased arrays." *2006 IEEE Lester Eastman Conference on High Performance Device*, Aug. 2006.

[43] E. Torkildson, B. Ananthasubramaniam, U. Madhow, and M. Rodwell, "Millimeter-wave MIMO: wireless links at optical speeds (invited paper)." *Proc. of 44th Allerton Conference on Communication, Control and Computing*, Monticello, IL, Sep. 2006.

[44] iMEC R&D fact sheet "IMEC 60GHz wireless demo." pp. 1–2, Feb. 2009

[45] S. Ranvier, J. Kivinen, and P. Vainikainen, "Millimeter-wave MIMO radio channel sounder." *IEEE Trans. Instrum. Meas.*, Vol. 56, No. 3, pp. 1018–1024, 2007.

[46] D. M. Pozar, "Input impedance and mutual coupling of rectangular microstrip antennas." *IEEE Trans. Antennas Propag.*, Vol. 30, pp. 1191–1196, Nov. 1982.

[47] S. Ranvier, J. Kivinen, and P. Vainikainen, "Development of a 60GHz MIMO radio channel measurement system." *Instrumentation and Measurement Technology Conference,* 2005. IMTC 2005. Proceedings of the IEEE Vol. 3, pp. 1878–1882, May 2005.

[48] K. Sulonen, P. Suvikunnas, L. Vuokko, J. Kivinen, and P. Vainikainen, "Comparison of MIMO antenna configurations in picocell and microcell environments." *IEEE J. Sel. Areas Commun.,* special issue on MIMO systems and Application, Vol. 21, No. 5, pp. 703–712, 2003.

[49] S. Ranvier, M. Kmec, R. Herrmann, and J. Kivinen, "Mm-wave wideband MIMO channel sounding." *XXVIIIth General Assembly of URSI,* New Delhi, 2005.

[50] M. J. Gans, "Channel capacity between antenna arrays—part I: sky noise dominates." *IEEE Trans. Commun.*, Vol. 54, No. 9, pp. 1586–1592, 2006.

[51] M. J. Gans, "Channel capacity between antenna arrays—part II: amplifier noise dominates." *IEEE Trans. Commun.*, Vol. 54, No. 11 pp. 1983–1992, 2006.

[52] I. J. Guptha and A. K. Ksienski, "Effect of the mutual coupling on the performance of the adaptive arrays." *IEEE Trans. Antennas Propag.*, Vol. 31, No. 5, pp. 785–791, 1983.

[53] W. Rotman, "EHF dielectric lens antenna for multibeam MIL-SATCOM applications." in Dig. I982 Int. IEEE-APSIURSI Symp., Albuquerque, NM pp. 132–135, Jun. 1982.

[54] K. Iizuka, M. Mizusawa, S. Urasaki, and J. Ushigome, "Volume-type holographic antenna." *IEEE Trans. Antennas Propag.*, Vol. 23, pp. 807–810, Nov. 1975.

[55] P. Bhartia, K. V. S Rao, and R. S. Tomar, *Millimetre-Wave Microstrip and Printed circuit antennas*, London: Artech House, 1991.

[56] T. Sventenson, and A. Ranheim, "Mutual coupling effects on the capacity of the multielement antenna system." *Acoustic, Speech and Signal Processing,* 2001, Proceedings (ICASSP'01), 2001.

[57] J. W. Wallace and M. A. Jensen, "Mutual coupling in MIMO wireless systems: a rigorous network theory analysis." *IEEE Trans. Wireless Commun.*, Vol. 3, pp. 1317–1325, 2004.

[58] S. Krusevac, P. Rapajic, and R. Kennedy, "Method for MIMO channel capacity estimation for electromagnetically coupled transmit antenna elements." *AusCTW* 2004, pp. 122–126, Feb. 2004.

[59] S. M. Krusevac, R. A. Kennedy, and P. B. Rapajic, "Effect of signal and noise mutual coupling on MIMO channel capacity." *Wireless Personal Communications*, Vol. 40, No. 3, pp. 317–328, Aug. 2006.

[60] S. M. Rytov, Yu. A. Krastov, and V.I. Tatarskii, *Principles of Statistical Radio Physics 3, Elements of the Random Fields*, Berlin: Springer, 1987.

[61] A. T. Starr, *Electric Circuit and Wave Filters*, London: Pitman, 2nd edition, p. 78, 1946.

[62] S. Krusevac, P. B. Rapajic, R. A. Kennedy, and P. Sadeghi, "Mutual coupling effect on thermal noise in multi-antenna wireless communication systems." *Communications Theory Workshop, 2005. Proceedings. 6th Australian*, pp. 209–214, Feb. 2005.

[63] S. Krusevac, P. Rapajic, and R. Kennedy, " Mutual Coupling effect on thermal noise in multi-antenna wireless communication systems." Aus CTW 2005, pp. 209–214.

[64] R. Q. Twiss, "Nyquist's and Thevenin's generalized for nonreciprocal linear networks." *J. Applied Physics*, Vol. 26, pp. 559–602, May 1955.

[65] Moon et al., "Flat-plate MIMO array antenna with isolation element." US patent application US2007/0069960.

[66] K. Huang and Z. Wang, "Millimetre-wave circular polarized beam-steering antenna array for gigabit wireless communications." *IEEE Trans. Antennas Propag.*, Vol. 54, No. 2, pp. 743–746, 2006.

6

ADVANCED DIVERSITY OVER MIMO CHANNELS

In this chapter, we present new aspects of advanced diversity over multiple-input-multiple-output (MIMO) channels for millimeter wave wireless systems. The use of diversity minimizes the errors occurring in data reception when the dynamic wireless channel is under deep fading. If we supply to the receiver several replicas of the same information signals transmitted over independently fading channels, the probability of having fadings for all signals is reduced definitely. As mentioned in Section 5.1, diversity could be realized in space, time, frequency, and polarization domains, which are defined as space diversity, time diversity, frequency diversity, and polarization diversity.

This chapter is organized as follows. Section 6.1 introduces the main concept of advanced diversity over MIMO channels. The inherent reliability issue of millimeter wave wireless systems is identified and the potential advantages of using diversity and MIMO are illustrated. Section 6.2 describes one specific solution to the issue by implementing spatial diversity and fulfilling the link budget requirement for high rate applications (beyond Gbps). In addition, the idea of integrating time diversity together with space diversity is presented to improve efficiency. This diversity-integration could be used for high-definition television (HDTV) applications. Section 6.3 discusses the concepts of receiver antenna selection and transmitter antenna diversity, and the combination of the both techniques is also discussed. A specific example to implement frequency diversity together with space diversity is illustrated. Finally, the dynamic

Millimeter Wave Communication Systems, by Kao-Cheng Huang and Zhaocheng Wang

allocation of antennas, frequency, and modulation schemes for millimeter wave systems are summarized in Section 6.4.

6.1 POTENTIAL BENEFITS FOR MILLIMETER WAVE SYSTEMS

We know from Chapter 5 that MIMO can increase capacity and also realize diversity [1, 2]. In addition, space and time diversity can be achieved by using efficient space-time coding [3–13]. The layered space-time code is introduced in [12] and the space-time trellis code is discussed in [9]. For millimeter wave wireless systems, narrow beam antennas are preferred to fulfil the strict link budget requirement introduced by high rate applications (beyond Gbps). Since the wavelength λ is short for millimeter wave signals, the received signal strength is very sensitive to obstacles between the transmitter and the receiver. When the size of the obstacles is comparable with the wavelength λ and the obstacle intercepts the line-of-sight (LOS) link between the transmitter and the receiver using narrow-beam antennas, the received signal level is reduced sharply. Diversity is one of the key techniques used by millimeter wave wireless systems to solve this issue. It refers to the existence of two or more signal paths that fade independently. This happens when the radio channel consists of several paths that are sufficiently separated in space, frequency, time, or polarization. The key idea is that, if several paths have their own dynamic channel coefficients that are statistically independent, it is almost impossible that they will all fade at the same time, so the probability is low when the overall received signal strength is below the required detection threshold.

High rate applications need huge bandwidth. A wireless system that makes use of large available bandwidth and small antenna size is known to be the millimeter wave solution, which allows a high rate (e.g., beyond Gbps) wireless data transmission. In a wireless system that experiences frequency-selective fading, a channel equalizer, such as a linear, decision feedback, or maximum likelihood sequence estimation (MLSE) equalizer, is required. In the case of a high data rate transmission, the symbol duration is correspondingly short such that the multipath channel delay spread may be over tens of the symbol periods. As a consequence, the equalizer becomes complex and needs a lot of processing power.

Furthermore, a well-known solution uses the orthogonal frequency division multiplexing (OFDM) technique for resisting against multipath interference in wireless communications. The OFDM modulation scheme has already been implemented, for example, for transmissions in wireless local area network (LAN) systems [14]. However, as an OFDM signal is the sum of a large number of subcarriers, it tends to have a high peak-to-average power ratio (PAPR). Because of high PAPR produced by the inherent linear modulation of the OFDM technique, the overall power consumption of a power amplifier (PA) is very high compared with other multiplexing techniques. Since the power consumption of a millimeter wave system is already high compared with conventional wireless systems, the problem is exaggerated by the use of the OFDM method. Another disadvantage is that OFDM demodulation requires complex units for carrying out high-speed fast Fourier transform (FFT) and other signal processing methods.

An alternative solution to reduce the complexity of equalizer is to use MIMO and diversity techniques. Since the antenna size of millimeter wave wireless systems is small, several pairs of small-size sharp-beam antennas could be used for both the transmitter and the receiver to keep a reliable wireless link, wherein each pair of sharp-beam antennas can be steered to match the direction of its corresponding strong reflection path. Depending on the steering resolution, the strong reflection path can be matched and other reflection paths can be disregarded. This results in the channel delay spread being shortened, and the required equalization function becomes simple. On the other hand, as very few communication paths or ideally only the strongest reflection path is received, the overall received power is drastically reduced. This drawback has to be compensated for by the introduction of additional antenna gains obtained from the sharp-beam antennas.

In addition, different data might be transmitted at the same time using different pairs of narrow-beam antennas from the transmitter to the receiver to further increase the data rate. Since the bandwidth provided by millimeter wave systems is high compared with other traditional wireless systems [15], the required transmitted power could be further reduced by selecting the bandwidth having good channel transfer function. In addition, dynamic modulation scheme selection can be used to achieve the optimum trade-off between data rate and power consumption. Furthermore, when there are enough margins for link budget from the system point of view, a simple and noncoherent detection scheme without local oscillator (LO) and mixer can be considered to reduce the high power consumption introduced by the LO.

6.2 SPATIAL AND TEMPORAL DIVERSITY

Millimeter wave communication systems exploit dimensions of time, frequency, and space to maximize the data rate and system capacity. Designs of space-time coding, equalization, adaptive antennas, and RAKE receiver techniques all rely on accurate characterization of the propagation channel. For millimeter wave systems with high data rate and directional antennas, the small-scale fading is more complex to characterize than for omnidirectional and narrowband wireless systems. More advanced channel models that include time dispersion and angular dispersion have to be addressed.

Conventional wireless systems commonly use a wide/omnidirectional beam antenna at the transmitter (Tx) side and a wide/omnidirectional beam antenna at the receiver (Rx) side. One example is shown in Figure 6.1, whereby the path with a cross is the transmission path being blocked, and there are many reflective objects between the transmitter and the receiver.

According to the characteristics of millimeter wave signals, when the size of the obstacle is comparable with wavelength λ, it can block the wireless transmission between the transmitter and the receiver. When a LOS wireless link is blocked by the obstacle, as shown in Figure 6.1, the transmitted signals are reflected by several reflective objects and received by the receiver, and this may cause a plurality of NLOS transmission paths P1, P2, P3, and P4. When the data rate is high (e.g., over Gbps) and symbol period is short, a channel delay spread might be over tens of symbol periods, which leads to severe

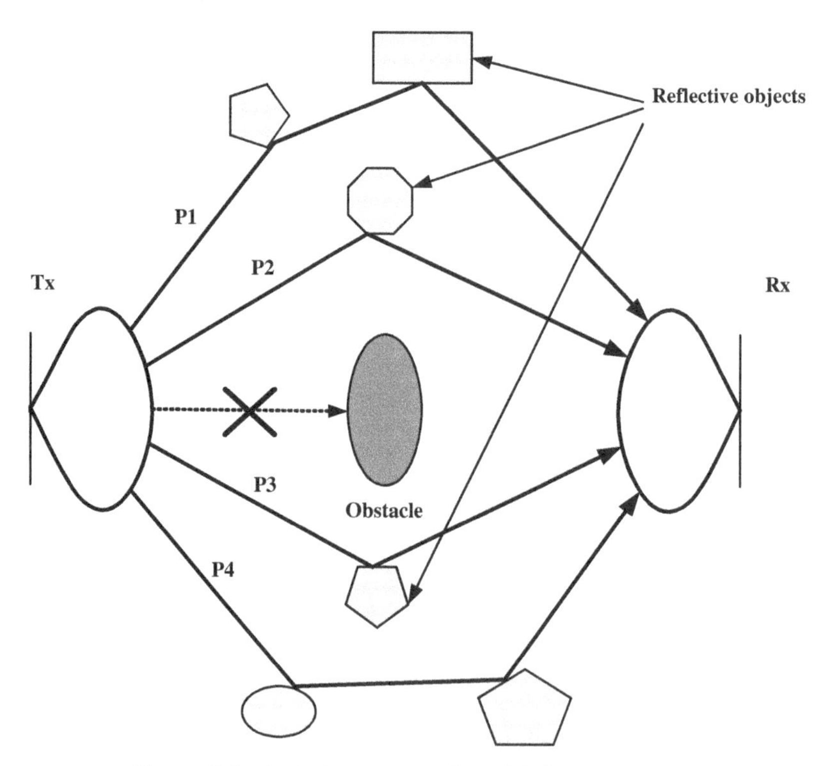

Figure 6.1. A wireless system using wide beam antennas

intersymbol interference (ISI) due to deep frequency selective fading. In addition, the signal strength at the Rx side may degrade sharply when there are obstacles between the Tx and Rx sides. As a consequence, the link budget requirement cannot be satisfied for high rate applications.

In order to meet the link budget requirement, Figure 6.2 shows a wireless communication system using a steerable narrow-beam antenna at the Tx side and a steerable narrow-beam antenna at the Rx side, whereby the paths with crosses are transmission paths being blocked. Because of the inherent high antenna gains from narrow antenna radiation patterns from both Tx and Rx sides, the link budget requirement is achieved and the complexity of the equalizer is reduced due to the decrease of multipath components.

In such communication systems the narrow-beam antennas provide a high antenna gain and, additionally, a tracking of the best or strongest transmission paths is carried out so that only the best or strongest transmission paths are used for actual communication. Both the narrow-beam antennas are steered to an optimum position (optimum antenna beam direction) where the best LOS signals or strongest NLOS reflection signals can be transmitted and received. As a result, only a very small number of reflection signals reach the receiver (in the example of Figure 6.2 only the signal corresponding to transmission path P2 is radiated by the Tx antenna and received by the Rx antenna). Therefore, the

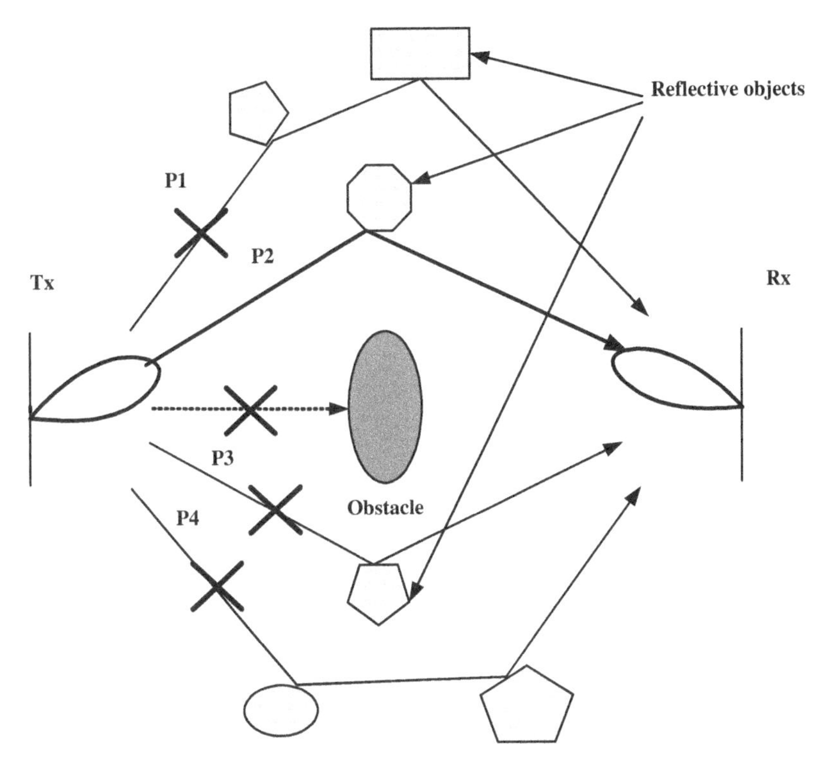

Figure 6.2. A wireless system using a pair of narrow beam antennas

channel delay spread is shortened dramatically, the complexity of the baseband circuit can be reduced, and low power consumption for baseband processing can be achieved. In addition, the signal strength can be kept strong due to the high antenna gains, and the link budget requirement can be easily satisfied even for high rate applications.

One example of high rate applications is HDTV. Although the link budget is fulfilled, the wireless link might not be reliable. Due to changes of the surrounding environment (e.g., moving or emerging objects and/or obstacles) and small wavelength of millimeter wave signals, the transmission paths are easily interrupted, even by small obstacles, and therefore video data may be lost and the quality of a video signal as perceived by a viewer is degraded. The interruption persists during a period (antenna search and switch period) in which an operable communication is determined and switched to find alternative reflection paths.

One solution to this problem is the adoption of spatial diversity. Several pairs of sharp-beam antennas from both the Tx and Rx sides are used, where each pair of sharp-beam antennas is steered to its corresponding strong reflection path. Each reflection signal can be assumed to be independent and can be treated as going through a relative frequency nonselective slow fading channel. The probability of all of the signals becoming weak is small and the path diversity gain can be acquired.

Each pair of sharp-beam antennas is treated as one finger of the RAKE receiver described in spread spectrum wireless systems, which can be adjusted based on the dynamic wireless environment. From the Tx side, only one Tx RF chain is used and connected with several Tx sharp-beam antennas. From the Rx side, each Rx sharp-beam antenna can be connected to one specific Rx chain directly or preferably by the switch network, where the number of required Rx chains can be reduced and is equivalent to the diversity order. The proposed system is illustrated in Figure 6.3. If one strong LOS or NLOS reflection path is intercepted or blocked, another unblocked link could still be alive to support the wireless link and viewers would not feel the interruption of HDTV signal. At the same time, another alternative path could be searched for using a beam steering algorithm. More detailed information about beam steering will be discussed in Chapter 7.

For HDTV user scenarios, an unequal error protection (UEP) method is applied to video information bits according to an importance level of the bits such that more important bits are provided with more protection for transmission error recovery [16–19]. Unequal protection is achieved by the way of using asymmetric coding and/or asymmetric constellation mapping. The method is applied to uncompressed video with 24 bits per pixel (i.e., 8 bits per pixel color component such as red, green, or blue), whereby for each component the four bits of lower numerical significance correspond to

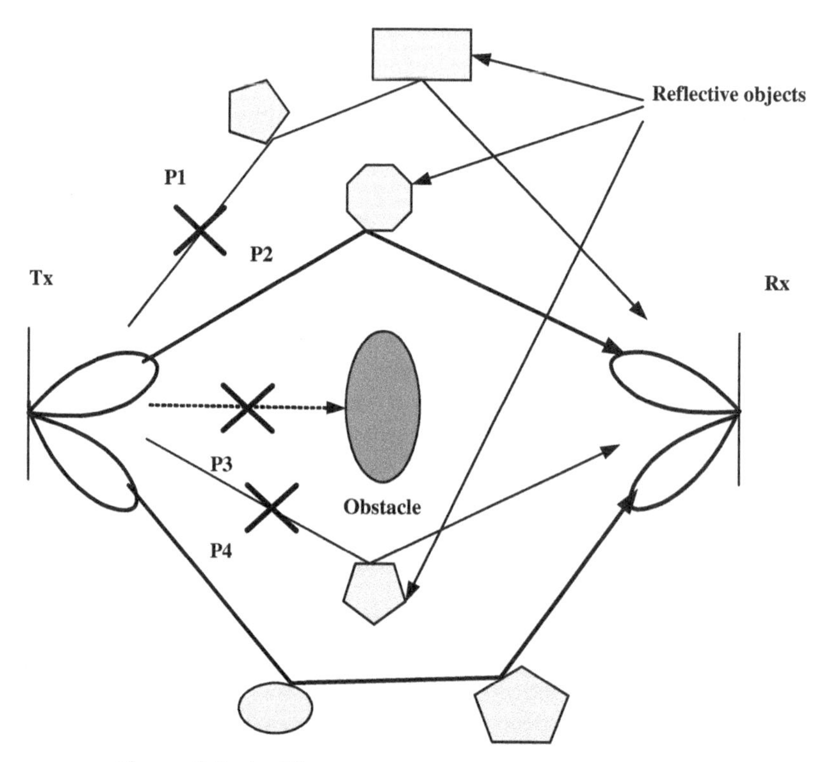

Figure 6.3. A millimeter wave system using spatial diversity

a low importance level and the four bits of higher numerical significance correspond to a high importance level. Such UEP technology, however, cannot solve the issue of degraded quality of millimeter wave signals since both the visually important bits of a pixel and the visually unimportant bits of a pixel are lost during the antenna search and switch period.

The problem could be solved by the combination of spatial diversity and temporal diversity. HDTV video data are split into at least two classes, and each class is further split into one or more blocks. The blocks are transmitted over communication channels. For at least some blocks, a different number of communication channels are used depending on the class of the blocks. Different communication channels correspond to different transmission time slots and/or different transmission paths.

HDTV video data today is typically represented by 24 bits per pixel. Uncompressed video may, however, also be represented with more or less than 24 bits per pixel. For example, it may be represented by 30 bits per pixel, corresponding to 10 bits per color component or by 36 bits per pixel, corresponding to 12 bits per color component. The color components can be treated differently. The green color component is considered visually more important than the red and the blue color components, and the red color component is treated as visually more important than the blue color component. An advantage of treating color components differently over treating them equally is that the perceived signal quality is increased when the data rate/amount of data is fixed or that the data rate/amount of data is reduced when the perceived signal quality is kept unchanged. For example, any of the bits in $n\_g = 6$ to 3 out of 8 bits of the green pixel component may be considered as visually important, as well as $n\_r = 4$ to 1 out of 8 bits of the red pixel component and $n\_b = 2$ to 0 out of 8 bits of the blue pixel component, whereby $n\_g > n\_r$, $n\_g > n\_b$ and $n\_r > n\_b$. Hereby, the remaining bits of each pixel component are considered as visually unimportant. One example is to use $n\_g = 4$, $n\_r = 2$, and $n\_b = 0$, which leads to a ratio of $6/24 = 1/4$ of the important bits in a pixel to the total number of bits in a pixel. This provides a very good tradeoff between additional resources needed and improvement of perceived robustness of transmission. For other than 8-bit quantizations, it is proposed that the ratio of the visually important bits to the total number of bits per pixel component is the same as in the example of the 8-bit quantization. Therefore, the ratio r_g for the green pixel component may be in the range of 6/8 to 3/8, the ratio r_r for the red pixel component in the range of 4/8 to 1/8, and the ratio r_b for the blue color component in the range of 2/8 to 0, where $r\_g > r\_r$, $r\_g > r\_b$, and $r\_r > r\_b$. The bits that are considered to be visually important are the bits of a relatively high numerical significance including the most significant bit (MSB), and the bits that are considered visually unimportant are the bits of a relatively low numerical significance such as the least significant bit (LSB).

One example of using spatial diversity to obtain the UEP effect is to transmit visually important bits over more than one pair of narrow-beam antennas and transmit visually unimportant bits over only one pair of narrow-beam antennas [18, 19]. As a consequence, the diversity gain of visually important bits is better than that of visually unimportant bits. Good trade-off between the video quality, the power consumption of millimeter wave systems, and the throughput could be achieved. The idea is summarized in Figure 6.4, whereby the P2 and P4 paths are used to transmit visually important

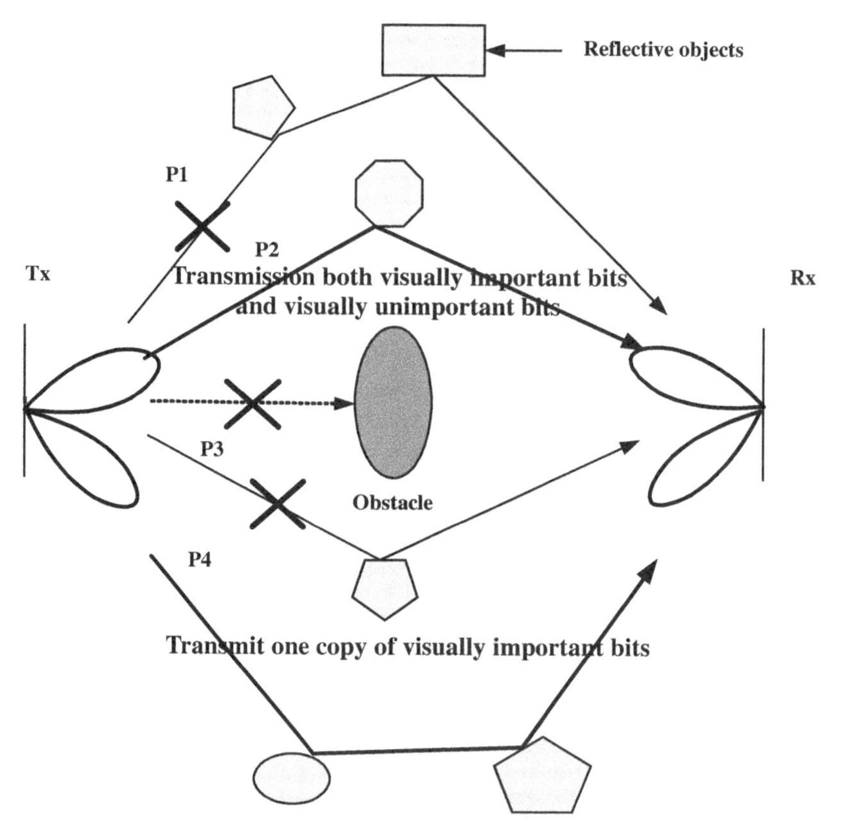

Figure 6.4. A millimeter wave system using spatial diversity for UEP

bits generated from HDTV signals and only the P2 path is used to transmit visually unimportant bits.

Another example using spatial diversity and temporal diversity to obtain the UEP effect is to transmit both visually important bits and visually unimportant bits over the same number of pairs of narrow-beam antennas. However, visually important bits are transmitted over more than one time slot in order to achieve temporal diversity gain against visually unimportant bits radiating over only one time slot. Sub-optimum trade-off between the video quality, the power consumption, and the data rate could be achieved. The idea is summarized as follows. In this case, for example, the P2 and P4 paths of Figure 6.4 are used to transmit both visually important bits and visually unimportant bits to acquire the spatial diversity gain. Let A1, A2, A3, ... define the sequences/blocks of visually important information bits, while B1, B2, B3, ... represent the sequences/blocks of visually unimportant bits. Two time slots are used to transmit one block of visually important bits to obtain additional temporal diversity gain. Figure 6.5(a) shows the regular pattern of temporal diversity and Figure 6.5(b) shows the irregular pattern of temporal diversity. During the usage of regular pattern, the visually important blocks of A1, A2, A3, ... and visually unimportant blocks of B1, B2, B3, ... are

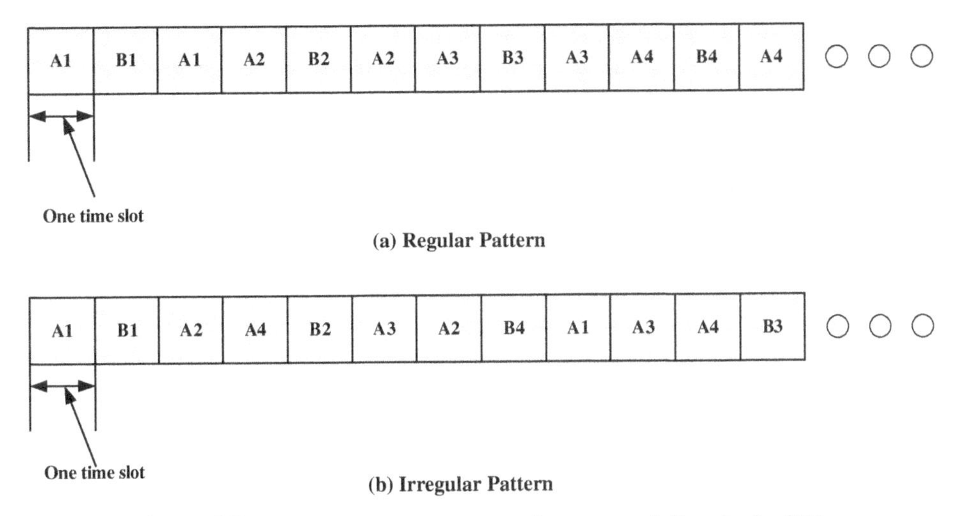

Figure 6.5. A millimeter wave system using temporal diversity for UEP

transmitted orderly, for example, A1 is transmitted earlier than A2, A2 is transmitted earlier than A3, and so on. It is apparent that the irregular pattern has higher temporal diversity gain than the regular pattern. However, the complexity of the irregular pattern is slightly higher.

6.3 SPATIAL AND FREQUENCY DIVERSITY

In many millimeter wave based user scenarios, temporal diversity is either not applicable or could not be efficiently implemented due to the strictly delay constraints of the data being transmitted. In these cases, looking back to the spatial domain and frequency domain is a natural option.

The MIMO technique can improve the capacity and reliability of millimeter wave wireless systems. However, the multiple RF chains associated with multiple antennas are expensive in terms of size, power, and cost. Antenna selection is a low-cost and low-complexity alternative choice that could capture many of the advantages of MIMO systems [20–22]. One example is described in Figure 6.2, where only one RF chain including LO, PA, and other millimeter wave components is connected with one transmission antenna selected from M_T ($M_T > 1$) sharp-beam antenna candidates, and only one RF chain including LO, mixer, and other millimeter wave components is connected with one receiver antenna selected from M_R ($M_R > 1$) sharp-beam antenna candidates. Selecting one transmitter antenna from M_T candidates in order to track the strongest LOS or reflection path is called transmitter antenna selection. Selecting one receiver antenna from M_R candidates in order to track the strongest LOS or reflection path is called receiver antenna selection.

Conventional receiver diversity via several antennas is illustrated in Figure 6.6, whereby the receiver sees different versions of the radiated signals, experienced over

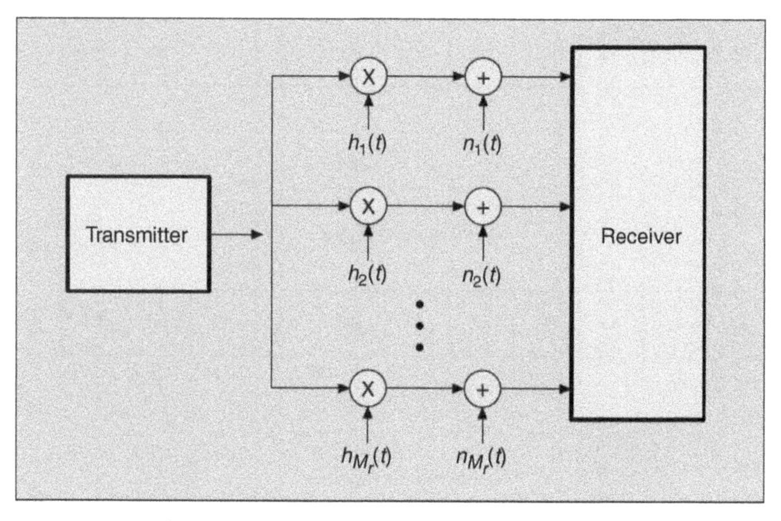

Figure 6.6. Receiver diversity [22] (© 2004 IEEE)

different fading channels with coefficients $h_i(t)$ $(1 \leq i \leq M_r)$ and noise $n_i(t)$ $(1 \leq i \leq M_r)$. To exploit diversity, these signals must be combined in a gainful manner. Diversity combining could be classified into three categories. Selection diversity chooses the path with the highest received signal strength indicator (RSSI) or signal-to-noise ratio (SNR) and performs a detection based on the signal from the selected antenna. Maximum ratio combining (MRC) takes decisions based on an optimal linear combination of the received signals from M_r receiver antennas. Equal gain combining (EGC) simply adds the signals after they have been co-phased. Due to the high cost of millimeter wave components like LO, mixer, PA, low noise amplifier (LNA), selection diversity has attracted a lot of attention.

Figure 6.7 describes a general concept of receiver antenna selection, where a single receiver antenna is chosen based on RSSI or SNR measurements. Since only one RF branch is available, we face a dilemma: we need to know all the branch SNRs for optimal selection, but how can we know all the SNRs simultaneously when there is only one RF

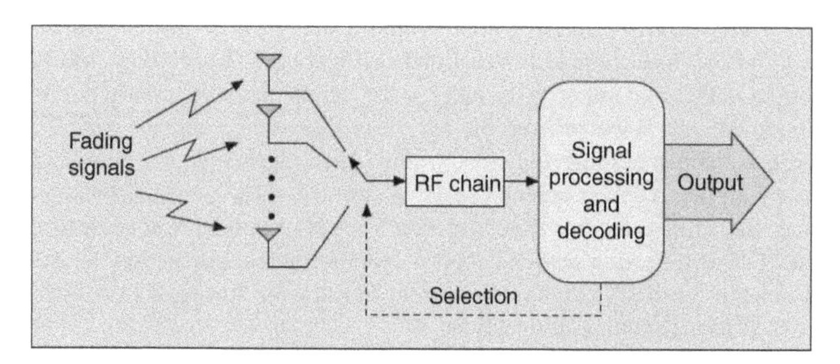

Figure 6.7. Receiver antenna selection [22] (© 2004 IEEE)

chain? There are several ways to address this problem based on the quasi-stationary nature of the channel gains. For example, one may use a training signal in a preamble to transmit data. During this period, the receiver scans the antennas, finds the antenna with the highest channel gain, and selects it for receiving the next incoming data burst. There are many practical considerations to implement the antenna selection. For example, the optimal choice must be based on the SNR or RSSI of the received signals, but in practice it is simpler to use an envelope detector and select the branch with the highest signal plus noise level, as shown in Figure 6.7. The feedback from the receiver to the transmitter is not necessary for realizing receiver antenna selection.

Information about how to implement receiver antenna diversity for millimeter wave systems is further detailed in Figure 6.8, whereby sharp-beam antennas for both the transmitter and the receiver can be realized by a single pin or rod antenna [23] or by phase array antennas [24]. Firstly, an omnidirectional/wide beam antenna is used in the Tx side to radiate the channel sounding signals to the receiver. Secondly, a sharp-beam antenna is used at the Rx side and rotated to different antenna beam directions step by step as shown in Figure 6.8. The channel conditions from those beam directions are estimated using the corresponding training signals/channel sounding signals. The criteria to judge the channel conditions include RSSI, SNR, cyclic redundancy check (CRC), and channel delay profile measurement. Based on the measurements, one antenna beam direction with the strongest received signals is chosen and connected with the RF chain. As mentioned previously, the preferred narrow-beam direction could be realized by either a single rod antenna or phase array antennas.

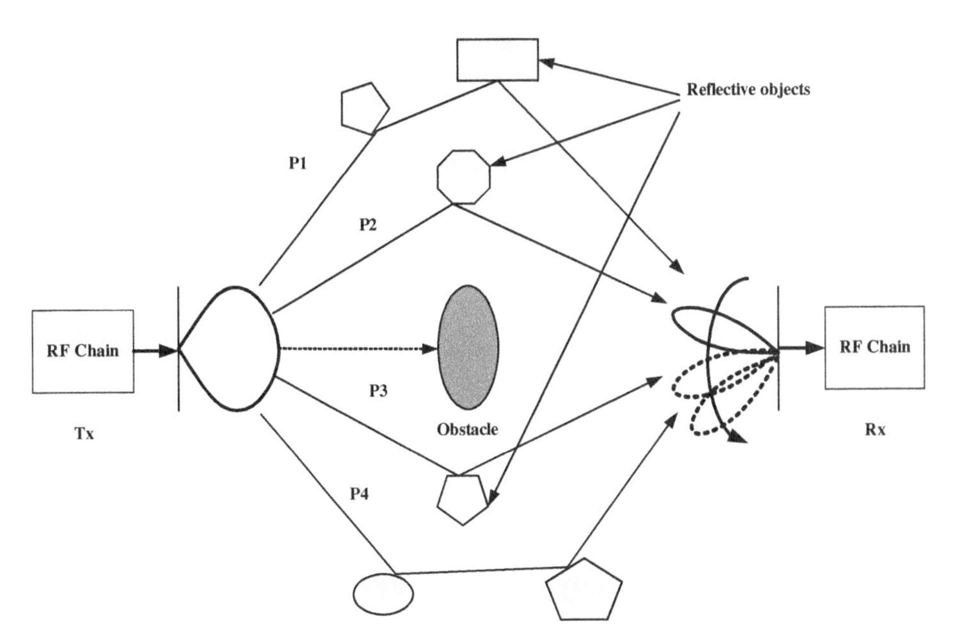

Figure 6.8. Receiver antenna diversity for millimeter wave systems

Transmitter antenna selection, different from receiver antenna selection, requires a feedback path from the receiver to the transmitter. This feedback rate is rather low, especially for single antenna selection. Therefore, a wide beam antenna, as shown in Figure 6.8, is adopted to improve the robustness for the feedback signals, while at the same time still fulfil the low rate link budget requirement. Aside from this difference, transmitter antenna selection is very similar to receiver antenna selection—the antenna is selected to provide the best channel conditions, and the criteria to judge the channel conditions are the same. More information about how to implement transmitter antenna diversity for millimeter wave systems is detailed in Figure 6.9, whereby sharp-beam antenna directions from both sides could be realized by using a single pin or rod antenna [23] or phase array antennas [24]. Firstly, the selected beam direction from the receiver antenna selection is kept unchanged from the Rx side to test the training signals/ channel sounding signals radiated from the transmitter. Secondly, a sharp-beam antenna is used at the Tx side and rotated to different antenna beam directions step by step as shown in Figure 6.9. The channel conditions at different antenna beam directions from the transmitter are estimated at the Rx side and sent back to the Tx side. Based on the feedback information, one beam direction from the Tx side with the strongest received channel sounding signals is selected. At the end, one RF chain is connected with the selected antenna at the Tx side for high rate data exchange.

If cost and size of millimeter wave components are not issues, more than one RF chain can be used for the transmitter, whose number is defined as L_T ($L_T > 1$, $L_T \ll M_T$) and more than one RF chain can be used for the receiver, whose number is defined as L_R

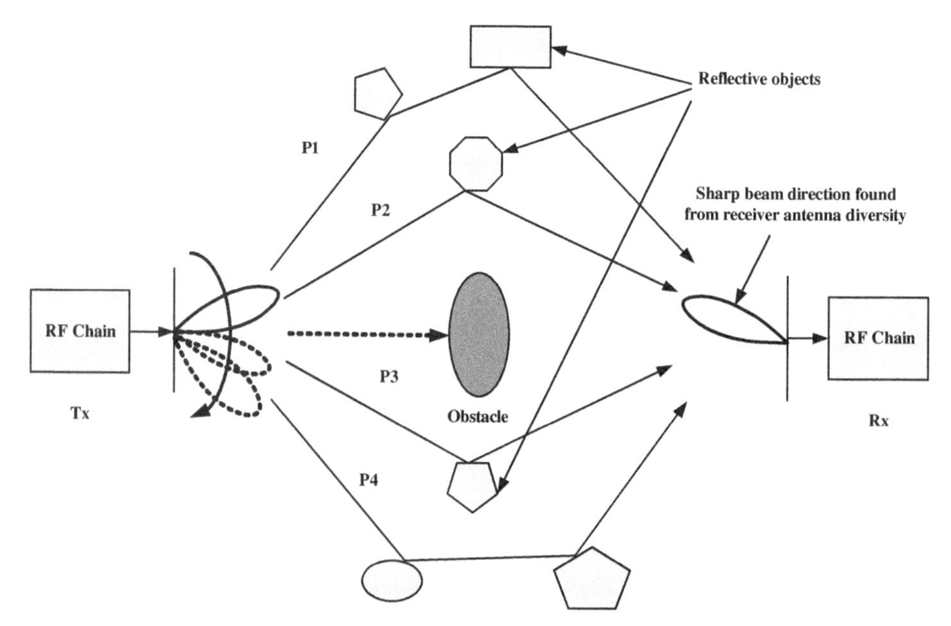

Figure 6.9. Transmitter antenna selection for millimeter wave systems

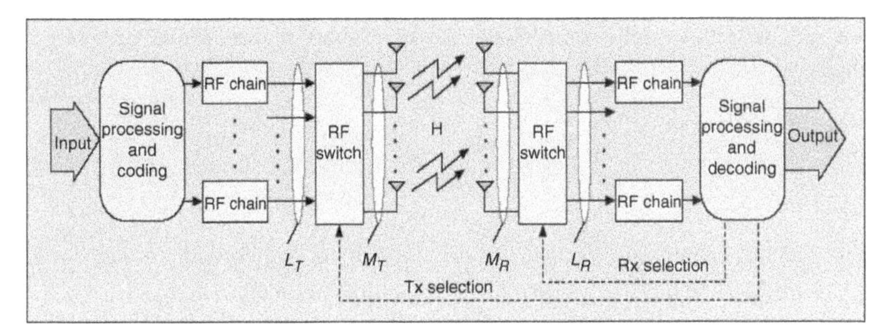

Figure 6.10. Antenna selection in MIMO [22] (© 2004 IEEE)

($L_R > 1$, $L_R \ll M_R$). The next step is to apply selection diversity simultaneously to both the transmitter and the receiver, as shown in Figure 6.10. In this scenario, there are M_T transmitter antennas and M_R receiver antennas. The Tx and Rx sides have L_T and L_R RF chains, respectively. Therefore, it is possible to transmit L_T parallel data streams, so a space-time code can be used to provide diversity [5, 10, 11]. Denote the overall $M_R \times M_T$ channel matrix by $\mathbf{H}$ and the $L_R \times L_T$ channel matrix corresponding to the selected antennas by $\hat{\mathbf{H}}$. Let us now consider an example of the orthogonal block space-time codes [5]. Those codes have a simple decoder and lead to an equivalent single-input-single-output (SISO) channel with the equivalent channel gain as follows

$$h_{eq} = \sqrt{\frac{1}{L_R}\sum_{i=1}^{L_R}\sum_{j=1}^{L_T}\left|\hat{h}_{ij}\right|^2} \tag{6.1}$$

where $\hat{h}_{ij}$ are the elements of $\hat{\mathbf{H}}$. The SNR of the equivalent channel is proportional to the Frobenius norm of the selected channel matrix, that is, $\left\|\hat{\mathbf{H}}\right\|^2 = \sum_{ij}\left|\hat{h}_{ij}\right|^2$. As a consequence, joint transmitter and receiver diversity should select a subset of the rows and columns of $\hat{\mathbf{H}}$ to maximize the sum of the squared magnitudes of the selected channel transfer function coefficients, that is $\left\|\hat{\mathbf{H}}\right\|^2$. However, the optimum solution is difficult to implement. For example, successively choosing the best antennas from the receiver and then the best antennas from the transmitter will not necessarily result in an overall optimal choice. In fact, except for exhaustive search, no systematic solution to the optimal joint transmitter and receiver antenna selection is known currently. Sub-optimal joint selection algorithms have to be considered and are worthy of further investigation. Since millimeter wave components like LO, mixer, PA, and LNA are not cheap, at most times, there is only one RF chain available at the Tx side and only one RF chain available at the Rx side. It is realistic to successively choose the optimum antenna from the receiver and after that the optimum antenna from the transmitter, which could converge to the joint optimal solution.

Frequency diversity happens when the multiple paths are spread out sufficiently, relative to the sampling period, so that multiple copies of the same radiated data are received over different symbol periods. At the *n*th symbol period, the received signals are expressed as

$$y(n) = \sum_{l=0}^{L-1} h_l x(n-l) + w(n) \tag{6.2}$$

whereby $x(n)$ is the transmitted signal, $w(n)$ is the noise component with zero mean and variance of σ^2, L is the number of symbol-spaced multiple paths, and $h_0, \ldots, h_{L-1}$ are the symbol-spaced channel tap coefficients for the different paths that are modeled as statistically independent. Assuming that $SNR = \frac{1}{\sigma^2}$ and optimum coherent detection is applied, the average detection probability is

$$P_e = E\left[Q\left(\sqrt{2 \times SNR \times \left(\sum_{l=0}^{L-1} |h_l|^2\right)}\right)\right] \tag{6.3}$$

where

$$Q(x) = \frac{1}{\sqrt{2\pi}} \int_x^{\infty} e^{-\frac{x^2}{2}} dx$$

It is clear that full diversity gain is achieved when frequency diversity is utilized and SNR is relatively high. As mentioned before, in millimeter wave systems, an original scheme of frequency selection diversity together with spatial diversity is proposed. As illustrated in Figure 6.3, after the combination of transmitter antenna selection and receiver antenna selection, path 2 and path 4 are selected as the two strongest candidates for high rate data exchange. Assume that the corresponding channel transfer functions for P2 and P4 are as shown in Figure 6.11.

If we consider the whole available millimeter wave bandwidth from 59 GHz to 65 GHz [15], path P4 is stronger than path P2, which is manifested by the relationship

$$\int_{f=59\text{ GHz}}^{65\text{ GHz}} |H_2(f)|^2 df > \int_{f=59\text{ GHz}}^{65\text{ GHz}} |H_1(f)|^2 df \tag{6.4}$$

Based on the criteria of the previous ideas of joint transmitter and receiver antenna selection, path P4 and its corresponding antenna beam directions from both Tx and Rx sides would be chosen. However, the data rate of millimeter wave applications might be limited, for example to less than 2 Gbps, and only the limited bandwidth of $B = f_2 - f_1$ is required (see Figure 6.11(a)). For this limited bandwidth, we have different relationship as follows:

$$\int_{f=f_1}^{f_2} |H_2(f)|^2 df < \int_{f=f_1}^{f_2} |H_1(f)|^2 df \tag{6.5}$$

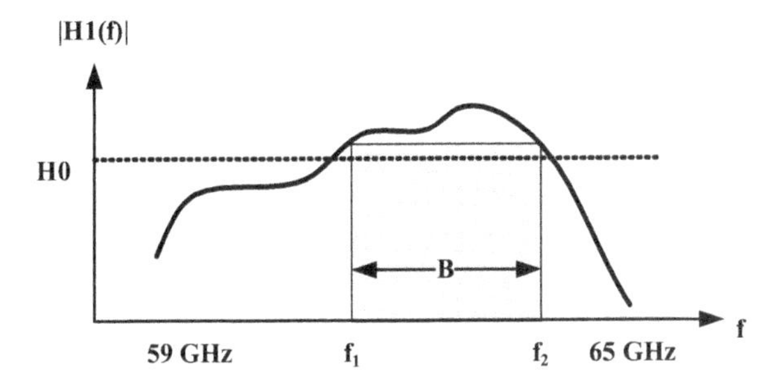

(a) Channel transfer function of path P2

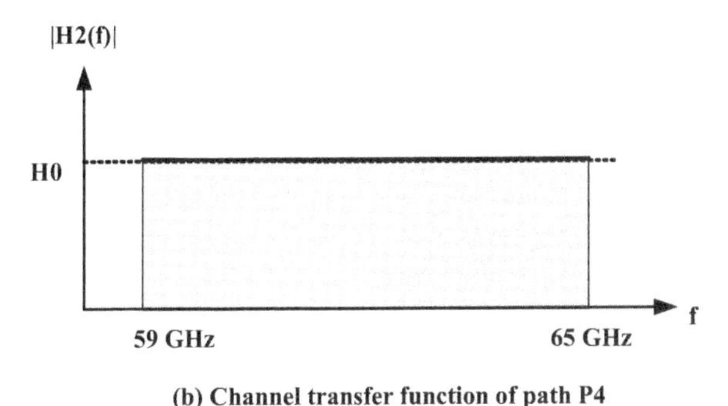

(b) Channel transfer function of path P4

Figure 6.11. Channel transfer functions for path P2 and P4

Therefore, when frequency selection diversity is applied together with spatial diversity using joint transmitter and receiver antenna selection, P2 is chosen together with operating bandwidth from f_1 to f_2. As a consequence, the optimum diversity gain can be achieved and the power consumption can be reduced using the combination of space and frequency selection diversity when there is only one RF chain for the transmitter and the receiver, respectively.

6.4 DYNAMIC SPATIAL, FREQUENCY, AND MODULATION ALLOCATION

We have learned that it is possible to use a pair or several pairs of sharp-beam antennas for both the transmitter and the receiver of a wireless communication. Each pair of sharp beam antennas is steered to match the direction of its corresponding strong reflection path, an original method to further reduce the power consumption of millimeter wave wireless systems when data rate is kept unchanged is proposed using dynamic spatial and frequency allocation [25]. This process is summarized as follows:

- Establish at least one communication path between at least one pair of narrow beam antennas for the transmitter and the receiver.

- Select at least one part of the bandwidth available on the communication path.
- Transmit data between the pair of antennas over the selected part of the bandwidth.

To increase the data rate, the MIMO technique is used, whereby different data can be transmitted over different pairs of narrow-beam antennas. The multi-path propagation is furthermore a benefit for the user as it may be used to improve the capacity and/or reduce the power consumption. The remaining issue is how to select a part of the available bandwidth according to the channel transfer function of the communication path. The method can comprise selecting at least one part of the bandwidth for which the channel transfer function of the communication path is above a threshold. The threshold can be constant, frequency dependent, or time dependent. The transmission power could be concentrated on the selected part of the bandwidth.

Combined with a dynamic modulation allocation scheme, the above method can be extended to transmit data over the communication path with a modulation scheme that is selected depending on the channel transfer function of the communication path.

Several pairs of antennas are used to transmit the data between the transmitter and the receivers, as shown in Figure 6.3, whereby it is possible to reduce the multipath channel delay spread and the fluctuations of the frequency response of the communication path due to narrow-beam antenna radiation patterns from both Tx and Rx sides. The narrow-beam antenna is characterized by its aperture and by its half-power beam width (HPBW), which defines the angle within which the power radiated is above one-half of what it is in the most preferential direction. When the aperture or HPBW of the narrow-beam antenna is reduced, the delay spread can be shortened. On the other hand, the communication path has to be maintained even if a station moves and even if an obstacle appears such that the complexity of the steering mechanism and of the tracking algorithm may increase. For practical reasons, HPBW should be kept at a reasonable value and the fluctuations of the channel frequency response may appear. Furthermore, in millimeter wave circuits there is generally a fluctuation of the frequency response that is caused by design tolerance and mismatching. The actual channel transfer function of each candidate wireless path corresponding to one pair of sharp-beam antennas for both Tx and Rx sides is sometimes not flat, which is illustrated in Figure 6.12. In this scenario, the channel transfer functions are defined as H1(f), H2(f), and H3(f), respectively.

The channel transfer functions H1(f) and H3(f) exhibit a frequency selective fading over the available bandwidth, which extends, for example, from 59 GHz to 65 GHz [15]. H2(f) has a constant value H0 over the whole bandwidth and thus exhibits a frequency nonselective fading.

Since the entire available bandwidth is not completely used for transmitting data, only portions of the available bandwidth are used wherever the corresponding channel transfer function is good. Figure 6.13 illustrates the data being transmitted from the transmitter to the receiver over the bandwidth B1 of the first communication path delimited by frequencies f_1 and f_2, over the bandwidth B2 of the second communication path delimited by frequencies f_3 and f_4, as well as over the bandwidth B3 of the third communication path delimited by frequencies f_5 and f_6. The bandwidth B1 is the portion of the available bandwidth where the channel transfer function H1(f) is equal or above a predefined threshold value H0,1. Likewise, the bandwidths B2 and B3 are chosen such

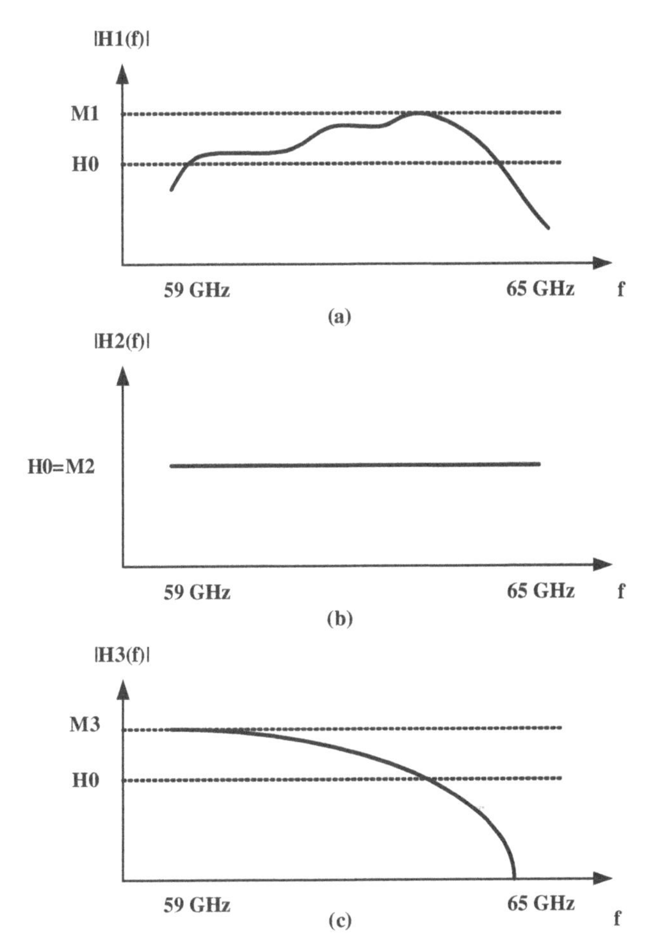

Figure 6.12. Channel transfer functions of different wireless links using different pairs of sharp-beam antennas

that the channel transfer functions H2(f) and H3(f) of the respective communication paths are equal or above the threshold values H0,2 and H0,3, respectively. The threshold values H0,1 H0,2, and H0,3 are either common to all communication paths or chosen individually. These values can be constants or may vary with frequency or time.

In the case that a channel transfer function is flat or relatively flat, as shown in Figure 6.13(b), the whole available bandwidth may be selected for transmitting data. The boundary frequencies f_3 and f_4 of the selected bandwidth B2 then correspond to the limits of the available bandwidth, which are 59 GHz and 65 GHz in this case [15].

The threshold values H0,1 H0,2, and H0,3 may vary depending on the amount of data to be transmitted, or on the used modulation scheme, or even on a preferred size for the resulting selected bandwidths B1, B2, and B3. In the last case, the threshold values are modified such that the selected bandwidths B1, B2, and B3 being above the thresholds have the given or predetermined sizes. It is also possible to set the threshold values H0,1

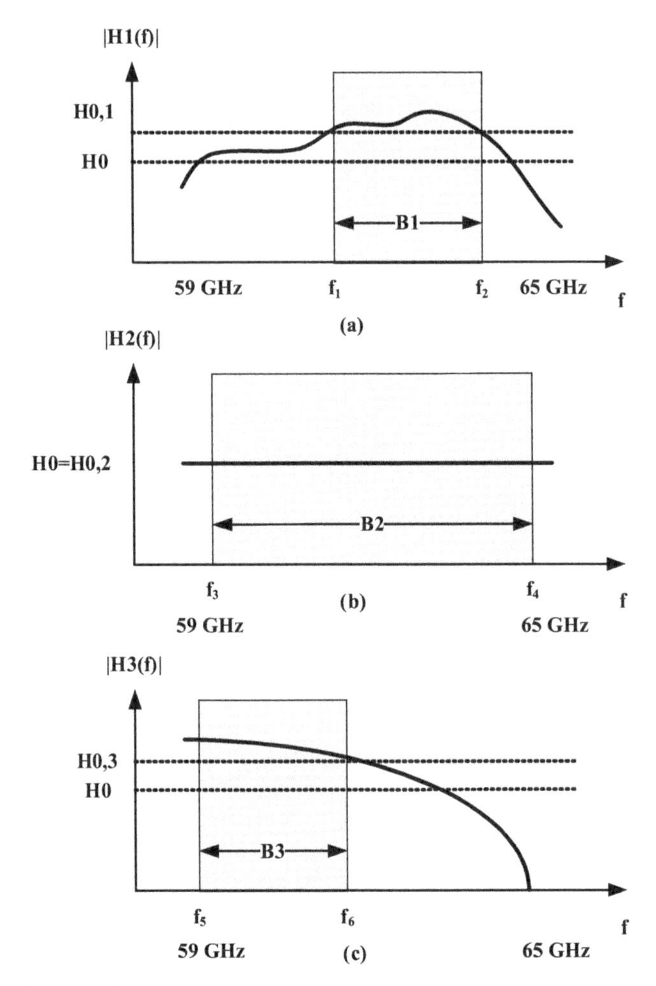

Figure 6.13. Proposed method with dynamic frequency selection

H0,2, and H0,3 depending on the maximal values M1, M2, and M3 of the corresponding channel transfer function, as shown in Figure 6.12. As an example, it is possible to transmit data via the first communication path only over frequencies that have a channel transfer function between the maximal value M1 and a lower value M1 – Vp, whereby Vp is a predetermined value.

The selected bandwidth B1 for transmitting data can consist of several separated portions or frequency ranges of the total available bandwidth if, for example, two separated regions of the channel transfer function H1(f) are above the threshold value H0,1.

The transmission power is adapted to the size of the selected bandwidths B1, B2, and B3. If the selected bandwidths B1, B2, and B3 are smaller than the wanted bandwidth, it is proposed to increase the transmission power PB1, PB2, and PB3 allocated to the selected bandwidths B1, B2, and B3, respectively.

Different modulation schemes (e.g., QPSK, 16APSK, 16-QAM, 64-QAM, 64APSK, or higher-order modulation schemes) are dynamically selected for each communication path depending on the RSSI or SNR measurement results in order to keep the data rate unchanged, and this is shown in Figure 6.14. Even if data are transmitted only over the B1 part of the available bandwidth, the data rate can be maintained constant by increasing the number of constellation points of the modulation scheme.

The modulation scheme may also be selected dynamically for each communication path depending on the channel transfer function in order to maximize the capacity. The number of constellation points of the modulation scheme are then optimized in accordance with the channel conditions in the selected bandwidth B1, B2, and B3. In

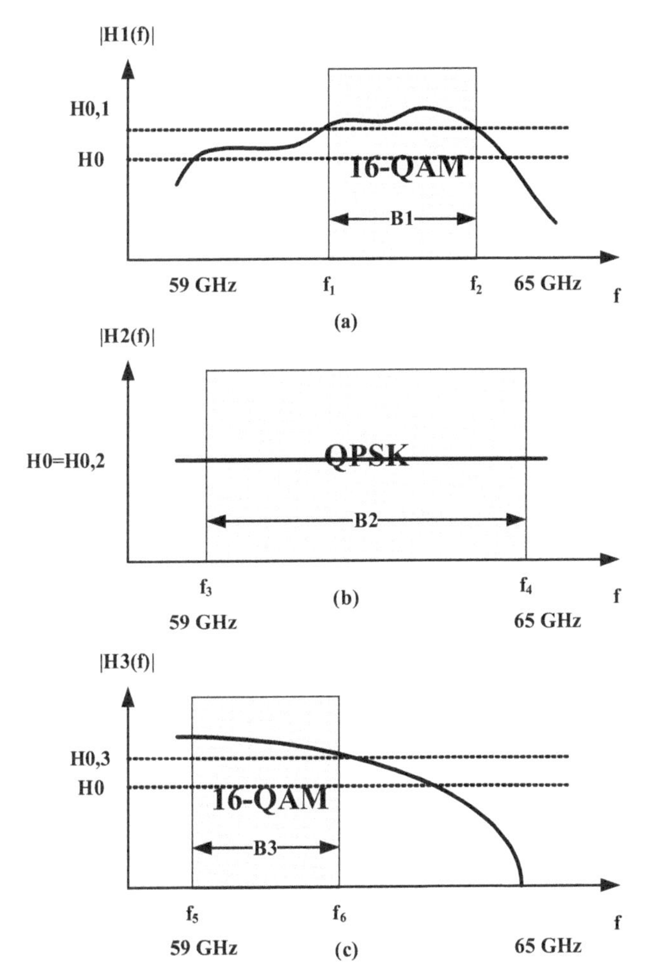

Figure 6.14. Proposed method with dynamic frequency and modulation scheme selection

Figure 6.14, the modulation scheme QPSK is used for transmitting data over the second communication path. In the selected bandwidths B1 and B3 for transmitting data over the first and third communication paths, the values of the respective channel transfer functions $H_1(f)$ and $H_3(f)$ are higher than $H_2(f)$ in the selected bandwidth B2 of the second communication path. Accordingly, the modulation scheme of the first and third communication paths can be increased to 16-QAM without the loss of quality or at least can maintain the data rate at a constant level. Based on the dynamic frequency and modulation scheme allocation, the overall power consumption for millimeter wave wireless systems can be reduced while the same data rate is maintained.

Figure 6.15 summarizes the solution using dynamic allocation across space, frequency, and modulation dimensions. The narrow-beam antennas of the transmitter have a HPBW of 20°, a horizontal or azimuthal beam control range of 80°, as well as a vertical or elevational beam control range of 80°. When determining the communication paths for data transmission between the transmitter and the receiver, a list of candidate wireless paths is established and a list of candidate antenna positions can be deduced. In the case where the available number of candidate antenna positions is

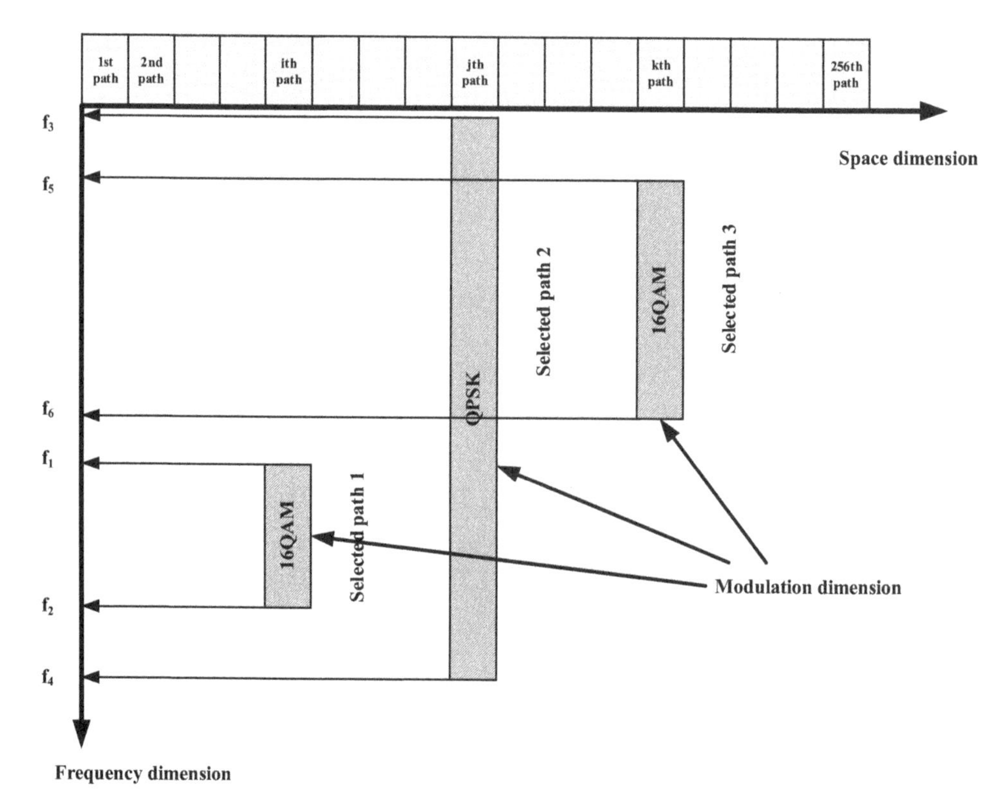

Figure 6.15. Proposed scheme with dynamic space, frequency, and modulation scheme selection

16 (4 × 4) for both the transmitter and the receiver, the number of total candidate wireless paths will be 256 (16 × 16). Figure 6.15 shows a two-dimensional space-frequency grid that is calculated when determining the candidate antenna positions and the corresponding candidate wireless paths from 1 to 256. The available bandwidth is from 59 GHz to 65 GHz. Three candidate paths with the strong channel conditions are assumed to be chosen based on the results of a RSSI or SNR measurement. Since fluctuations of the channel transfer functions H1(f), H2(f), and H3(f) of the three communication paths exist and good channel transfer functions suitable for data transmission might only be found within a part of the available bandwidth, data may only be transmitted between the frequencies f_1 and f_2 for the first communication path, between the frequencies f_3 and f_4 for the second communication path, and between the frequencies f_5 and f_6 for the third communication path. On top of this two-dimensional space-frequency grid, another dimension is added based on modulation, whereby various modulation schemes are selected based on actual channel conditions and the required data rate. As a result, the whole radiated power can be reduced and the power consumption of PA could be decreased.

Furthermore, a MIMO channel model is built up due to the usage of several pairs of narrow-beam antennas from both Tx and Rx sides, where the matrix of channel transfer function based on Figure 6.15 can be defined as

$$\mathbf{H} = \begin{bmatrix} h_{11} & h_{12} & h_{13} \\ h_{21} & h_{22} & h_{23} \\ h_{31} & h_{32} & h_{33} \end{bmatrix} \tag{6.6}$$

where h_{ij} defines the channel transfer function from the transmitter antenna corresponding to the ith selected path from the list to the receiver antenna corresponding to the jth selected path from the list. Differing from the conventional MIMO systems, the features of this model are summarized as follows:

- Due to the usage of sharp-beam antennas at both Tx and Rx sides, $h_{ii}(1 \leq i \leq 3)$ is much greater than $h_{ij}(i \neq j$ and $1 \leq i \leq 3,\ 1 \leq j \leq 3)$ and
- $h_{ii}(1 \leq i \leq 3)$ could be reasonably assumed to be quasi-static.
- In a conventional MIMO system, all transmitter and receiver antennas use the same bandwidth and the same carrier frequency. Since our proposed MIMO solution uses different pairs of sharp-beam antennas, which might occupy different bandwidths and spectrums, $h_{ij}(i \neq j)$ are further reduced.

In conclusion, the cross-interference caused by $h_{ij}(i \neq j)$ between different pairs of sharp-beam antennas becomes much smaller due to the usage of sharp-beam antennas and different bandwidths, and they can be easily handled and cancelled using the state-of-the-art algorithms already adopted in the conventional MIMO systems [1–13]. A good choice for a matrix of channel transfer function is a diagonal matrix.

REFERENCES

[1] A. J. Paulraj, D. A. Gore, R. U. Nabar, and H. Bölcskei, "An overview of MIMO communications – a key to gigabit wireless." *Proc. IEEE*, Vol. 92, No. 2, pp. 198–218, 2004.

[2] G. L. Stüber, J. R. Barry, S. W. Mclaughlin, and Ye (Geoffrey) Li, "Broadband MIMO-OFDM wireless communications." *Proc. IEEE*, Vol. 92, No. 2, pp. 271–294, 2004.

[3] S. M. Alamouti, "A simple transmit diversity technique for wireless communications." *IEEE J. Sel. Areas Commun.*, Vol. 12, No. 8, pp. 1452–1458, 1998.

[4] B. L. Hughes, "Differential space–time modulation." *IEEE Trans. Info. Theory*, Vol. 46, No. 7, pp. 2567–2578, 2000.

[5] X. B. Liang, "Orthogonal designs with maximal rates." *IEEE Trans. Info. Theory*, Vol. 49, No. 10, pp. 2468–2503, 2003.

[6] B. M. Hochwald and T. L. Marzetta, "Unitary space-time modulation for multiple communications in Rayleigh flat fading." *IEEE Trans. Info. Theory*, Vol. 46, No. 2, pp. 543–564, 2000.

[7] B. M. Hochwald and T. L. Marzetta, "Systematic design of unitary space-time constellations." *IEEE Trans. Info. Theory*, Vol. 46, No. 6, pp. 1962–1973, 2000.

[8] D. Cui and A. M. Haimovich, "Performance of parallel concatenated space-time codes." *IEEE Commun. Lett.*, Vol. 5, No. 6, pp. 236–238, 2001.

[9] V. Gulatiand and K.R. Narayanan, "Concatenated codes for fading channels based on recursive space-time trellis codes." *IEEE Trans. Wireless Commun.*, Vol. 2, No. 1, pp. 118–128, 2003.

[10] V. Tarokh, H. Jafarkhani, and A. R. Calderbank, "Space–time block codes from orthogonal designs." *IEEE Trans Info. Theory*, Vol. 45, No. 5, pp. 1456–1467, 1999.

[11] V. Tarokh, A. F. Naguib, N. Seshadri, and A. R. Calderbank, "Space-time codes for high data rate wireless communication: performance criteria in the presence of estimation errors, mobility and multiple paths." *IEEE Trans. Commun.*, Vol. 47, No. 2, pp. 199–207, 1999.

[12] X. Yan and G. B. Giannakis, "High-rate space-time layered OFDM." *IEEE Commun. Lett.*, Vol. 6, No. 5, pp. 187–189, 2002.

[13] N. Sharma and C. B. Papadias, "Improved quasi-orthogonal codes through constellation rotation." *IEEE Trans. Commun.*, Vol. 51, No. 3, pp. 332–335, 2003.

[14] IEEE Draft STD 802.11n, Part II: Wireless LAN Medium Access Control (MAC) and Physical Layer (PHY) Specifications, Sep. 2009.

[15] S. A. Khan, A. N. Tawfik, C. J. Gibbins, and B. C. Gremont, "Extra-high frequency line-of-sight propagation for future urban communications." *IEEE Trans. Antenna Propag.*, Vol. 51, No. 11, pp. 3109–3121, 2003.

[16] H. Xu, V. Kukshya, and T. S. Rappaport, "Spatial and temporal characteristics of 60GHz indoor channels." *IEEE J. Sel. Areas Commun.*, Vol. 20, No. 3, pp. 620–630, 2002.

[17] H. Ha and Cc Yim, "Layer-weighted unequal error protection for scalable video coding extension of H.264/AVC." *IEEE Trans. Consum. Electron.*, Vol. 54, No. 2, pp. 736–744, 2008.

[18] Z. Wang, F. Hohl, R. Stirling-Gallacher, and Q. Wang,"Improving video robustness using spatial and temporal diversity." European patent EP2109317.

[19] T. Gan, L. Gan, and K. K. Ma, "Reducing video-quality fluctuations for streaming scalable video using unequal error protection, retransmission, and interleaving." *IEEE Trans. Image Processing*, Vol. 15, No. 4, pp. 819–832, 2006.

[20] I. Bahceci, T. M. Duman, and Y. Altunbasak, "Antenna selection for multiple antenna transmission systems: performance analysis and code construction." *IEEE Trans. Info. Theory*, Vol. 49, No. 10, pp. 2669–2681, 2003.

[21] S. D. Blostein and A. Nosratinia, "Multiple antenna systems: their role and impact in future wireless access." *IEEE Commun. Mag.*, pp. 94–101, Jul. 2003.

[22] S. Sanayei and A. Nosratinia, "Antenna Selection in MIMO Systems." *IEEE Commun. Mag.*, pp. 68–73, Oct. 2004.

[23] K. Huang and Z. Wang, "V-band patch-fed rod antennas for high data-rate wireless communications." *IEEE Trans. Antenna Propag.*, Vol. 54, No. 1, pp. 297–300, 2006.

[24] K. Huang and Z. Wang, "Millimeter-wave circular polarized beam-steering antenna array for gigabit wireless communications." *IEEE Trans. Antenna Propag.*, Vol. 54, No. 2, pp. 743–746, 2006.

[25] Z. Wang and U. Masahiro,"Wireless communication method and system." European patent EP1841092.

7

ADVANCED BEAM STEERING AND BEAM FORMING

In this chapter, we present the concepts of advanced beam steering and beam forming for millimeter wave wireless systems. When a millimeter wave system employs a narrow-beam antenna with a regular radiation pattern, beam steering is preferred to track the strong line-of-sight (LOS) or non-line-of-sight (NLOS) reflection paths and thereafter for setting up a reliable wireless link for high rate applications. On the other hand, when phase array antennas are utilized, the amplitude and/or the phase of each patch element is calculated based on the beam forming principle to build up the wireless link with the best channel quality. At this time, the generated antenna beam might have regular or irregular antenna radiation patterns. The frame structure to support the function of beam steering and beam forming is discussed. The concept of exchanging the antenna identification information between the transmitter and the receiver to facilitate beam steering and beam forming is presented.

This chapter is organized as follows. Section 7.1 introduces narrow-beam antennas with or without regular antenna radiation patterns. Section 7.2 summarizes various frame structures to enable beam steering and beam forming with reduced overhead and hardware complexity. Section 7.3 discusses the concept of beam steering using regular antenna radiation patterns. Section 7.4 presents the idea of exchanging antenna identification information between the transmitter and the receiver, which is critical to accelerating the setup of a reliable wireless link for millimeter wave high rate

Millimeter Wave Communication Systems, by Kao-Cheng Huang and Zhaocheng Wang

applications. Section 7.5 analyzes the concept of beam forming, where the method of calculating the coefficients for each patch element is described.

7.1 THE NEED FOR BEAM-STEERING/BEAM-FORMING

Modern millimeter wave communication systems are dedicated to consumer electronics applications and therefore have to meet strict requirements including small size, light weight, and low production cost. Moreover, there are numerous challenging constraints on the antenna design, such as high gain and wide bandwidth, while maintaining a small size.

To build a millimeter wave link, a single patch antenna is one solution to these requirements, but a standard microstrip antenna has the drawbacks of low antenna gain and narrow impedance bandwidth. To address these issues, there are solutions such as a single-layer, multi-resonant microstrip patch antenna on a relatively thin substrate [1], and a stacked patch on a multi-layer substrate to increase the substrate thickness [2]. Furthermore, an additional layer with parasitic elements is added to the layer containing the main patch element in order to increase the patch size and, thus, obtain the better radiation efficiency. In a millimeter wave communication system, the transmitting power can be increased either by a power amplifier or a high gain antenna. In general, 60 GHz power amplifiers consume high DC power so an antenna-based solution is more cost-effective. To increase the antenna gain, one of the solutions is to add a horn or rod on top of a patch antenna [3]. An example is illustrated in Figure 7.1 (a), whereby a rod antenna is fed by a patch and held by a waveguide. The rod antenna consists of a cylindrical part and a tapered part. The rod is fed by a patch, which is sequentially energized by a microstrip line connected to a coaxial connector. Figure 7.1 (b) shows the corresponding regular antenna radiation pattern with different height.

When a narrow-beam antenna is adopted for both the transmitter and the receiver, beam directions from both sides are rotated either mechanically or electrically in order to find the strongest wireless link based on the measured channel quality information, which is shown in Figure 7.2. During the rotation, the half-power beam width (HPBW) of the main beam or the main radiation pattern is kept unchanged. The algorithm to control the rotation, handle the coordination between the transmitter and the receiver, and successfully exchange the control signalling is called beam steering, which will be discussed in Section 7.2. At the end of beam steering, the strong LOS/NLOS path can be aligned and the link budget requirements can be satisfied due to the inherent antenna gain from both narrow-beam antennas.

As an alternatives to horn or rod antennas on top of the patch antennas, it is possible to adopt arrays of multiple patch elements [4, 5], although the antenna size and the losses due to the long feeding network are increased in this case. One example of phase array antennas is shown in Figure 7.3. From the transmitter (Tx) side, the baseband and RF signal are identical for all the patch antenna elements. The coefficients W_i for $i = 1, \ldots, M$ are adjusted to realize beam forming, where the coefficient amplitude for each patch element can be changed and its corresponding phase can be shifted. For millimeter wave circuits, the feeding loss for phase array antennas is relatively high. To improve the

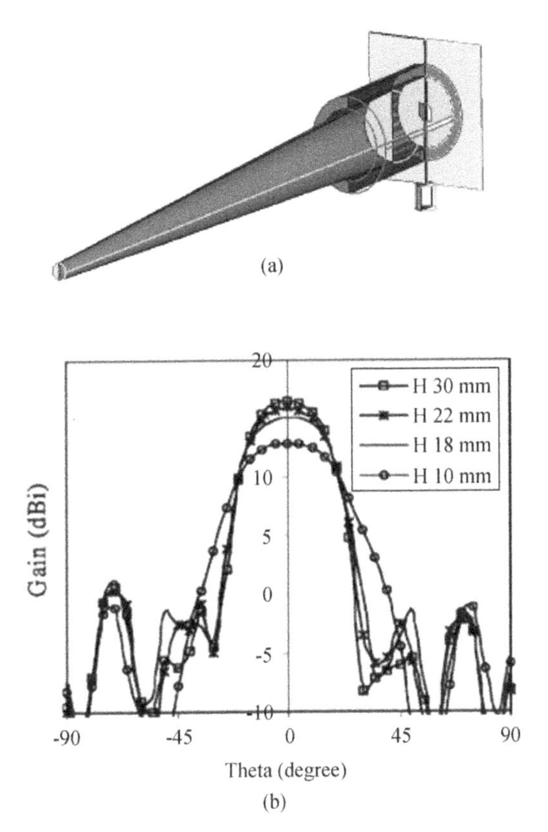

Figure 7.1. (a) One example of a narrow-beam antenna and (b) its radiation pattern [3] (©2006 IEEE)

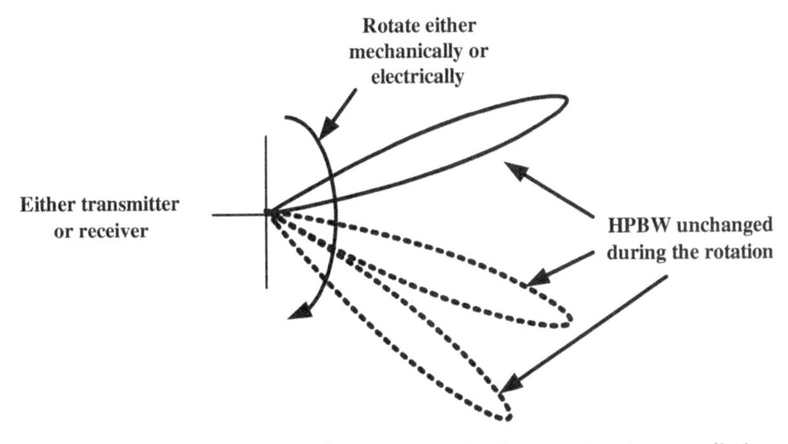

Figure 7.2. Beam steering of a narrow-beam antenna having a main antenna radiation pattern

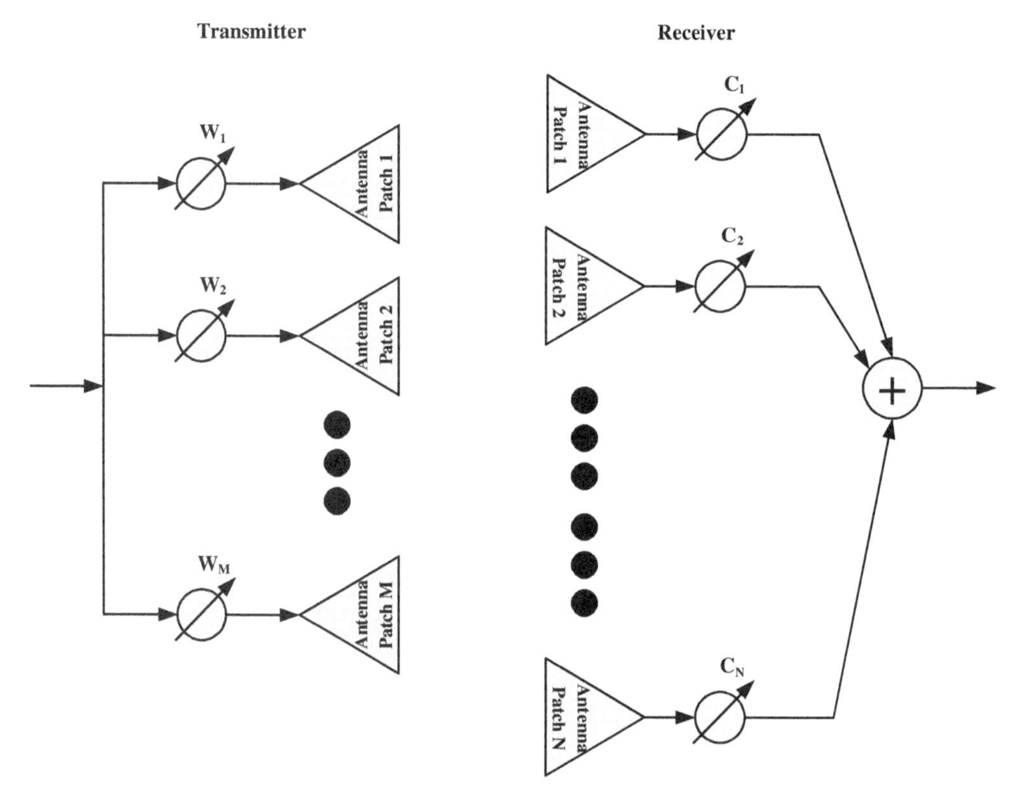

Figure 7.3. System model of phase array antennas

performance, each patch antenna could have its own embedded power amplifier to reduce the effect of the feeding loss. From the receiver (Rx) side, the multiple received RF signals are combined from all the patch antenna elements. The coefficients C_i for $i = 1, \ldots, N$ are adjusted to obtain the maximum received signal-to-noise ratio (SNR), whereby the coefficient amplitude for each patch element can be changed and its corresponding phase can be shifted. As with the transmitter, each patch element could have its own embedded low noise amplifier before the combiner to reduce the effect of the feeding loss and fulfil the link budget requirements.

At the Tx side, multiple phase shifters with coefficients $W_i \quad i = 1, \ldots, M$ receive the output from the mixer working at the millimeter wave range. A de-multiplexer is included to control which phase shifters receive the signal. These phase shifters are either quantized phase shifters or complex multipliers. The phase and/or magnitude of the currents in each patch element are adjusted to produce a desired beam pattern, as illustrated in Figure 7.4. The beam pattern might not have a single regular beam direction. An irregular beam pattern is generated in order to maximize the received SNR under dynamic wireless conditions.

With respect to the Rx side, a phase array antenna with a plurality of patch elements receives the signals and provide them to the phase shifters having coefficients

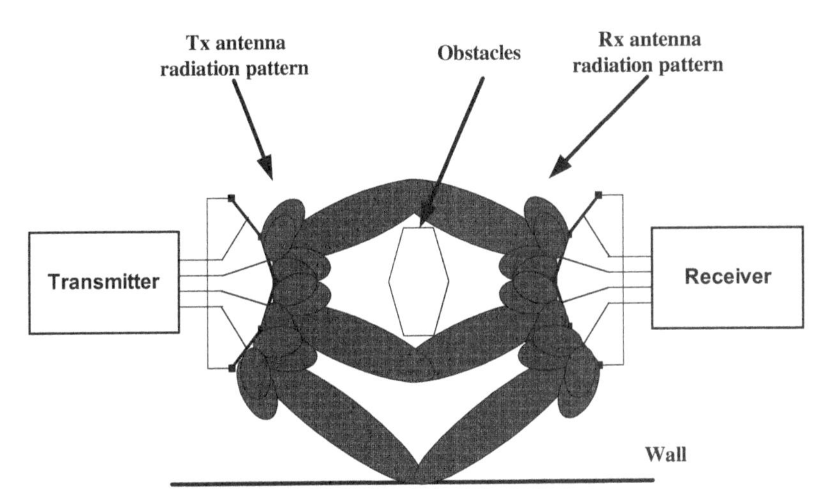

Figure 7.4. One example of a phase array antenna with an irregular antenna radiation pattern

$C_i \quad i = 1, \ldots, N$. As discussed above, phase shifters comprise either quantized phase shifters or complex multipliers. Phase shifters receive the signals from antennas, which are combined to form a single-line feed output as shown in Figure 7.3. A multiplexer is used to combine the signals from different elements and output through the single feed line. As mentioned before, the phases and magnitudes of the patch elements in the phased array antenna are adjusted to produce a desired beam pattern as illustrated in Figure 7.4. The beam pattern might have more than one regular beam direction. An irregular beam pattern is generated to maximize the received SNR.

For millimeter wave circuits, complex multipliers are difficult to implement from a design point of view; quantized phase shifters with very limited amplitude or phase selections are preferred, especially for 60 GHz based applications [6]. The algorithm to find the optimum coefficients $W_i \quad i = 1, \ldots, M$ and $C_i \quad i = 1, \ldots, N$ for quantized phase shifters from the transmitter and the receiver is called beam forming, which will be discussed in Section 7.5. The idea of beam forming is not concentrated on the search for an antenna radiation pattern having good shape, but instead concentrated on maximizing the received SNR under a dynamic wireless environment. The corresponding antenna radiation patterns from both Tx and Rx sides might have strange beam shapes, which are illustrated in Figure 7.4.

7.2 ADAPTIVE FRAME STRUCTURE

Sharp-beam antenna radiation patterns having high antenna gains are required to fulfil the link budget requirements for high-rate (beyond Gbps) millimeter wave applications. Beam steering is used to steer the antenna radiation pattern from both Tx and Rx sides so that they align with the strongest LOS or NLOS reflection path. One communication path corresponds to one specific narrow antenna beam direction from the transmitter and one specific narrow antenna beam direction from the receiver. Beam forming, on the other

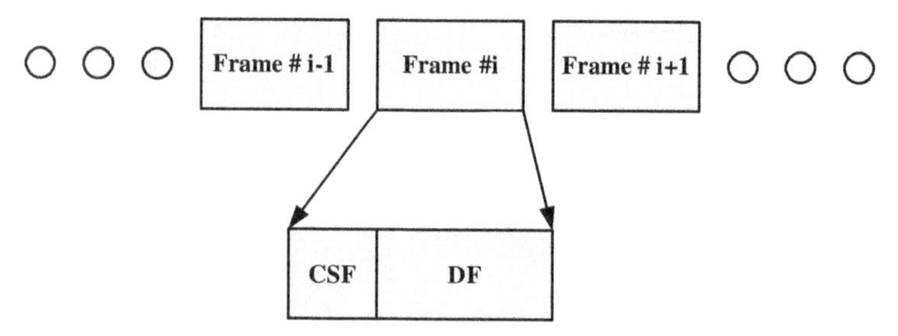

Figure 7.5. Frame structure to enable beam steering or beam forming

hand, refers to the adjustment of the coefficients of W_i for $i = 1,\ldots,M$ at the transmitter and the coefficients C_i for $i = 1,\ldots,N$ at the receiver.

A physical frame structure, as illustrated in Figure 7.5, is proposed to enable the function of beam steering or beam forming, whereby each consecutive frame is composed of two different types of frames, defined as the channel sounding frame (CSF) and the data frame (DF).

In Figure 7.6, the CSF comprises two parts, a training sequence Tc for aligning the directions of sharp-beam antennas from both Tx and Rx sides to the candidate communication paths and for performing time and frequency synchronisation, and a preamble sequence Pc enabling the estimation of channel quality information for the candidate communication paths. The DF comprises three part, one a training sequence Td for aligning the directions of sharp-beam antennas from both Tx and Rx sides to a currently used transmission path, and for carrying out time and frequency synchronisation; another a preamble sequence Pd enabling the estimation of channel quality information for the currently used transmission path; and the third a data part for normal high rate transmission from the transmitter to the receiver.

For high-rate applications like wireless high-definition television (HDTV), the transmitter and the receiver are stationary most of the time. It is not necessary to change a currently used transmission path frequently. For the physical frame structure shown in Figure 7.5, the overhead introduced by CSF (used for realizing beam steering or beam forming) is rather high. One possible solution is to transmit one CSF between several DF frames to reduce the overhead. This adaptive physical frame structure is illustrated in Figure 7.7.

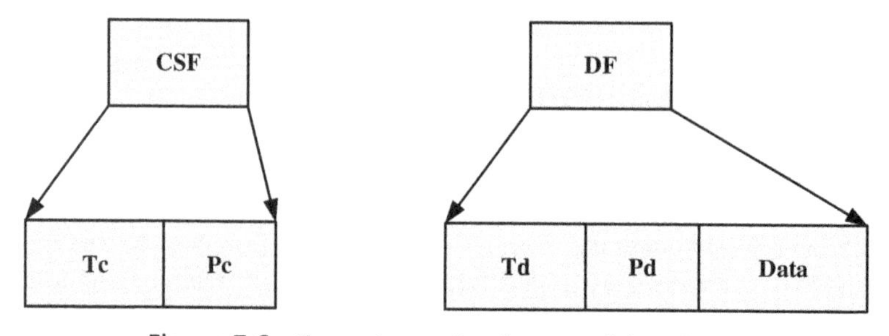

Figure 7.6. Channel sounding frame and data frame

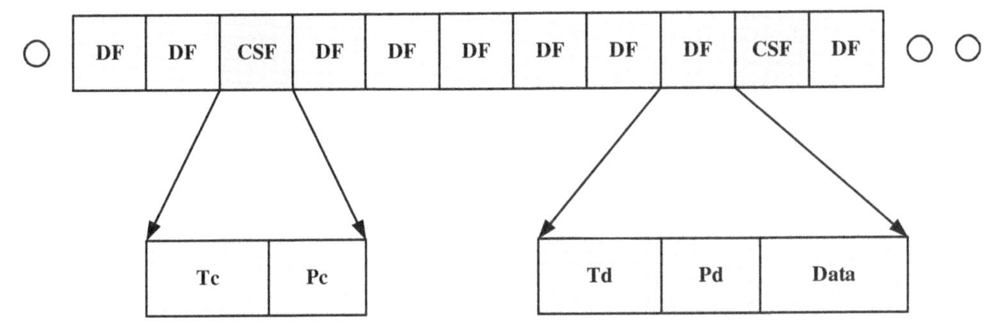

Figure 7.7. Adaptive frame structure to reduce the CSF overhead

However, if the channel quality of the currently used transmission path deteriorates, the frequency of CSFs might be increased in order to quickly find alternative candidate transmission paths with better channel quality. In other words, the number of CSFs transmitted from the transmitter to the receiver can be adapted depending on the estimated channel quality for the currently used transmission path. Additionally, the frequency of the CSFs in relation to the DFs can be dynamically adjusted based on additional features and/or parameters of the transmitter and the receiver, for example, the movement of one of the devices such as acceleration, rotation, and so forth. One example for an increased frequency of the transmission of CSFs is schematically illustrated in Figure 7.8.

In Figure 7.8, there exist different requirements between DF right after CSF and DF right after another DF. For DF right after CSF, its training sequence Td is longer since, within that period, the sharp-beam antennas from both Tx and Rx sides have to be switched from the position corresponding to one possible candidate transmission path

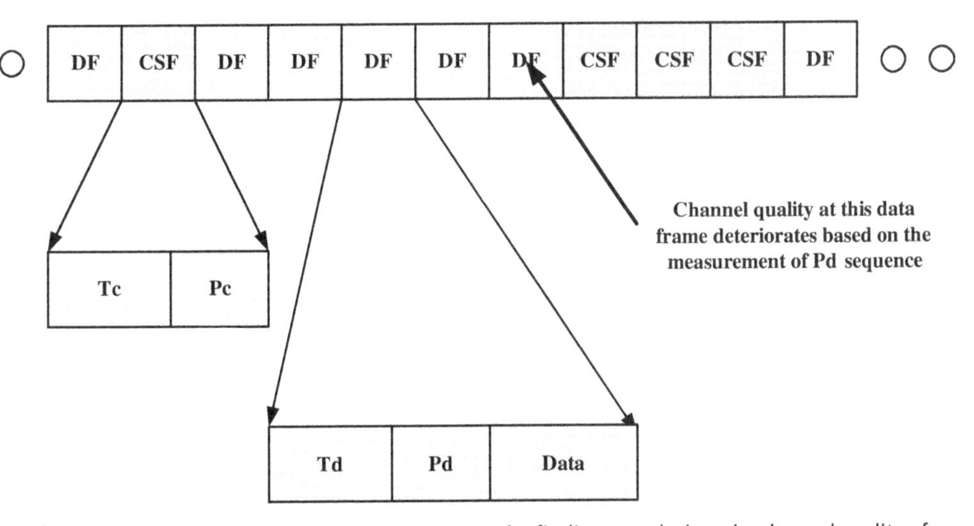

Figure 7.8. Adaptive frame structure to improve the finding speed when the channel quality of the currently used transmission path deteriorates

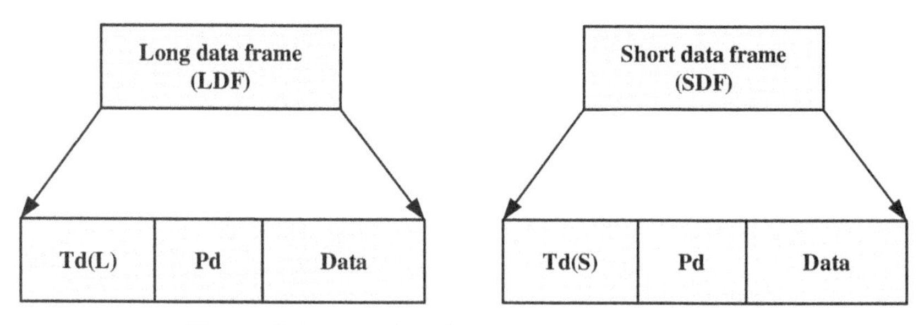

Figure 7.9. Long data frame and short data frame

back to the position relating to the currently used transmission path. For a DF immediately after another DF, such switching is not necessary. The switching of the transmission paths means adjustments of some parameters of millimeter circuits, for example, the automatic gain controller (AGC). Since the AGC circuit needs time to converge, a longer time is required for a DF immediately after a CSF. Therefore, in order to further reduce the overhead introduced by beam steering or beam forming, the data frame is divided into two categories, long data frame (LDF) and short data frame (SDF), illustrated in Figure 7.9. LDF is defined as the DF immediately after CSF and SDF is the DF immediately after another DF.

An LDF comprises a training sequence Td(L), a preamble sequence Pd, and a data part. The training sequence Td(L) is used to enable the antenna switching for both the transmitter and the receiver, and then to enable synchronization, that is, timing or carrier frequency recovery for the currently used transmission path. A SDF similarly comprises a training sequence Td(S), a preamble sequence Pd, and a data part. The training sequence Td(S) is, however, only used to enable synchronization. Therefore, the length of Td(S) is shorter than that of Td(L). If a re-synchronization is not necessary, Td(S) could be omitted. An improved frame structure can be seen in Figure 7.10, where most of the time only the transmission of an SDF is required. The overhead introduced by the training sequences is therefore reduced.

7.3 ADVANCED BEAM STEERING TECHNOLOGY

Spatial diversity is achieved using several pairs of sharp-beam antennas from both Tx and Rx side, as shown in Figure 6.3. Each pair of sharp-beam antennas is steered to its corresponding strong reflection path. Each reflection signal can be assumed independent and can be treated as going through a frequency non-selective slow fading channel. The probability of all signals becoming weak is small and the path diversity gain can be acquired. If one strong LOS or NLOS reflection path is blocked, another unblocked link could still be alive to support the wireless transmission and viewers would not notice the interruption of HDTV signals, while at the same time another alternative transmission path can be searched using either the beam steering or the beam forming algorithm.

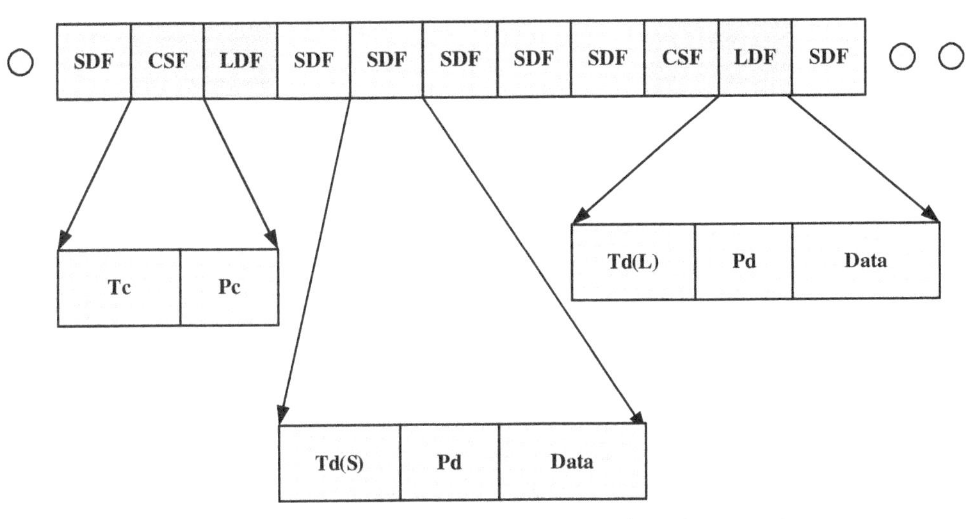

Figure 7.10. Proposed frame structure to reduce the overhead from training sequences

The beam steering algorithm is applied when there is one strong and regular main lobe for antenna radiation pattern [3, 7]. To find the strongest reflection paths, it is necessary to search all the possibilities when sharp HPBW antennas from both the transmitter and the receiver have many two-dimensional choices. For example, if the scanning range is 100° and the HPBW of the sharp-beam steering antenna is 20°, the number of two-dimensional choices from each side is 25 (5×5) and the total number of choices for both of Tx and Rx sides is 625 (25×25). For each choice, channel quality information like received signal strength indicator (RSSI), SNR, bit error rate (BER), frame error rate (FER), and channel delay profile estimation need to be measured, stored, and compared. After that, several best choices will be used for high data rate (HDR) communication. The corresponding wireless link is defined as the HDR link. It is apparent that the complexity of estimating and processing the channel quality information for all the 625 candidate transmission paths is rather high.

To reduce the complexity of the exhausted search–based sharp-beam antenna solution, wide-beam antennas could be added, which are used for low data rate control signaling exchange between the transmitter and the receiver. The corresponding wireless link is defined as the low data rate (LDR) link. In addition, the antennas are also used to accelerate the search for the strongest LOS or NLOS reflection path, where the number of candidate choices is reduced from 625 to 50. The initial search of the strongest LOS or NLOS reflection path is called the acquisition stage. When this is completed, sharp-beam antennas from both the transmitter and the receiver are steered to that strongest path to set up one HDR for high-rate applications and the beam steering algorithm goes to the tracking stage. During the tracking period, the beam steering algorithm continues to search other alternative LOS or NLOS strong reflection paths and update the candidate path table. When the currently used HDR link is blocked by moving obstacles, the beam steering algorithm will choose another strong reflection path from the table to set up an alternative HDR wireless link. The proposed acquisition and tracking algorithms are illustrated in Figure 7.11 and summarized in detail as follows:

Step 1. The transmitter sends the control signal-like beacons using a wide beam antenna and waits for the response from the receiver.

Step 2. Initially the receiver is configured to use a wide-beam antenna. When it wants to communicate with the transmitter, it sends back the control signal using that antenna. The LDR wireless link is then established. Since the data rate is low, the channel distortion introduced by the wide-beam antennas is negligible and the link budget requirement can be satisfied.

Step 3. When the link is set up, the transmitter is then configured to use both narrow-beam and wide-beam antennas simultaneously. The wide-Beam antenna is used for demodulating the control signal from the corresponding receiver. The beam direction of the narrow-beam antenna is steered step-by-step. At the Rx side, the wide-beam antenna is adopted, and for each narrow-beam direction from the Tx side, channel quality information such as RSSI, SNR, BER, FER, and channel delay profile are measured, stored, and compared. This information is then sent back to the transmitter using the LDR wireless link, and is stored and compared at the transmitter. At the end, the strongest narrow-beam direction from the transmitter is identified.

Step 4. The beam position of narrow HPBW steering antenna from the Tx side remains unchanged.

Step 5. At the corresponding Rx side, switch the wide-beam antenna of the receiver to another narrow HPBW steering antenna. The beam direction of the narrow-beam antenna is rotated step-by-step as in Step 3 to identify the strongest narrow-beam direction.

Step 6. The strongest narrow-beam directions from both the Tx and Rx sides are finally identified and used for the setup of one HDR wireless link for high rate applications. The acquisition stage is finished and the beam steering algorithm goes to the tracking stage.

Step 7. The frame structure to enable the beam steering algorithm has been discussed in Section 7.2. During the DF period, the previously selected narrow-beam directions from both the transmitter and the receiver are used for the HDR wireless link. During the CSF period, different beam directions of narrow-beam antennas from both the transmitter and the receiver are continuously checked according to Steps 3 to 6, and the second strongest path is found and the second pair of narrow HPBW steering antennas is used to track it. At this time, the required control signaling is exchanged using either wide-beam antennas or narrow-beam antennas used for the HDR wireless link. A wide-beam antenna can still be configured and used for improving the search speed for alternative strong wireless paths. Finally, several strong paths are chosen for constructing several HDR wireless links, which can be used for increasing capacity or achieving

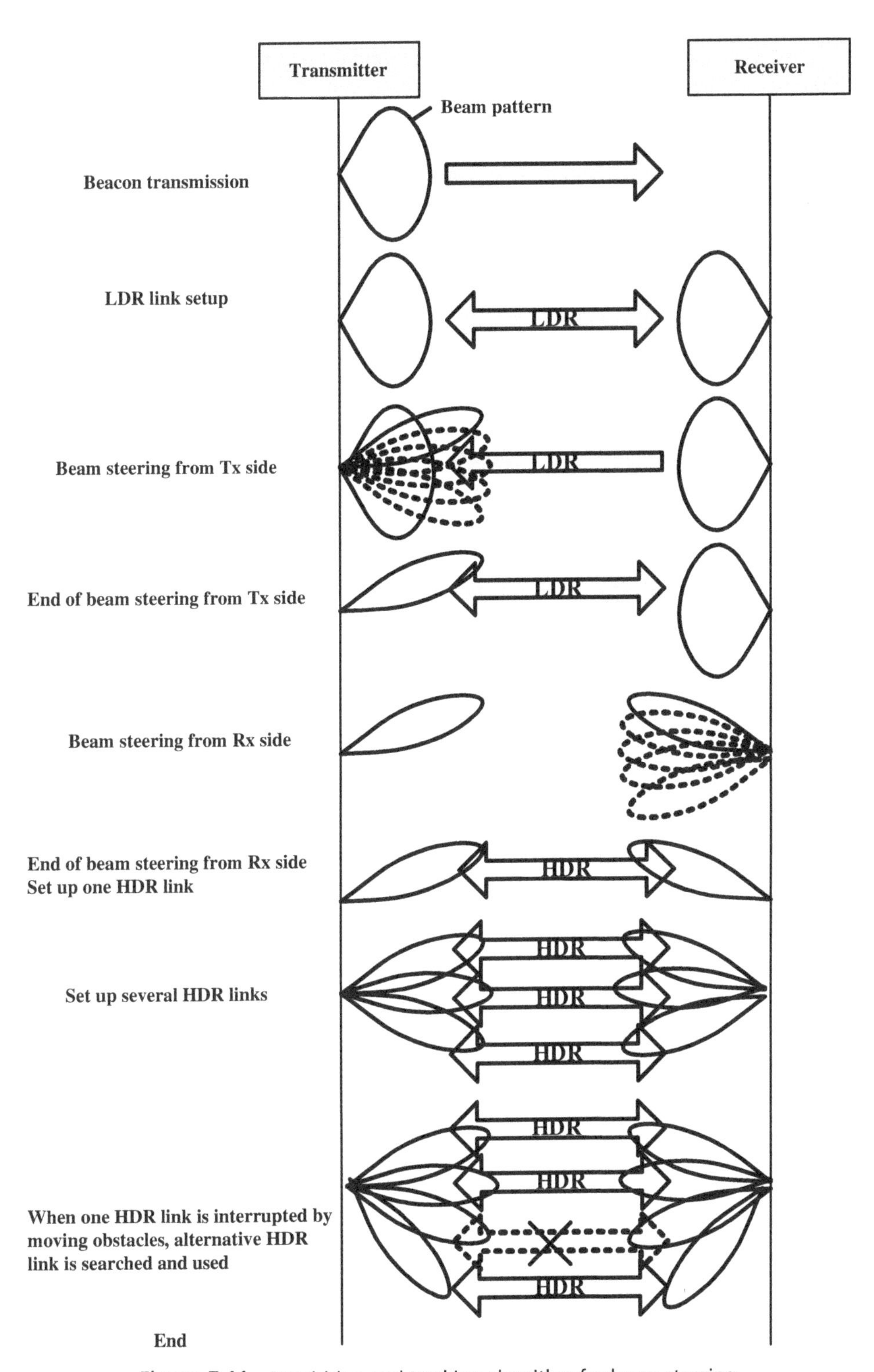

Figure 7.11. Acquisition and tracking algorithm for beam steering

additional spatial diversity gain to improve the robustness of millimeter wave wireless systems.

Step 8. Other strong paths not used for the HDR wireless link are stored in the candidate path table. Due to the movement of the transmitter, the receiver, the reflective objects, or the obstacles between the transmitter and the receiver, one HDR wireless link might disappear and another strong candidate path taken from the candidate path table might be used to build up another HDR wireless link, as shown in Figure 7.11. The candidate path table is updated continuously. When one new strong reflection path is found using CSF, it is added into the candidate path table. Similarly if a strong reflection path becomes weak due to the movement of the transmitter, the receiver, the reflective objects, or the obstacles, it is deleted from the table.

Step 9. Steps 7 and 8 are repeated until the high rate communication between the transmitter and the receiver is done. When there is only one HDR wireless link and it deteriorates, and there are no other strong paths from the candidate path table, the beam steering algorithm will go from the tracking stage back to the acquisition stage, and Steps 2 to 6 are repeated.

As already explained, if the scanning range is 100° and the HPBW of the narrow-beam antenna is 20°, narrow-beam antennas configured at both the transmitter and the receiver are steerable into 25 different positions in order to achieve spatial diversity gain, which is critical for millimeter wave high-rate applications. These 25 different positions for the transmitter and the receiver are schematically drawn in the channel candidate tables, as illustrated in Figure 7.12.

Each of the narrow-beam antennas can be steered into five different positions along a first axis denoted as the x-axis, and can be steered into another five different positions along a second axis denoted as y-axis. Thereby, each antenna can be steered into total 25 (5×5) different positions in a two-dimensional plane. The channel quality of each position at both Tx and Rx sides is measured. When one good transmission path between the transmitter and the receiver is found, it is stored into a candidate path table for the transmitter and into a candidate path table for the receiver, respectively. Each of the squares in Figure 7.12(a) and Figure 7.12(b) corresponds to a different position for both the transmitter and the receiver and thereby to a different transmission path.

A channel quality estimation method is discussed in [8]. The antennas of the transmitter and of the receiver are steered into a first position and then CSF signals are sent. In a next step, the position of the narrow-beam antenna of the transmitter is changed and again CSF signals are sent. This goes on until the narrow-beam antenna of the transmitter has run through all the different positions. The receiver is then, depending upon the received CSF signals, able to decide which of the positions of the narrow-beam antenna of the transmitter provides the best transmission characteristics. The narrow-beam antenna of the transmitter will then be steered into that position. The same procedure repeats for the narrow-beam antenna of the receiver, namely the position of the narrow-beam antenna of the receiver is changed, thereby enabling an estimation of the

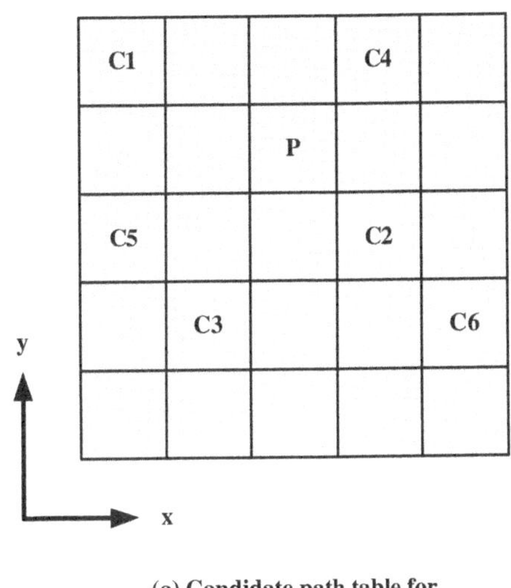

(a) Candidate path table for the transmitter

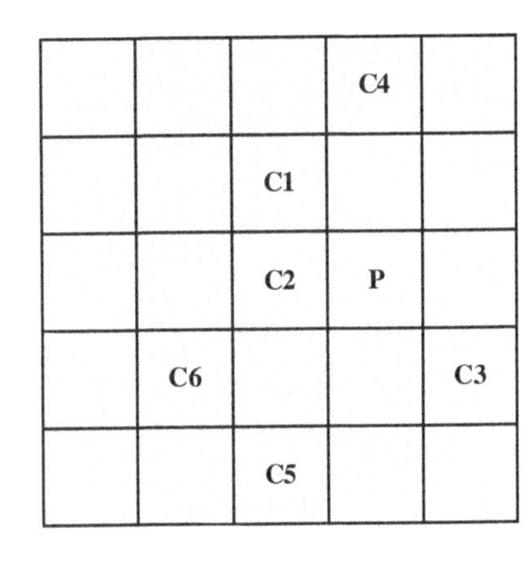

(b) Candidate path table for the receiver

Figure 7.12. Candidate path tables based on the channel quality information

best position of the narrow-beam antenna of the receiver. Once this is done, the strongest transmission path, namely the present transmission path, is identified and stored in the candidate path tables. The entire procedure may be repeated until all the candidate transmission paths are found and stored.

It is clear that the transmitter and the receiver have the knowledge about the present and next transmission paths and their steering positions. Furthermore, some feedback control signaling about the present and candidate transmission paths from the receiver to the transmitter and vice versa are provided.

The channel quality measurement is based on the transmitted CSF signals. After channel quality estimation of the first transmission path, the narrow-beam antennas are steered to further positions, thereby covering every possible transmission path. Depending on the channel quality information, a present transmission path having the best channel quality will be selected. This is schematically denoted by letter P in the path candidate table in Figure 7.12. Furthermore, several candidate transmission paths C1 to C6, which can be used to form alternative HDR wireless links, are stored in the candidate path tables, as shown in Figure 7.12, in case the performance of present transmission path P deteriorates.

It is preferable to use a CSF with a relatively short period in order to reduce the overhead for beam steering. The problem, however, arises because the accuracy of the channel quality estimation cannot be guaranteed. As already explained, when the CSF period is short, the resolution and accuracy of the channel quality estimation are reduced due to insufficient sounding samples. There might be many selected transmission paths between the transmitter and the receiver, which were initially identified to fulfil the requirements to become a candidate transmission path, for example, their channel quality might be above the predefined threshold, but some of these paths might not be good enough for actual high rate applications (e.g., HDTV signals). Therefore, when the present transmission path is interrupted by moving obstacles, it is difficult for the beam steering algorithm to choose the right candidate transmission path from the candidate path tables and switch the beam directions at both Tx and Rx sides to another reliable HDR wireless link, and viewers might feel the deterioration of HDTV signals.

As mentioned previously, after steering the narrow-beam antennas to different positions corresponding to different transmission paths, the candidate path tables are generated. From all these possible transmission paths, a present transmission path P is selected, and several further transmission paths C1 to C6, whose channel quality measurements are all above a predefined threshold, are selected as candidate transmission paths. One may repeat a transmission of CSF signals with a longer duration than previous CSF from the transmitter to the receiver and test only the candidate transmission paths C1 to C6 once more.

Because a longer CSF is used, the channel quality estimation is more accurate. In addition, since the number of candidate transmission paths of C1 to C6 is much smaller than the total available transmission paths of 625 (25×25), the additional overhead introduced is negligible. For example, if after checking C1 to C6 using a longer CSF, the channel quality measurements of C1, C2, and C3 are judged to be sufficiently good for HDR wireless links, while those of C4, C5, and C6 are not, only the candidate transmission paths C1, C2, and C3 are then identified and stored in the candidate path tables. Thus, the candidate tables are updated from those of Figure 7.12 to those shown in Figure 7.13. Therefore, when the present transmission path is interrupted by moving obstacles, it is possible for the beam steering algorithm to choose the right candidate transmission path from the candidate path tables of both the transmitter and the receiver.

The method to refine the candidate path tables will now be summarized in the flow chart in Figure 7.14. The process starts in step S0 with the need for setting up a

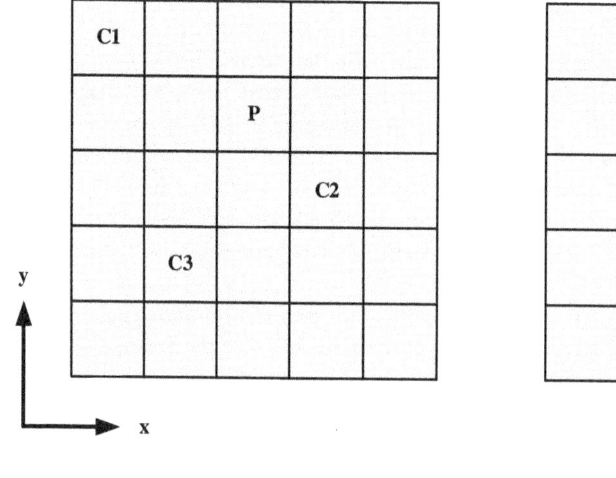

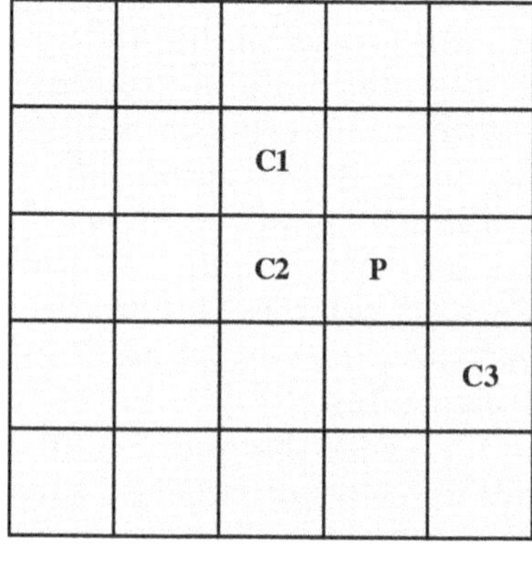

Figure 7.13. Candidate path tables with a reduced number of transmission paths

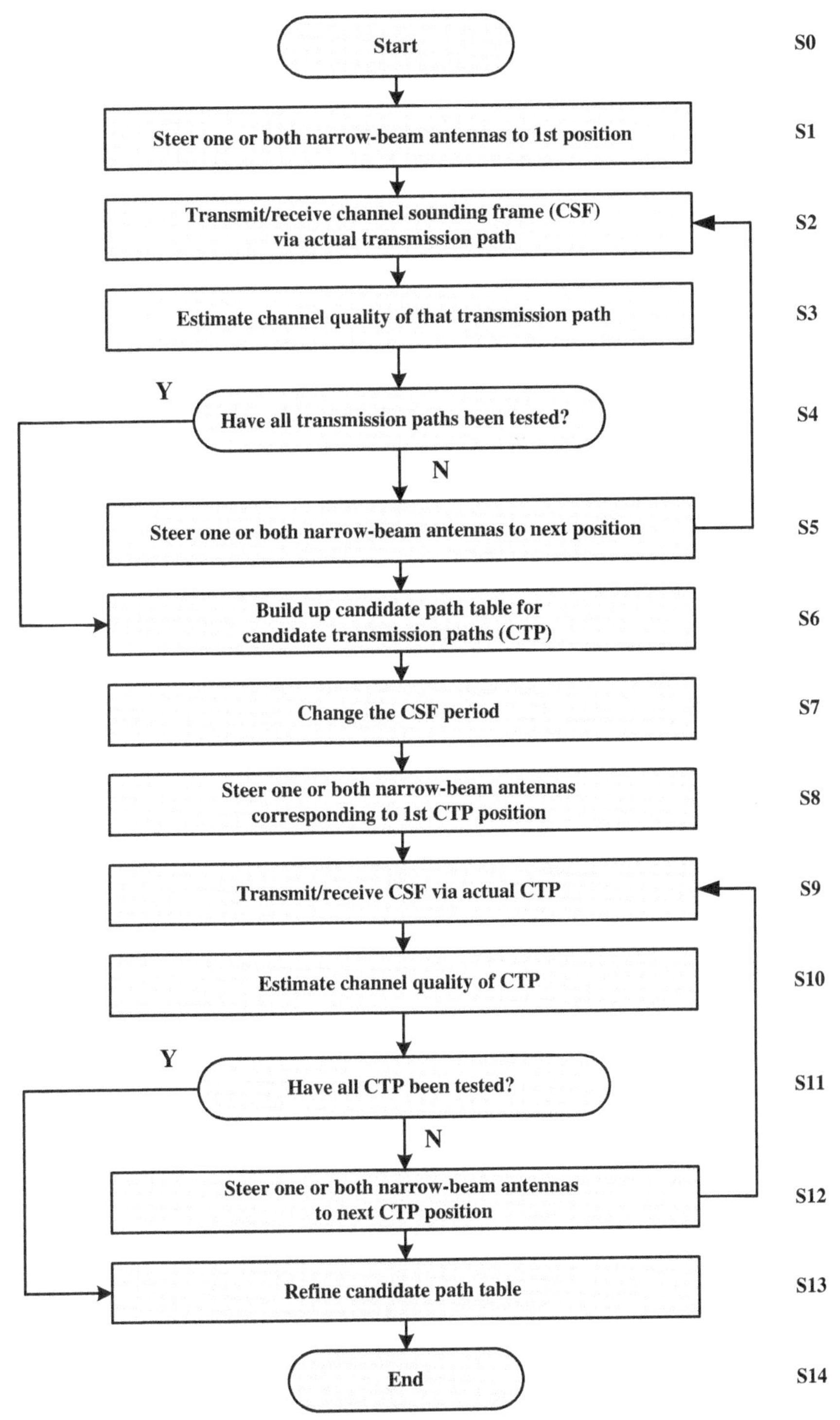

Figure 7.14. Flowchart of beam steering algorithm

communication. In step S1, one narrow-beam antenna from either Tx or Rx or both narrow-beam antennas of Tx and Rx are steered to the first position corresponding to the first transmission path. In step S2, a signal comprising the CSF frame is transmitted from the transmitter to the receiver, and the channel quality is estimated based on the received CSF in step S3. In step S4, a check is done to see whether all the possible transmission paths have been tested. If it is decided that not all or a insufficient number of transmission paths have been tested, then in the next step S5 one or both narrow-beam antennas are steered to the next position corresponding to the next transmission path. The process then goes back to step S2, where the CSF signal is transmitted via the new transmission path.

If it is determined in step S4 that all or a sufficient number of transmission paths have been tested, the process continues directly to step S6, where a candidate path table is built with several candidate transmission paths (CTP) identified from the tested transmission paths.

In step S7, the period of CSF is increased. In step S8, one or both narrow-beam antennas are steered to the first position corresponding to the first CTP selected in step S6. In step S9, the updated CSF is transmitted via the path, and the channel quality of the selected CTP is estimated based on the received new CSF in step S10.

In step S11, a check is done to see whether all the CTP paths have been tested. If not, then in step S12 one or both narrow-beam antennas are steered to the next position, corresponding to the next CTP path out of the selected CTP paths. The process goes back to step S9 with a transmission of the signal comprising the updated CSF. Otherwise, in step S13 the path candidate table is refined with a reduced number of CTP paths. The process ends at step S14 when the communication between the transmitter and the receiver is terminated.

7.4 ADVANCED ANTENNA ID TECHNOLOGY

The beam steering algorithm mentioned previously is used to realize spatial diversity, whereby several pairs of sharp-beam antennas from the transmitter and the receiver are used to build up several independent wireless paths, as shown in Figure 6.3. Each path can be assumed to be independent and can be treated as going through a frequency nonselective slow fading channel. The probability of all signals becoming weak is small and the diversity gain can be achieved. If one strong LOS or NLOS reflection path is intercepted, another unblocked link could still be alive to support the wireless link, while at the same time another alternative path can be searched for using the beam steering algorithm. How to exchange the control signaling between the transmitter and the receiver to enable the beam steering is addressed in this section.

During the beam steering period, when a narrow-beam antenna is used from the Tx side, it has to be steered to different positions using the corresponding control unit. At each of these positions, a *link request signal* is generated and transmitted by the antenna. The term "*link request signal*" is used to cover all kinds of possible control signalings (such as *beacon signals* and the like), which are initially transmitted from the transmitter to identify the presence of the transmitter, request the establishment of a communication, and so on. At the beginning of the link establishment procedure, the narrow-beam antenna is steered to a specific position and sends out a *link request signal*. Specifically, this *link request signal* comprises information related to the current position of the narrow-beam direction [9]. Furthermore, the narrow-beam antenna is steered to different positions, where in each position a corresponding *link request signal* is generated. The details of this process are discussed in Figure 7.15, where the *link request signal* is called a "*beacon signal*" and the information relating to the current position of the narrow-beam antenna is called Ant_ID. For simplicity, the number of available narrow-beam antenna positions is limited to 4.

As shown in the upper portion of Figure 7.15, the narrow-beam antenna transmits a *link request signal* (or *beacon signal*) from various antenna positions, which are identified by Ant_ID = 1, 2, 3, and 4. This sequence of various antenna positions can either be a random pattern or a predefined pattern. If the sequence is a predefined pattern,

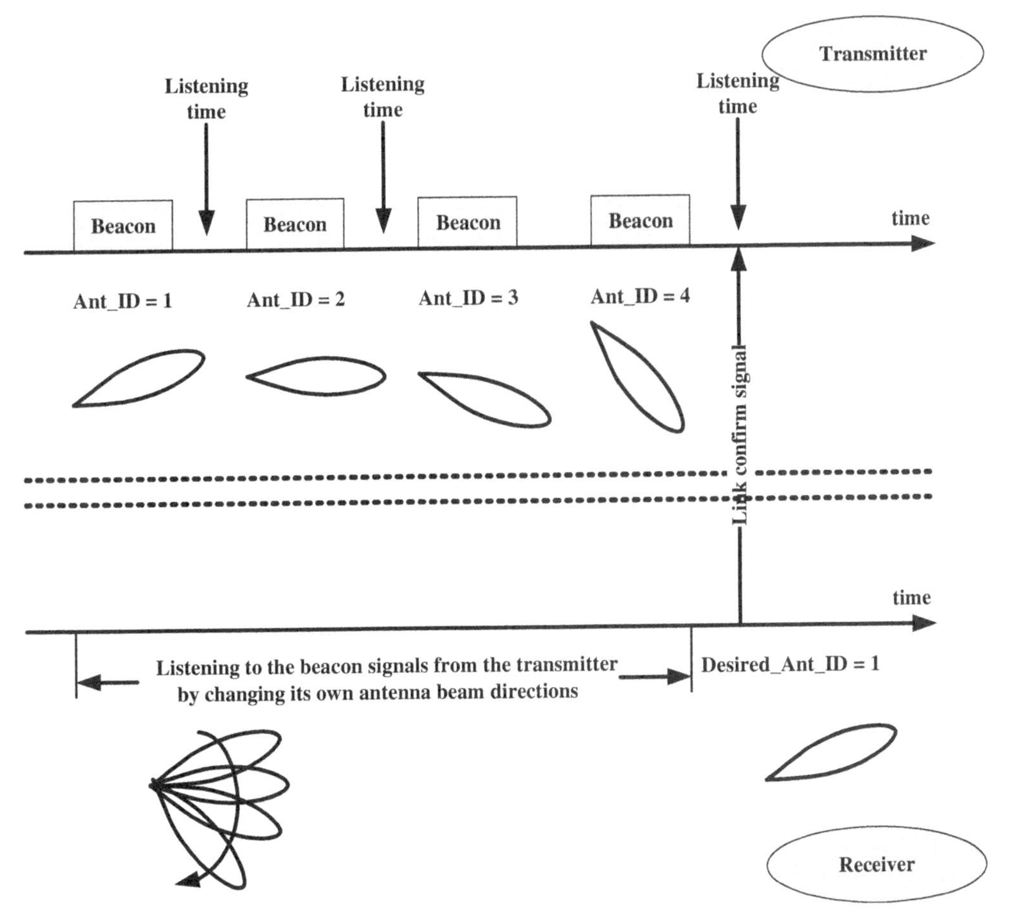

Figure 7.15. Antenna_ID exchange between the transmitter and the receiver

this information can be transmitted or made known to the receiver to accelerate the link establishment procedure.

After the transmission of each *beacon signal*, the narrow-beam antenna from the transmitter remains at the same position for a certain time (the listening time indicated in Figure 7.15), during which an *answer signal* (or *link confirm signal*) from the receiver can be acquired, listened to and checked. If one reliable wireless link is found based on channel quality information such as RSSI, SNR, BER, FER, and/or channel delay profile estimation, the receiver can process the received *link request signal* and send the corresponding *link confirm signal* back to the transmitter.

The *link confirm signal* comprises the information about the position of the narrow-beam antenna of the transmitter from which the corresponding *link request signal* was received. As can be seen in Figure 7.15, while the transmitter radiates *link request signals* from various positions of the narrow-beam antenna, the receiver changes the position of its own narrow-beam antenna to various positions in order to listen or receive *beacon signals* from the transmitter. When a *beacon signal* is acquired, the quality of the corresponding wireless channel is measured, and if the channel quality is considered to be sufficiently good, a *link confirm signal*, which also comprises the information about the position of the antenna radiating the *link confirm signal*, is sent back to the transmitter. This, as explained, takes place while the narrow-beam antenna of the transmitter is still at the same position.

In Figure 7.15, this position is named Desired_Ant_ID = 1, signifying that the antenna position 1 of the narrow-beam antenna has a sufficient quality in the corresponding transmission path. Alternatively, instead of answering immediately as soon as a *link request signal/beacon signal* over a transmission path with a sufficient channel quality is received by the receiver, it could adjust and change the position of its narrow-beam antenna a predefined number of times to try to receive several *link request signals* from the transmitter, and each *link request signal* is evaluated in terms of the channel quality. After this process, the positions of the narrow-beam antennas from both the transmitter and the receiver associated with a transmission path with the best transmission quality are sent back to the transmitter using *link confirm signal.*

Figure 7.16 visualizes a link establishment procedure between the transmitter and the receiver. In the first step, the transmitter sends *link request signals* (*beacon signals*) from different positions of its narrow-beam antenna according to a predefined or random pattern. At the same time, the receiver changes the position of its narrow-beam antenna also either on the basis of a random pattern or a predefined pattern attempting to receive a *link request signal.* In the second step, when the receiver has acquired one or more *link request signals* over a transmission path with a sufficient channel quality, a *link confirm signal* is sent back to the transmitter, which comprises information about the desired position of the narrow-beam antenna of the transmitter (Desired_Ant_ID). The transmitter then, in the third step, acknowledges the requested communication link, and a communication link is established. During the exchange of control signaling, video data, and others over the established communication link, the transmitter as well as the receiver may from time to time change the positions of their narrow-beam antennas to test alternative transmission paths in case the current transmission path breaks down or deteriorates.

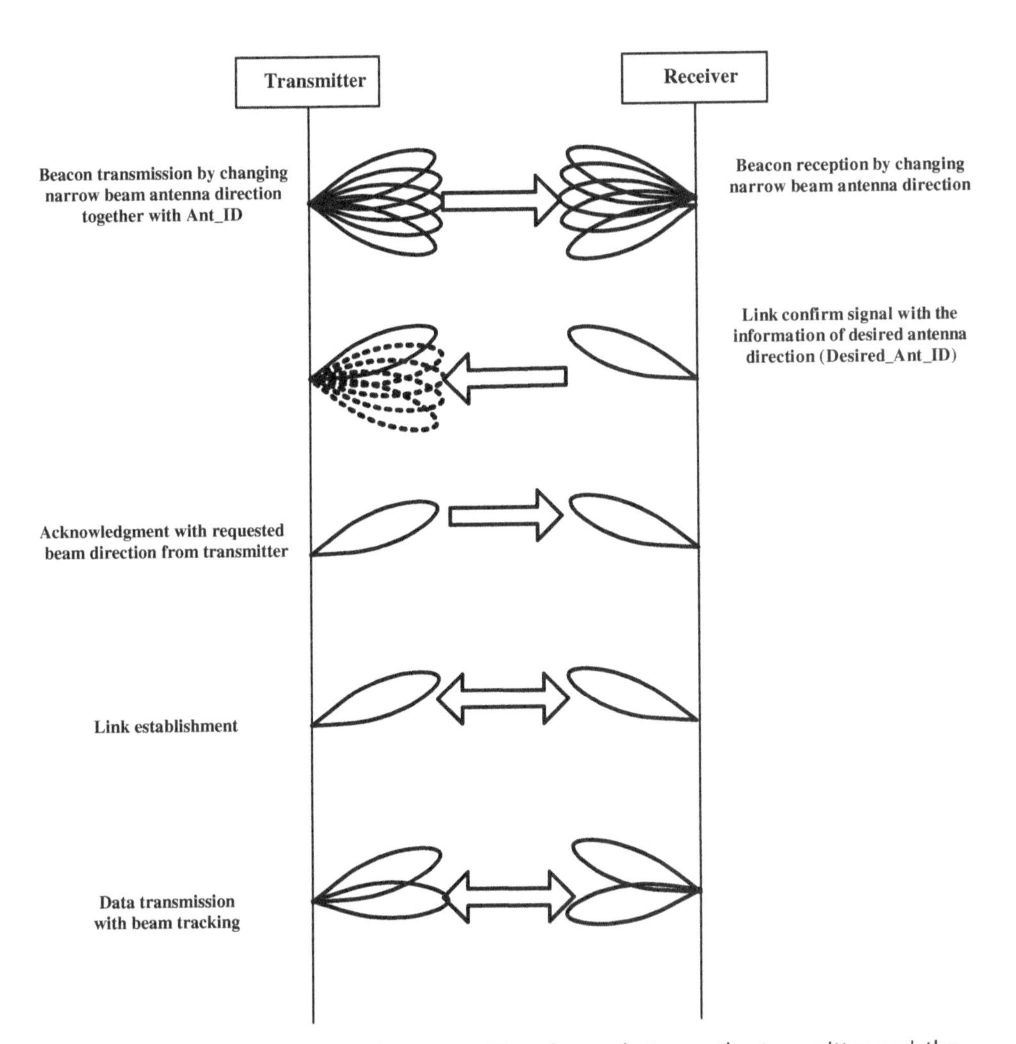

Figure 7.16. Sequence chart of Antenna_ID exchange between the transmitter and the receiver

7.5 ADVANCED BEAM FORMING TECHNOLOGY

The concept of beam forming is briefly introduced in Section 7.1. More detailed information about the conventional beam forming using a regular main antenna beam can be found in [10–15]. In the case of millimeter wave wireless systems, the function of beam forming does not only consider the antenna radiation pattern but focus on maximizing the received SNR under dynamic wireless conditions. The corresponding transmitter and receiver both might have non-conventional antenna radiation patterns.

One example is illustrated in Figure 7.4. In this section, the system model of millimeter wave based beam forming is assumed to be bi-directional. As shown in Figure 7.17, two devices are exchanging the information using millimeter wave based wireless links. Both the device 1 (DEV1) and the device 2 (DEV2) might use the same physical antennas for transmission and reception, which is referred to as an antenna symmetric system (ASS); otherwise it is referred to as an antenna asymmetric system (AAS). It is clear that the ASS is a special case of AAS. Consider a wireless link between DEV1 and DEV2, as shown in Figure 7.17, where DEV1 has M_T transmitter antennas and M_R receiver antennas; and DEV2 has N_T transmitter antennas and N_R receiver antennas. The developed model makes no assumptions regarding the antenna arrangements, whether these are 1-D or 2-D antenna arrays or any other arrangements. A cyclic prefixed OFDM system with a FFT length of N subcarriers and a cyclic prefixed single carrier (SC) system with a burst length of N can have the same system model. It is clear that the same beam forming concept can be applied to the both systems. We further assume that the cyclic prefix is longer than any multipath delay spread between any pair of antenna elements.

Considering an OFDM symbol (or a burst block from the single carrier system) radiated from DEV1 to DEV2, which can be expressed as

$$x(t) = \sum_{n=0}^{N-1} s_n \delta(t - nT_c) \tag{7.1}$$

where T_c is the sample (chip) duration, $\{s_n\}_{n=0}^{N-1}$ are the complex data, $s_n \in C$ depending on the modulation scheme (QPSK, 16QAM, 64QAM or others), and

$$\delta(t) = \begin{cases} 1 & 0 \le t \le T_c \\ 0 & \text{others} \end{cases} \tag{7.2}$$

At the transmitter of DEV1, each s_n is processed by the following beam forming vector

$$\mathbf{W}_1 = [\,w(1,1) \quad w(1,2) \quad \cdots \quad w(1,M_T)\,]^T \tag{7.3}$$

and then transmitted into a MIMO channel with a frequency domain channel state information (CSI) $\mathbf{H}^{1\to 2}(n) \in \mathbf{C}^{N_R \times M_T}$ at the nth subcarrier for $0 \le n < N$, which is expressed by

$$\mathbf{H}^{1\to 2}(n) = \begin{bmatrix} h_{1,1}^{1\to 2}(n) & h_{1,2}^{1\to 2}(n) & \cdots & h_{1,M_T}^{1\to 2}(n) \\ h_{2,1}^{1\to 2}(n) & h_{2,2}^{1\to 2}(n) & \cdots & h_{2,M_T}^{1\to 2}(n) \\ \vdots & \vdots & \ddots & \vdots \\ h_{N_R,1}^{1\to 2}(n) & h_{N_R,2}^{1\to 2}(n) & \cdots & h_{N_R,M_T}^{1\to 2}(n) \end{bmatrix} \tag{7.4}$$

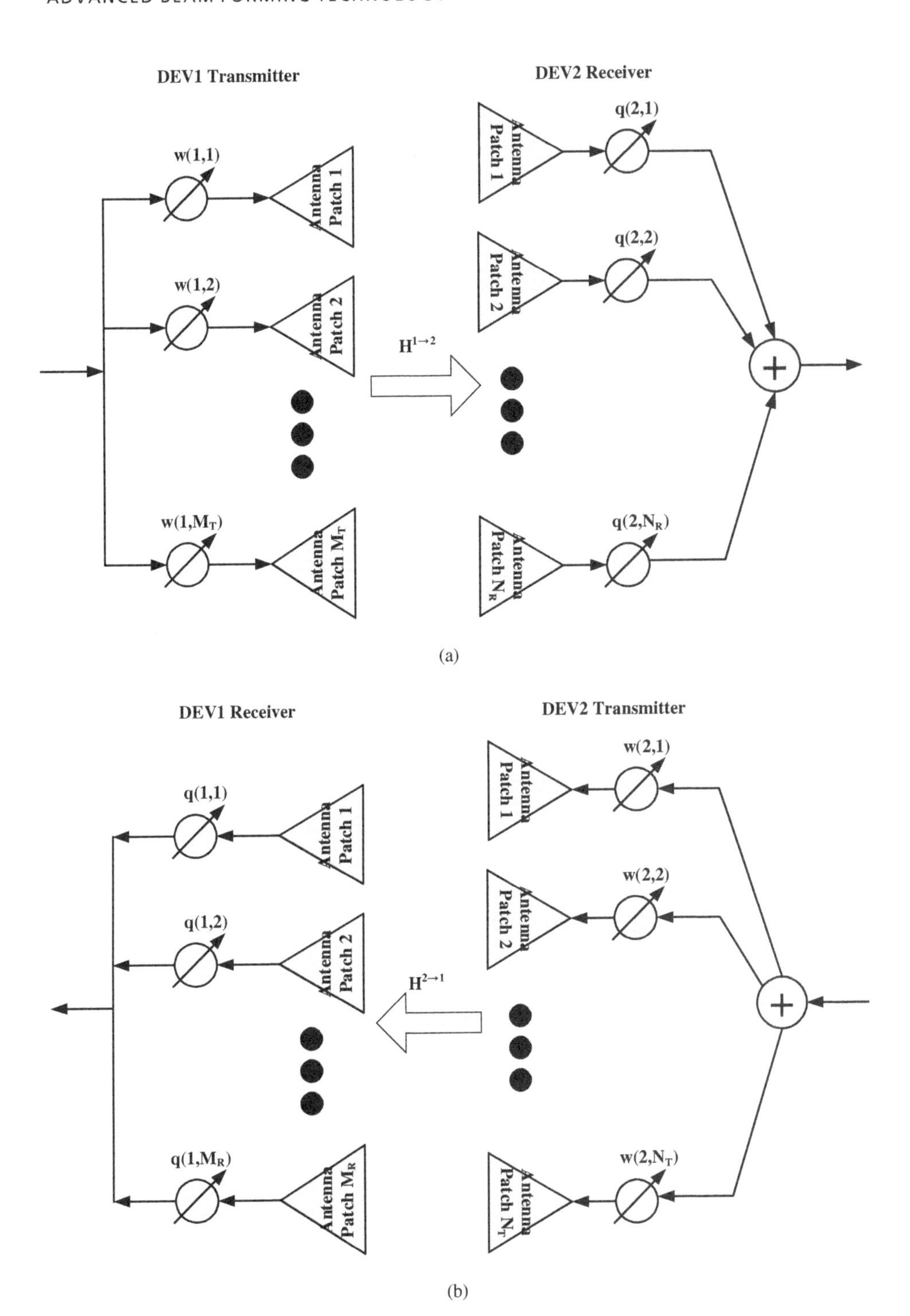

Figure 7.17. System model of beam forming: (a) transmitter, (b) receiver

where $h_{i,j}^{1\to 2}(n)$ denotes the channel transfer function from the *j*th transmitter antenna of DEV1 to the *i*th receiver antenna of DEV2.

At the receiver of DEV2, the received signal is processed through the combiner vector

$$\mathbf{Q}_2 = \begin{bmatrix} q(2,1) & q(2,2) & \cdots & q(2,N_R) \end{bmatrix}^T \tag{7.5}$$

The equivalent channel between the transmitter of DEV1 and the receiver of DEV2 is a single input single output (SISO) channel, with a frequency response at the the *n*th subcarrier ($0 \le n < N$) given by

$$p_n = \mathbf{Q}_2^H \mathbf{H}^{1\to 2}(n)\mathbf{W}_1 \quad \textit{for} \quad 0 \le n \le N-1 \tag{7.6}$$

The discrete received signal becomes

$$y_n = p_n s_n + b_n \quad \textit{for} \quad 0 \le n \le N-1 \tag{7.7}$$

whereby $\{s_n\}_{n=0}^{N-1}$ is the OFDM data symbol (or FFT of the single carrier data burst) and $\{b_n\}_{n=0}^{N-1}$ is the additive white Gaussian noise vector with variance of N_0.

Similarly, at the transmitter of DEV2, the data symbol stream is processed by the following beam forming vector

$$\mathbf{W}_2 = \begin{bmatrix} w(2,1) & w(2,2) & \cdots & w(2,N_T) \end{bmatrix}^T \tag{7.8}$$

and then transmitted into a MIMO channel with a frequency domain CSI of $\mathbf{H}^{2\to 1}(n) \in \mathbf{C}^{M_R \times N_T}$ at the *n*th subcarrier ($0 \le n < N$) given by

$$\mathbf{H}^{2\to 1}(n) = \begin{bmatrix} h_{1,1}^{2\to 1}(n) & h_{1,2}^{2\to 1}(n) & \cdots & h_{1,N_T}^{2\to 1}(n) \\ h_{2,1}^{2\to 1}(n) & h_{2,2}^{2\to 1}(n) & \cdots & h_{2,N_T}^{2\to 1}(n) \\ \vdots & \vdots & \ddots & \vdots \\ h_{M_R,1}^{2\to 1}(n) & h_{M_R,2}^{2\to 1}(n) & \cdots & h_{M_R,N_T}^{2\to 1}(n) \end{bmatrix} \tag{7.9}$$

where $h_{i,j}^{2\to 1}(n)$ denotes the channel transfer function from the *j*th transmitter antenna of DEV2 to the *i*th receiver antenna of DEV1.

At the receiver of DEV1, the receiver signal is processed through the combiner vector

$$\mathbf{Q}_1 = \begin{bmatrix} q(1,1) & q(1,2) & \cdots & q(1,M_R) \end{bmatrix}^T \tag{7.10}$$

The equivalent channel between the transmitter of DEV2 and the receiver of DEV1 is a SISO channel, with a frequency response at the *n*th subcarrier ($0 \le n < N$) given by

$$l_n = \mathbf{Q}_1^H \mathbf{H}^{21}(n)\mathbf{W}_2 \quad \text{for} \quad 0 \le n \le N-1 \tag{7.11}$$

For the both OFDM and SC wireless systems, when the wireless link from the transmitter of DEV1 to the receiver of DEV2 is considered, the SNR over the *n*th subcarrier is expressed as follows:

$$\mathrm{SNR}_n^{1\rightarrow 2} = \frac{E_s|p_n|^2}{N_0} = \frac{E_s\left|\mathbf{Q}_2^H\mathbf{H}^{1\rightarrow 2}(n)\mathbf{W}_1\right|^2}{N_0} \quad for \quad 0 \le n \le N-1 \tag{7.12}$$

where E_s is the energy per symbol, while when considering the wireless link from the transmitter of DEV2 to the receiver of DEV1, the SNR is given by

$$\mathrm{SNR}_n^{2\rightarrow 1} = \frac{E_s|l_n|^2}{N_0} = \frac{E_s\left|\mathbf{Q}_1^H\mathbf{H}^{2\rightarrow 1}(n)\mathbf{W}_2\right|^2}{N_0} \quad for \quad 0 \le n \le N-1 \tag{7.13}$$

We now define an effective SNR (ESNR) as a mapping from the instantaneous subcarriers to an equivalent SNR that takes into account the forward error correction. There are many ways to compute the ESNR, such as the mean of the SNR_ns over all the subcarriers, the quasi-static method, the capacity effective mapping, the convex metric method and so on.

The ultimate objective of a beam forming algorithm is to select the optimal beam former vectors $\mathbf{W}_1$ at DEV1 and $\mathbf{W}_2$ at DEV2, as well as the optimal combiner vectors $\mathbf{Q}_1$ at DEV1 and $\mathbf{Q}_2$ at DEV2 that maximize ESNR or other optimization criteria. Different ESNR methods can be adopted for SC and OFDM wireless systems. A minimum mean squared error (MMSE) SC equalizer tends to have an ESNR that can be estimated by the average of the SNR_ns over all the subcarriers, while an OFDM equalizer tends to have an ESNR that can be approximated by the geometric mean of the SNR_ns over the different subcarriers.

For an ASS-based millimeter wave wireless system, we have the following relationships between the beam former coefficients and the combiner coefficients from DEV1 and DEV2, respectively:

$$\mathbf{W}_1 = \mathbf{Q}_1 \tag{7.14}$$

and

$$\mathbf{W}_2 = \mathbf{Q}_2 \tag{7.15}$$

As a result, it is sufficient to consider only one direction of the wireless link, either from the transmitter of DEV1 to the receiver of DEV2 or from the transmitter of DEV2 to the receiver of DEV1. For an AAS-based millimeter wave wireless system, both links have to be considered.

For millimeter wave systems, in order to reduce the complexity and cost, quantized phase shifters are utilized to construct 2-D codebooks with limited sizes for DEV1 and DEV2. In the following, we consider separable 2-D antenna arrays. To form the 2-D codebook, we need only to specify codebooks for 1-D antenna arrays along the x-axis and y-axis. The simplest phased antenna array is an antenna array implementing the following phases 0° or 180° for each antenna element. The coefficients of either the

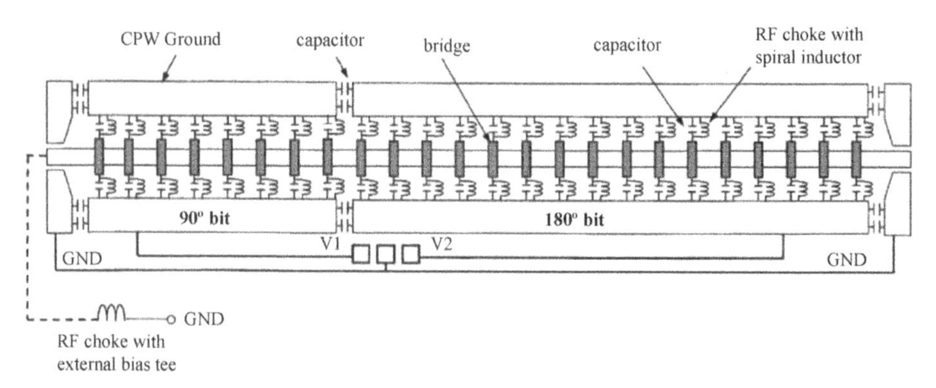

Figure 7.18. Diagram of the fabricated 60 GHz 2-bit (0°, 90°, 180°, 270°) phase shifter [16], (©2002 IEEE)

beam former or the combiner are selected from $\{+1, -1\}$. Using a quadrature transmitter with I & Q (in-phase and quadrature), this means that each antenna element will transmit or receive either I + Q (phase 0°), or −(I + Q) (phase 180°). A more flexible phased antenna array is an antenna array implementing the following phases, 0°, 90°, 180° or 270°, for each antenna element. An example of a low-loss 2-bit, (0°, 90°, 180°, or 270°), distributed phase shifter is shown in Figure 7.18. Using a quadrature transmitter means that each antenna element will transmit either I (phase 0°), or −I (phase 180°), or Q (phase 270°), or −Q (phase 90°). An equivalent set of signals would be: I + Q, I − Q, −I + Q, −I − Q. Such a system has the weights selected from the set $\{+1, -1, +j, -j\}$. The codebook of the beam former stores all the possible choices for the coefficients of each antenna element from the transmitter, and the codebook of the combiner stores all the possible choices for the coefficients of each antenna element from the receiver. The size of the codebooks depends on the quantized bit length of each antenna element and the total number of antenna elements in a phased array antenna.

The basic protocol for beam forming can be summarized as follows:

Step 1, Acquisition: DEV1 will train DEV2 by repeating transmission of beacon signals using $\mathbf{W}_1$ selected from a sufficient subset of the beam former codebook. After that DEV2 will acquire the CSI by using a sufficient subset of its combiner codebook.

Step 2, Estimation: DEV2 will estimate the optimal beam former vector $\mathbf{W}_1$ and optimal combiner vector $\mathbf{Q}_2$, which is related to the coefficients of quantized phase shifters.

Step 3, Feedback: DEV2 should feedback the optimal beam former $\mathbf{W}_1$ and possibly optimal combiner vector $\mathbf{Q}_2$ to DEV1.

Step 4, AAS Feedback: For an AAS-based system, repeat steps 1 to 3 with DEV1 replaced by DEV2 and DEV2 replaced by

DEV1 in order to acquire the optimal beam former vector $\mathbf{W}_2$ and optimal combiner vector $\mathbf{Q}_1$.

Step 5, Tracking: DEV1 tracks the beam former and the combiner vectors continuously by sending beacon signals using $\mathbf{W}_1$ selected from a sufficient subset of the beam former codebook at a lower rate compared with the acquisition phase.

Step 6, Update: DEV2 should feedback the optimal beam former $\mathbf{W}_1$ and possibly optimal combiner vector $\mathbf{Q}_2$ to DEV1 at a lower rate compared with the acquisition phase.

Step 7, AAS Update: For an AAS-based system, repeat steps 5 to 6 with DEV1 tracking the beam former vector $\mathbf{W}_2$ and the combiner vector $\mathbf{Q}_1$ and feeding them back to DEV2.

Assume that an ASS-based millimeter wave wireless system is adopted with the following number of quantized phase shifters:

$$M = M_T = M_R \tag{7.16}$$

and

$$N = N_T = N_R \tag{7.17}$$

In order to reduce the complexity of performing the necessary singular value decomposition (SVD), an adaptive method is proposed. In the remaining part of this section, instead of searching the optimum eigenvector, a suboptimum solution is chosen with reduced complexity by using a Hadamard matrix. An example of an 8×8 Hadamard matrix is given in Figure 7.19.

The flowchart of beam forming is summarized in Figure 7.20. It includes three elements a beam search process, a beam tracking process, and a beam steering state machine. Beam search and beam tracking are used to compensate for the time variations of the wireless channel and the possible obstruction to narrow-beams. When called, the beam search process finds the beam direction that maximizes the link budget between DEV1 and DEV2. The obtained parameters of $\mathbf{W}_1$, $\mathbf{Q}_1$, $\mathbf{W}_2$, and $\mathbf{Q}_2$ are then used for

$$\mathbf{H}_{8\times 8} = \begin{bmatrix} 1 & 1 & 1 & 1 & 1 & 1 & 1 & 1 \\ 1 & -1 & 1 & -1 & 1 & -1 & 1 & -1 \\ 1 & 1 & -1 & -1 & 1 & 1 & -1 & -1 \\ 1 & -1 & -1 & 1 & 1 & -1 & -1 & 1 \\ 1 & 1 & 1 & 1 & -1 & -1 & -1 & -1 \\ 1 & -1 & 1 & -1 & -1 & 1 & -1 & 1 \\ 1 & 1 & -1 & -1 & -1 & -1 & 1 & 1 \\ 1 & -1 & -1 & 1 & -1 & 1 & 1 & -1 \end{bmatrix}$$

Figure 7.19. 8×8 Hadamard matrix

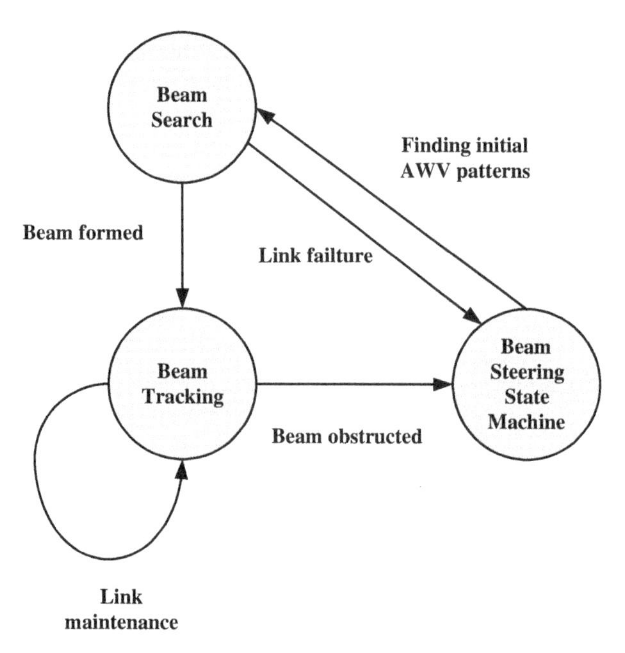

Figure 7.20. Beam forming flow chart

beam tracking. The beam tracking process tracks the beam versus small time variations of the channel transfer function for millimeter wave applications. The beam steering state machine uses an bad link (a bad link means that current beam direction is obstructed) detection mechanism, which can be based on payload or beam tracking results to detect whether the SNR of the current wireless link is below a desired threshold, and subsequently a new beam search is scheduled to find the next best beam direction.

Without the loss of generality, DEV1 is treated as the transmitter while DEV2 is treated as the receiver, and $\mathbf{W}_1$, $\mathbf{Q}_1$, $\mathbf{W}_2$, and $\mathbf{Q}_2$ are called antenna-array weight vectors (AWVs), which are complex weight vectors that can have phase and/or magnitude information. The beam steering state machine is used to find the initial AWV patterns for the transmitter and the receiver, respectively. Its principles are similar to those discussed in the Section 7.3. After acquiring the initial AWV patterns, the algorithm goes to the beam search process.

The beam search process includes several iterations, and each iteration has two steps. Step 1 is to find the AWV pattern for the transmitter, and Step 2 is to find the AWV pattern for the receiver. Several iterations are run to let AWV patterns for both the transmitter and the receiver converge to the optimum values under dynamic wireless conditions. During Step 1 of the first iteration, the transmitter radiates known symbol sequences over the air, which are used to estimate the resulting SISO channel transfer functions from the RF-modulated signal just before the coefficients of M antenna elements (see Figure 7.17) to the combined signal just after the coefficients of N antenna elements (see Figure 7.17). During this stage, the transmitter phased array antenna switches between AWV phase vectors derived from the M columns of

Hadamard matrix $\mathbf{H}_{M\times M}$ one at a time, spanning the entire space. The transmitter AWV vector includes M weight vectors. For each chosen transmitter AWV pattern, the received signal is correlated with the transmitted sequence. The vector of M complex-valued channel estimates is complex conjugated and multiplied by matrix $\mathbf{H}_{M\times M}$.

Angles of the complex-valued elements of this vector are then quantized into fixed point values, for example, typically about 2 to 4 bits for 60 GHz millimeter wave circuits, forming a vector of quantized phases. This vector is referred to as a maximum ratio combining (MRC)- based transmitter quantized phase shift (QPS) vector. Amplitudes of the complex-valued elements of this vector are also quantized into fixed point values, normally used to adjust the gain of power amplifier of the transmitter, forming a vector of quantized amplitudes. This vector is referred to as a MRC-based transmitter quantized amplitude (QA) vector. The quantized transmitter AWV vector is sent back to the transmitter via a reverse wireless channel, where it is used as the fixed transmitter AWV pattern for Step 2 of the first iteration.

During Step 2 of the first iteration, the transmitter AWV pattern is set equal to the coefficients calculated at Step 1. The receiver AWV pattern, on the other hand, is set equal to the N columns of the Hadamard matrix $\mathbf{H}_{N\times N}$ one at a time. Transmitting the same sequence and using the same correlation procedure, SISO channel transfer functions are estimated for each receiver AWV pattern.

Following the same procedure, one can find that the vector of N complex-valued channel estimates is complex-conjugated and multiplied by $\mathbf{H}_{N\times N}$. The generated vector is used as the fixed receiver AWV pattern for the next iteration in Step 1. Angles of the complex-valued elements of this vector are also quantized into fixed point values, for example, typically about 2 to 4 bits for 60 GHz millimeter wave circuits, forming a vector of QPS. Amplitudes of the complex-valued elements of this vector are then quantized into fixed point values, normally used to adjust the gain of low noise amplifier of the receiver, forming a vector of QA. To reduce the power consumption and complexity of millimeter wave circuits, the amplitude of the receiver AWV pattern could also be set to be a fixed value.

The same iteration procedure is repeated several times, where alternatively transmitter AWV patterns or receiver AWV patterns are set equal to the calculated quantized phase vectors from the previous iteration, while the AWV patterns for the opposite operation, that is, receiver or transmitter AWV patterns, are set equal to either the N columns of the $N \times N$ Hadamard matrix $\mathbf{H}_{N\times N}$, or the M columns of the $M \times M$ Hadamard matrix $\mathbf{H}_{M\times M}$, one at a time.

At the end of the iterations, the updated transmitter AWV and receiver AWV patterns are chosen for DEV1 and DEV2 to transfer high rate data using an irregular antenna beam pattern with the best link quality as shown in Figure 7.4.

When the beam search process fails, the initial AWV transmitter and receiver vectors might be wrong and the process goes back to the beam steering state machine. When the beam search process works, it goes to the beam tracking process. Firstly, the coefficients of the transmitter are set equal to the transmitter AWV patterns corresponding to the current beam direction (i.e., the transmitter AWV patterns are set to the current estimates), while the receiver AWV pattern is swept through the N columns of the $N \times N$ Hadamard matrix $\mathbf{H}_{N\times N}$ one at a time. From this operation, the MRC-based receiver QPS vectors and/or the QA vectors are calculated. The calculated parameters are then used as the fixed receiver AWV pattern for the next step, while the transmit AWV pattern is swept through the M columns of the M × M Hadamard matrix $\mathbf{H}_{M \times M}$ one at a time and the MRC-based transmit QPS vectors and/or the QA vectors are calculated. To improve the reliability, the transmitter AWV pattern that produces the strongest signal strength at the receiver is repeated more than once to allow the receiver to compensate for various phase inaccuracies inherent to the millimeter wave analog circuits, which are very common for the manufacturer of 60 GHz millimeter wave components.

When a wireless link is obstructed, the beam forming goes from the beam tracking process to beam steering state machine, where initial transmitter and receiver AWV patterns for both the transmitter and the receiver are estimated. Otherwise, the beam forming will stay at the stage of beam tracking, while high rate data exchange is realized using millimeter wave wireless systems.

REFERENCES

[1] K. Rambabu, M. Alam, J. Bornemann, and M. A. Stuchly, "Compact wideband dual-polarized microstrip patch antenna." *Proc. IEEE Antennas Propag. Soc. Symp*, Vol. 2, pp. 1955–1958, Jun. 2004.

[2] R. B. Waterhouse, D. Novak, A. Nirmalathas, and C. Lim, "Broadband printed millimeter-wave antennas." *IEEE Trans. Antennas Propag*, Vol. 51, No. 9, pp. 2492–2495, 2003.

[3] K. Huang and Z. Wang "V-band patch-fed rod antennas for high data-rate wireless communications." *IEEE Trans. Antennas Propag*, Vol. 54, No. 1, pp. 297–300, 2006.

[4] T. Seki, N. Honma, K. Nishikawa, and K. Tsunekawa, "Millimeterwave high-efficiency multilayer parasitic microstrip antenna array on teflon substrate." *IEEE Trans. Microw. Theory Tech.*, Vol. 53, No. 6, pp. 2101–2106, 2005.

[5] M. Al-Tikriti, S. Koch, and M. Uno, "A compact broadband stacked microstrip array antenna using eggcup-type of lens." *IEEE Micro. Wireless Compon Lett.*, Vol. 16, No. 4, pp. 230–232, 2006.

[6] CoMPA (Consortium of millimeter wave practical applications) proposal, IEEE 802.15-07-0693-01-003c, May 2007.

[7] K. Huang and D. Edwards, "60GHz multi-beam antenna array for Gigabit wireless communication networks." *IEEE Trans. Antennas Propag.*, Vol. 54, No. 12, pp. 3912–3914, 2006.

[8] M. Uno, Z. Wang, K. Huang, and V. Wullich,"Communication system and method." European patent EP1659813.

[9] M. Uno and Z. Wang,"Transmitting device, receiving device and method for establishing a wireless communication link." US patent US7551135.

[10] N. Herscovici and C. Christodoulou, "Analysis of beam-steering and directive characteristics of adaptive antenna arrays for mobile communications." *IEEE Antennas Prop. Mag.*, Vol. 43, No. 3, pp. 145–152, 2001.

[11] J. Bach Andersen, "Antenna arrays in mobile communications: gain, diversity, and channel capacity." *IEEE Antennas Propag. Mag.*, Vol. 42, No. 2, pp. 12–16, 2000.

[12] J. Kennedy and M. C. Sullivan, "Direction finding and smart antennas using software radio architectures." *IEEE Commun. Mag.* pp. 62–68, 1995.

[13] Alfred R. Lopez, "Performance predictions for cellular switched-beam intelligent antenna systems." *IEEE Commun. Mag.* pp. 152–154, 1996.

[14] D. Chizhik, "Slowing the time-fluctuating MIMO channel by beam forming." *IEEE Trans. Wireless Commun.*, Vol. 3, No. 5, pp. 1554–1565, 2004.

[15] W. Aerts, P. Delmotte, and G. A. E. Vandenbosch, "Conceptual study of analog baseband beam forming: design and measurement of an eight-by-eight phased array." *IEEE Trans. Antennas Propag.* Vol. 57, No. 6, pp. 1667–1672, 2009.

[16] H. T. Kim, J. H. Park, S. Lee, S. Kim, J. M. Kim, Y. K. Kim, and Y. Kwon, "V-band 2-b and 4-b low-loss and low-voltage distributed MEMS digital phase shifter using metal–air–metal capacitors." *IEEE Trans. Microw. Theory Tech*, Vol. 50, No. 12, pp. 2918–2923, 2002.

8

SINGLE-CARRIER FREQUENCY DOMAIN EQUALIZATION

In this chapter, we present the details of a single-carrier frequency domain equalizer (SC-FDE) with and without decision feedback equalization (DFE) and emphasize its strong points, compared to the orthogonal frequency division multiplexing (OFDM, see Section 2.5) technology. SC-FDE is a promising technique and it solves the power consumption issue of power amplifier (PA) in high rate (beyond gigabit/s) millimeter wave wireless systems, and lets the millimeter wave related products and devices become commercially viable.

This chapter is organized as follows. Section 8.1 compares the similarities and differences between SC-FDE and OFDM wireless systems. The issue of power consumption of PA in millimeter wave wireless systems is identified and the potential benefits in using the SC-FDE technique due to its inherently low peak-to-average power ratio (PAPR) to replace the OFDM technique are emphasized. Section 8.2 describes various frame structures to enable the reliable synchronization for burst transmission. Section 8.3 discusses channel estimation for SC-FDE. The new concept to adaptively track the dynamic wireless channels is presented. Section 8.4 analyzes the performance of SC-FDE based on the minimum mean squared error (MMSE) criterion. The low complexity suboptimal MMSE-based equalizer is illustrated. Section 8.5 analyzes the benefits of SC-FDE with DFE. The idea of eliminating the requirement of complex matrix inversion and supporting a multi-user scenario are also discussed.

Millimeter Wave Communication Systems, by Kao-Cheng Huang and Zhaocheng Wang

8.1 ADVANTAGES OF SC-FDE OVER OFDM FOR MILLIMETER WAVE SYSTEMS

SC-FDE gives the millimeter wave communication systems distinct advantages over its competitor, OFDM. It is known that OFDM suffers from a number of drawbacks, including a large PAPR, intolerance to amplifier nonlinearities, and high sensitivity to carrier frequency offsets [1]. These problems become extremely severe when radio frequency goes to the millimeter wave range. The OFDM technology does have a significant advantage of offering the robustness against frequency selective fading.

An alternative solution to ISI mitigation is the use of single-carrier (SC) modulation combined with frequency domain equalization (FDE). The complexity and performance of SC-FDE systems are comparable to that of OFDM while avoiding the above drawbacks associated with multi-carrier (MC) implementation. On the other hand, FDE does not represent an optimal solution to signal detection over ISI channels and SC systems cannot certainly offer the same flexibility as OFDM systems in the management of bandwidth and energy resources, both in single and multi-user communications. All these considerations have made a detailed analysis of SC-FDE techniques as well as a careful comparison between SC-FDE and OFDM much more critical.

The first MC scheme was proposed in 1966 [2], whereas the first SC-FDE approach in digital communication systems dates back to 1973 [3]. Despite the small time gap between their introductions, many efforts have been made by the scientific community in the study of MC solutions, but little attention has been paid to SC-FDE for many years. In the last decade, there has been a renewed interest in this area. The theoretical and practical gaps between these two solutions are becoming small, but the technical literature on MC communications is still by far much richer and broader than that on SC-FDE. In this chapter, we provide an overview of the principles as well as an up-to-date review of the latest research results for SC-FDE, with a particular focus on millimeter wave high rate applications.

The transmitter and receiver block diagrams of OFDM wireless systems are shown in Figure 8.1 (a) and (b), respectively.

The transmitter comprises a quadrature amplitude modulation (QAM) modulator, an inverse fast Fourier transformation (IFFT), a cyclic prefix insertion module, a radio frequency (RF) modulator, and an antenna. Firstly, an input signal to be modulated and transmitted is sent to the QAM module. The IFFT module applies an inverse fast Fourier transformation (FFT) transformation on the signal received from the QAM module. The cyclic prefix insertion module inserts a cyclic prefix into the signal received from the IFFT module. The RF module converts the signal received from the cyclic prefix insertion module into a modulated RF signal, which is then radiated by the antenna.

The receiver comprises an antenna, a RF receiver, a cyclic prefix removal module, a FFT module, a channel equalizer, a channel estimation module, and a QAM demodulator. The antenna converts the received electromagnetic signal into a RF electrical signal, which is then converted into a baseband signal by the RF receiver. After removing the inserted cyclic prefix by the cyclic prefix removal module, the received signal is then passed through the FFT module. The channel estimation module estimates the channel quality and other characteristics of the dynamic wireless channel between the transmitter

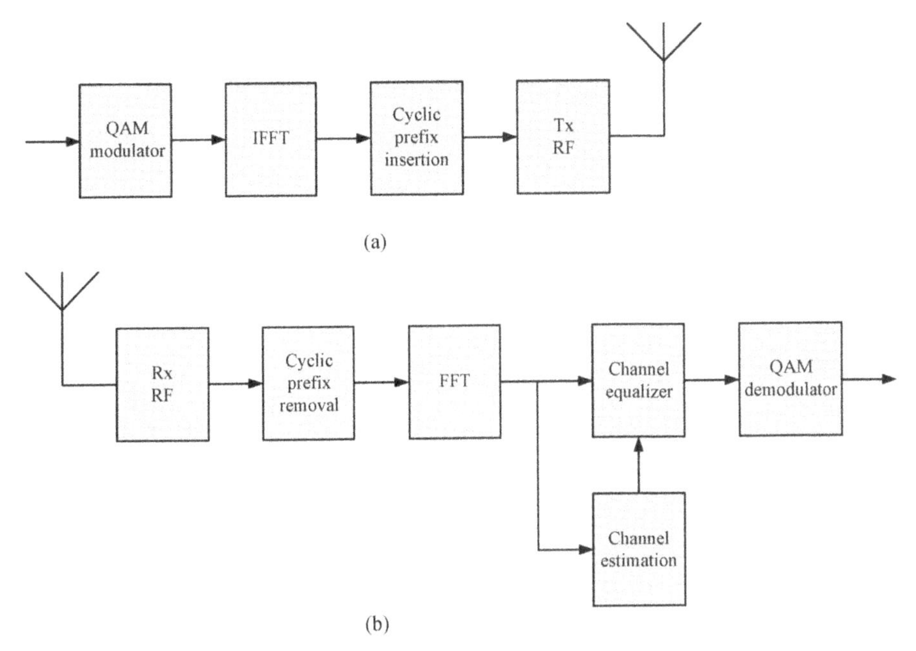

Figure 8.1. Block diagram of OFDM wireless systems: (a) transmitter, (b) receiver

and receiver, which are required by the channel equalizer to compensate for the frequency response of the wireless channel. The output of the channel equalizer is then passed to the QAM demodulator for recovering the transmitted data.

In contrast to the MC nature of OFDM wireless systems shown in Figure 8.1, the transmitter and receiver block diagrams of SC-FDE wireless systems are depicted in Figure 8.2 (a) and (b), respectively. The transmitter and receiver of Figure 8.2 operate based on single carrier.

The transmitter of SC-FDE wireless systems shown in Figure 8.2(a) appears to be similar to that of OFDM systems illustrated in Figure 8.1(a), except that it does not contain an IFFT module. Likewise, the receiver of SC-FDE systems depicted in Figure 8.2(b) looks very similar to that of Figure 8.1(b), except that it has an extra IFFT module between the channel equalizer and the QAM demodulator.

It is clear that the complexity of SC-FDE wireless systems depicted in Figure 8.2 is almost the same as that of OFDM wireless systems given in Figure 8.1. For a SC-FDE system, the transmitter and receiver can be merged in a single mobile device. Moreover, the whole transceiver can be integrated in a semiconductor chip, which also comprises additional modules to extend the functions for the customer.

The main advantages of SC-FDE systems are listed as follows:

- The energy of individual symbols is transmitted over the whole available frequency spectrum. Therefore, narrow band notches within the channel transfer function have only a small impact on the performance. For OFDM systems, however, the narrow band notches would degrade the performance of transmitted

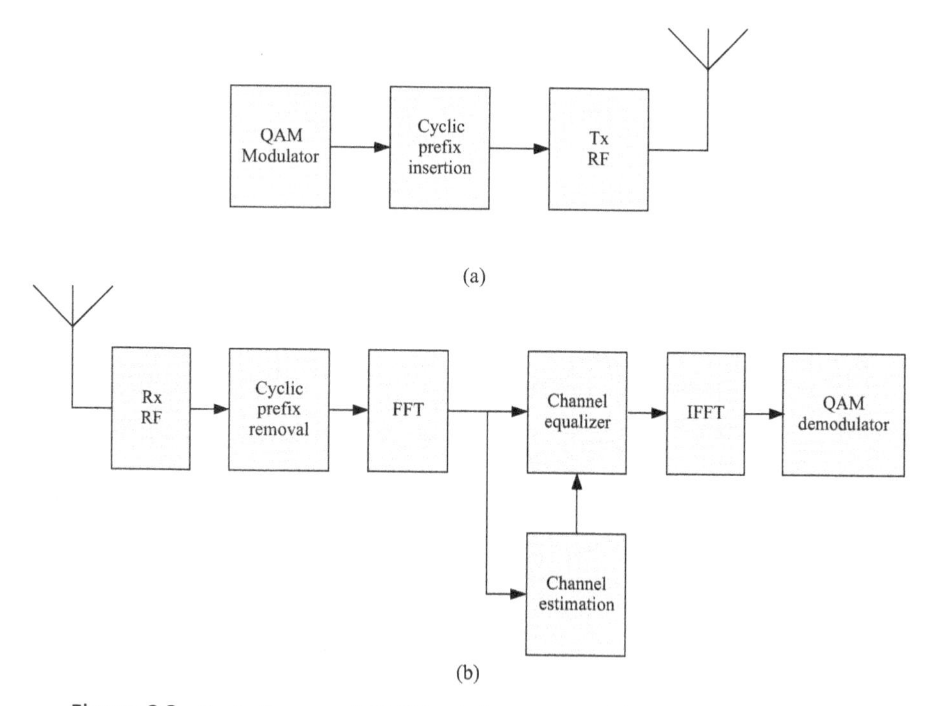

Figure 8.2. Block diagram of SC-FDE wireless systems: (a) transmitter, (b) receiver

symbols assigned over the relevant subcarriers. Of course, the diversity can be regained partly by utilizing a soft decision error control decoder with some performance penalty.

- Lower PAPR ratio for the radiated signal makes the PA from the transmitter (Tx) side more energy efficient and cheaper, especially for the millimeter wave wireless systems.
- SC-FDE is robust to the effect of phase noise, which makes the local oscillator (LO) simpler, especially for the millimeter wave wireless systems.
- A smaller number of analogue-to-digital converter (ADC) bits, significantly reduces the design and system cost for high rate applications.
- The carrier frequency error between the Tx side and the receiver (Rx) side can destroy the orthogonality between subcarriers and introduce the inter-subcarrier interference for OFDM systems. However, it has less effect on single-carrier systems with frequency domain equalization.

Therefore, SC-FDE is more suitable to the user scenarios where a simple transmitter with less power consumption is desired, while at the same time the receiver can be complex and have relatively high power consumption, such as high definition television scenarios.

In terms of power consumption and environment impact, SC-FDE with its low PAPR ratio is definitely an attractive technology for millimeter wave wireless systems. In the past, the design of PA is dominated by III-V semiconductors, which have high electron mobility, high breakdown voltage, and the availability of high quality-factor (Q) passives. Recently, cheap CMOS technology with downscaling of feature dimensions and feasibility of back-end integration attracts more and more attention. The details about the performance of CMOS PA can be found in [4, 5].

For a typical 60 GHz silicon radio, the 1-dB compression point $P_{1dB}(out)$ is 10–12 dBm approximately, while the saturation point of *Psat* is 16–17 dBm. Thus, the dynamic range of PA is limited. This is an important issue for a 60 GHz system.

An OFDM system has to back off considerably more than SC-FDE. If multiple power amplifiers are used, the power back-off is not necessary, but the power consumption increases with higher PAPR. For Class A amplifiers, the power consumption is proportional to the peak power. The higher the PAPR is, the higher the power consumption will be. Also, the cost increases almost proportionally with the power consumption. An example below demonstrates that a SC-FDE has lower PAPR than an OFDM system.

When PAPR is considered at $Pr\ (\text{PAPR} > \gamma) = 0.01$ and *Psat* is assumed to be 16 dBm, a PAPR comparison between SC-FDE and OFDM is shown in Figure 8.3 [6]. It is clear that PAPR gain (between SC-FDE and OFDM) for quadrature phase shift keying (QPSK), non-square 8-ary QAM and 16-ary QAM is about 4.7 dB, 3.6 dB, and 3.0 dB, respectively.

Performance comparisons between OFDM and SC-FDE wireless systems are summarized in [7–9]. When a nonlinear PA model is used and about 5.5 dB output

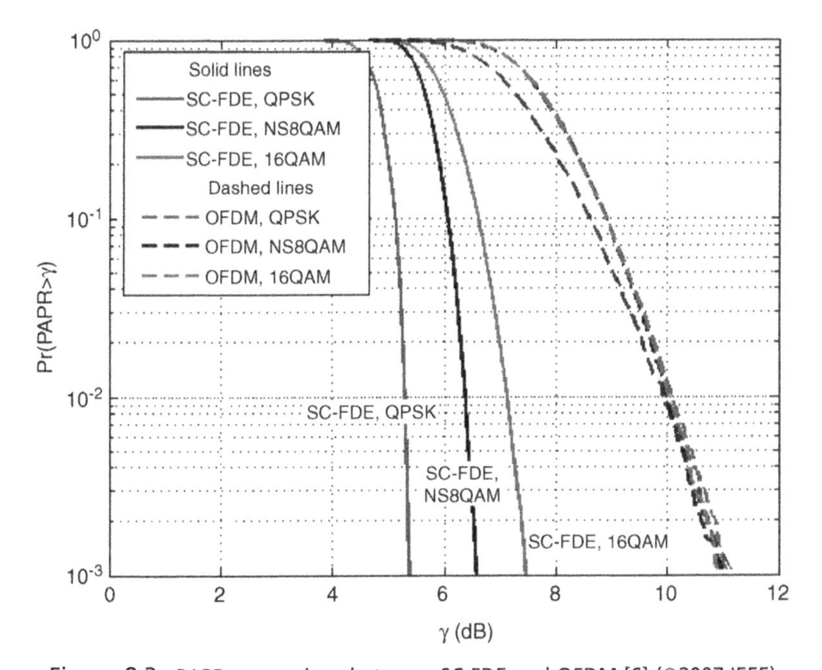

Figure 8.3. PAPR comparison between SC-FDE and OFDM [6] (©2007 IEEE)

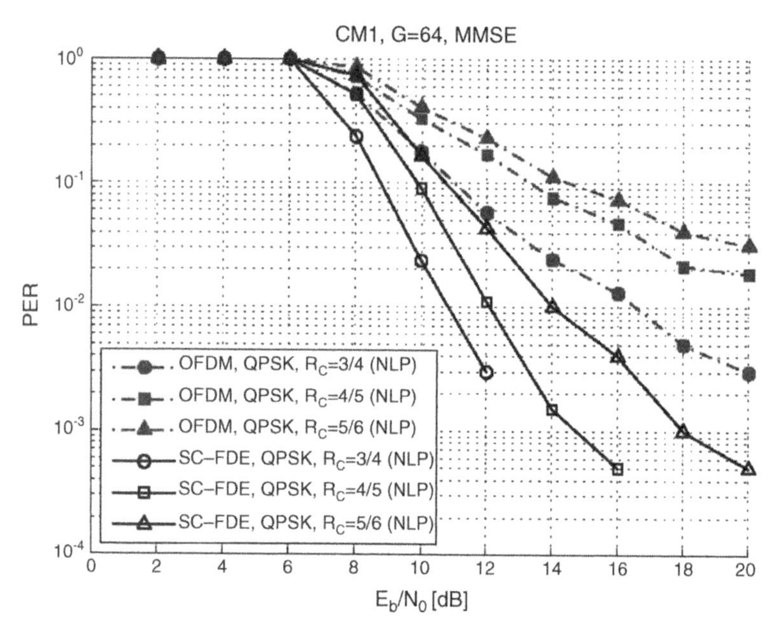

Figure 8.4. Performance comparison of OFDM and SC-FDE wireless systems [9] (©2006 IEEE)

back-off (OBO) is considered, the performance comparison between OFDM and SC-FDE wireless systems is illustrated in Figure 8.4, whereby *Rc* is the code rate. Detailed simulation conditions for Figure 8.4 are explained in [9]. It is concluded from Figure 8.4 that better packet error rate (PER) performance can be achieved when the SC-FDE technology is used for millimeter wave wireless personal area networks (WPAN).

8.2 PREAMBLE DESIGN

The basic frame structure of SC-FDE wireless systems is introduced in [10–16]. One example of frame structure of SC-FDE wireless systems is shown in Figure 8.5. A burst frame comprises a preamble, a header, and a packet payload. The preamble comprises a packet synchronization sequence (SYNC) field, a start of frame delimiter (SFD) field, and a channel estimation (CE) sequence field. The packet payload part comprises a plurality of basic structures, whereby one basic structure comprises one data frame plus one guard interval and could be chained successively with other basic structures. One basic structure is processed by the FFT module, whereby the FFT window (or FFT size) is, for example, as long as the length of a data frame plus the length of a guard interval.

Since the content of the guard interval is the same as the previous guard interval, based on the same principle regarding OFDM systems, the guard interval could be treated as the cyclic prefix of the next basic structure (comprising one data frame plus one guard interval). The inter-block interference introduced by the time-dispersive multi-path fading channel can be eliminated when the wireless channel delay is less than the length

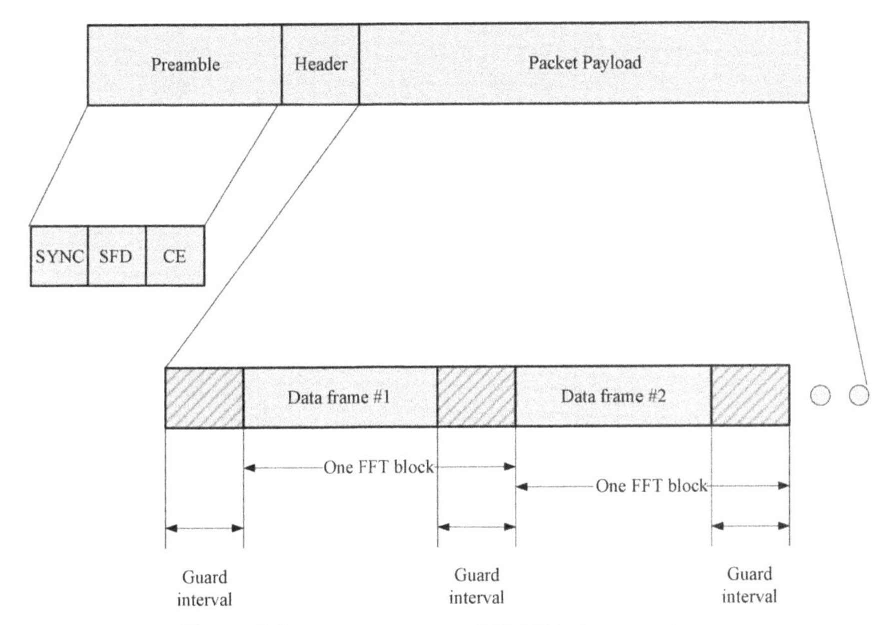

Figure 8.5. Frame structure of SC-FDE wireless systems

of guard interval. Thus, the guard interval provides a guard time to clearly separate the consecutive data frames from each other, so that the data of one data frame does not overlap with data of an adjacent data frame under the dynamic multi-path fading wireless channel environment. The guard interval can be filled with a known pseudo noise (PN) sequence or M sequence [15] in order to facilitate the coarse frame timing and carrier frequency recovery.

The preamble part described in Figure 8.5 is used for the start of frame indicator, coarse frame timing, coarse frequency offset estimation, channel estimation, and automatic gain controller (AGC). Different combinations of the sequences can be used to fill the preamble for different purposes, and some examples are discussed below.

Figure 8.6 shows an example of the burst frame structure comprising a preamble and a packet payload, which is similar to the payload structure of Figure 8.5. In Figure 8.6, the preamble consists of a single PN sequence. Furthermore, the PN sequence used to fill in the guard interval exhibits the similar correlation function to that of the preamble, as illustrated in Figure 8.6. The correlation function of a PN sequence can of course be obtained at the receiver by processing the received signal with the PN sequence using a correlator. The time duration D1 between the correlation peak of the preamble and the correlation peak of the first guard interval is much shorter than the time duration D2 between the correlation peaks of the two consecutive guard intervals. The values of D1 and D2 are known, and they can be predetermined and stored. Thus, by comparing the time duration of two consecutive correlation peaks with D1 and D2, the receiver can gain some valuable information regarding the starting position of the burst frame or the starting position of each data frame.

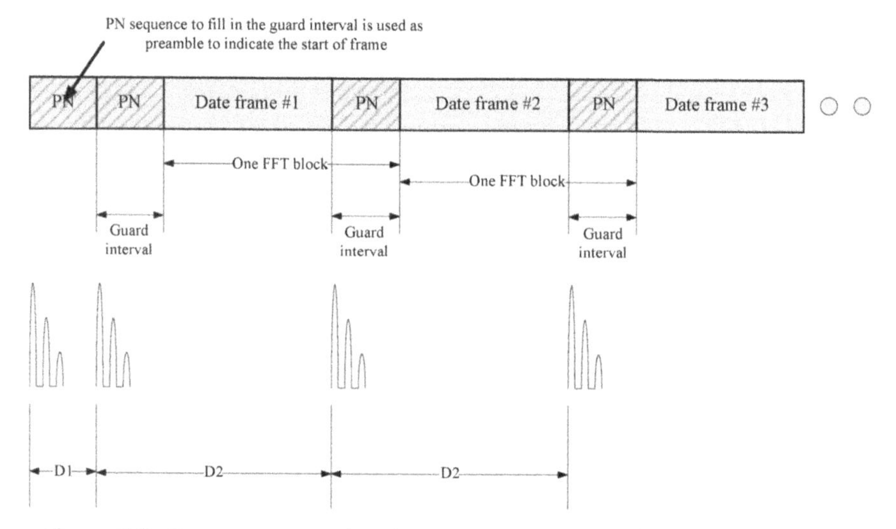

Figure 8.6. Frame structure using the PN sequence as preamble and guard interval

For the frame structure of Figure 8.6, the detection method can be summarized as follows:

- When the distance between the two autocorrelation peaks is roughly equal to D1, the coarse frame time is acquired, which indicates the beginning of the burst frame.
- When the distance between the two autocorrelation peaks is roughly equal to D2, the beginning of a data frame within the burst frame is identified.

Since the PN sequence for the preamble part is the same one as for the guard interval part, the same PN correlator can be used for the burst frame indicator as well as for processing the guard interval part. This reduces the hardware complexity.

The idea of detecting the beginning of frame position within the burst frame structure can be extended by adopting more than one PN sequence in the burst preamble to improve the detection reliability and to reduce the false alarm probability. Figure 8.7 shows an example of the burst frame comprising three PN sequences as the preamble.

When more than one PN sequence is used as a part of a burst preamble, sometimes a −PN sequence can be chosen instead of a PN sequence, where −PN stands for an inverse sequence of PN. Different combinations of plus and minus PN sequences can be used to transmit the control signaling for such burst applications.

The distances between two adjacent autocorrelation peaks for the three PN sequences of the preamble may be made different to bring further control signaling information. Figure 8.8 shows another example of a burst frame whereby the preamble comprises three PN sequences and a free zero space between the second PN sequence and the third PN sequence. The time duration D1 between the two autocorrelation peaks of the first and second PN sequences in the preamble is different from the time duration D3

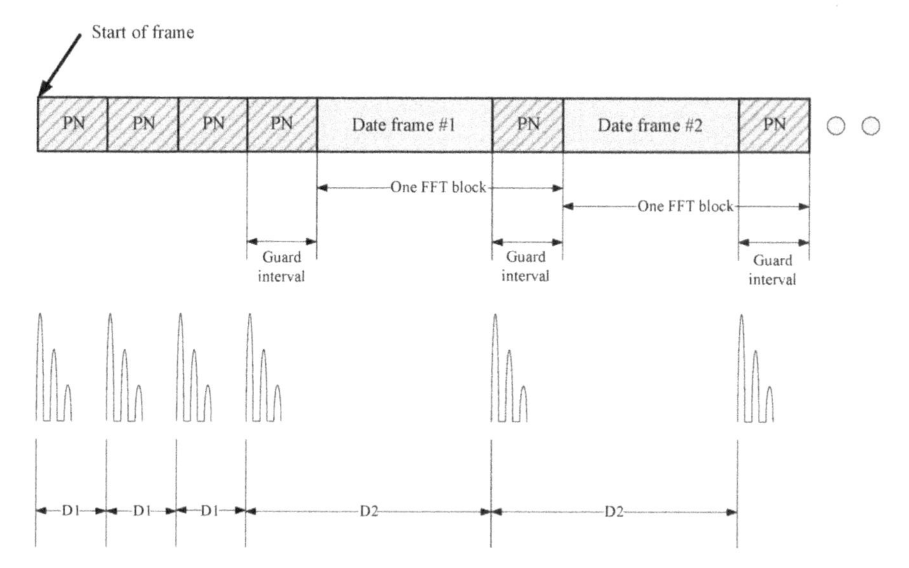

Figure 8.7. Frame structure using three PN sequences as preamble

between the two autocorrelation peaks of the second and third PN sequences in the preamble. As a result, more control signaling can be transmitted and processed with the same hardware complexity.

When constant envelope modulation schemes like offset quadrature phase shift keying (OQPSK), shaped offset quadrature phase shift keying (SOQPSK), and Feher offset quadrature phase shift keying (FOQPSK) are used for SC-FDE wireless systems to further reduce the PAPR ratio, the idea of inserting zero free space between PN sequences

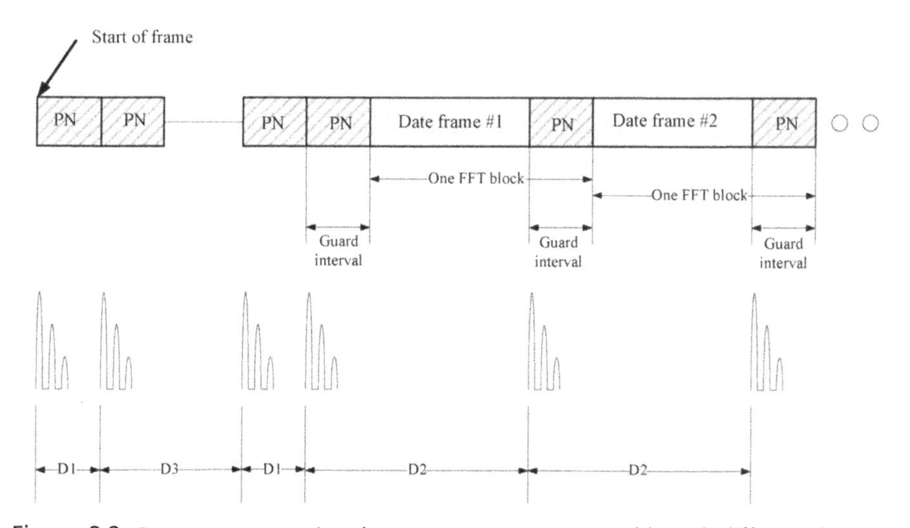

Figure 8.8. Frame structure using three PN sequences as preamble with different distances

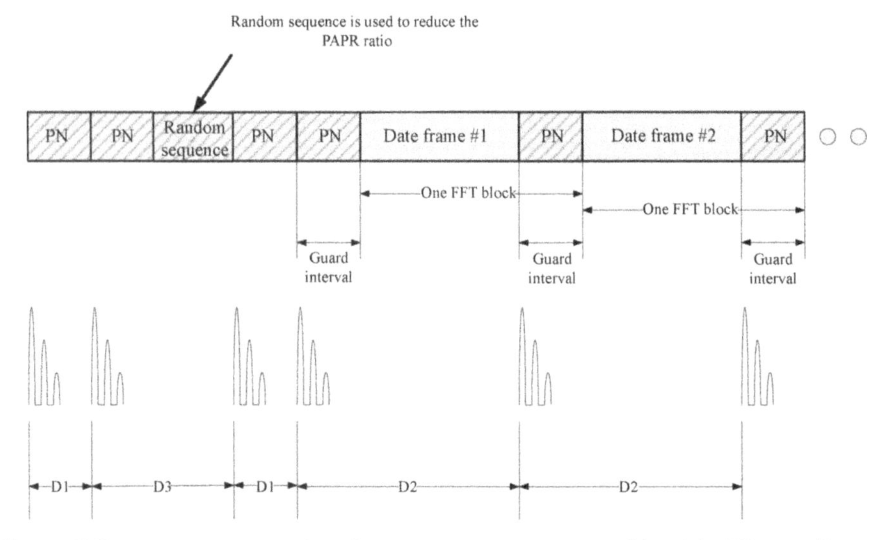

Figure 8.9. Frame structure using three PN sequences as preamble with different distances (insertion with random sequence)

has to be reconsidered. One alternative idea is to insert a random sequence between the second PN sequence of the preamble and the third PN sequence of the preamble. This preamble structure is illustrated in Figure 8.9.

8.3 ADAPTIVE CHANNEL ESTIMATION

The principle of channel estimation for OFDM systems is discussed in [17]. For SC-FDE wireless systems, channel state information (CSI) is typically estimated and tracked using time domain training sequences, which is presented in [18–26]. An alternative solution using training sequence as cyclic prefix/guard interval is discussed in [27]. To improve the performance, the MMSE-based algorithm is preferred, where the estimation of noise variance is required [28].

The block diagram of channel estimation is illustrated in Figure 8.10, where the FFT size of the SC-FDE system is N, the length of the channel delay profile is Len_CDP, and the length of the guard interval is T_G.

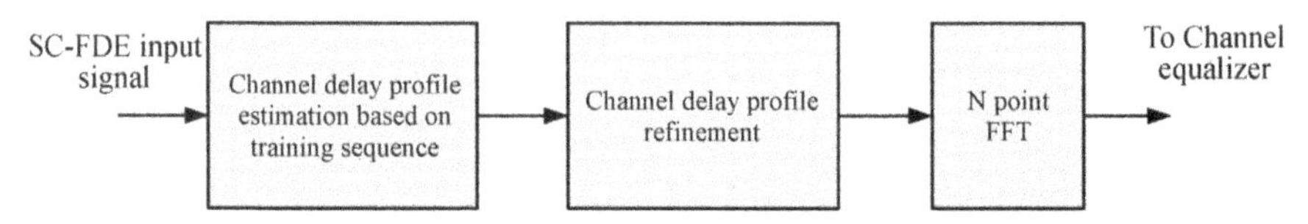

Figure 8.10. Block diagram of channel estimation for SC-FDE wireless systems

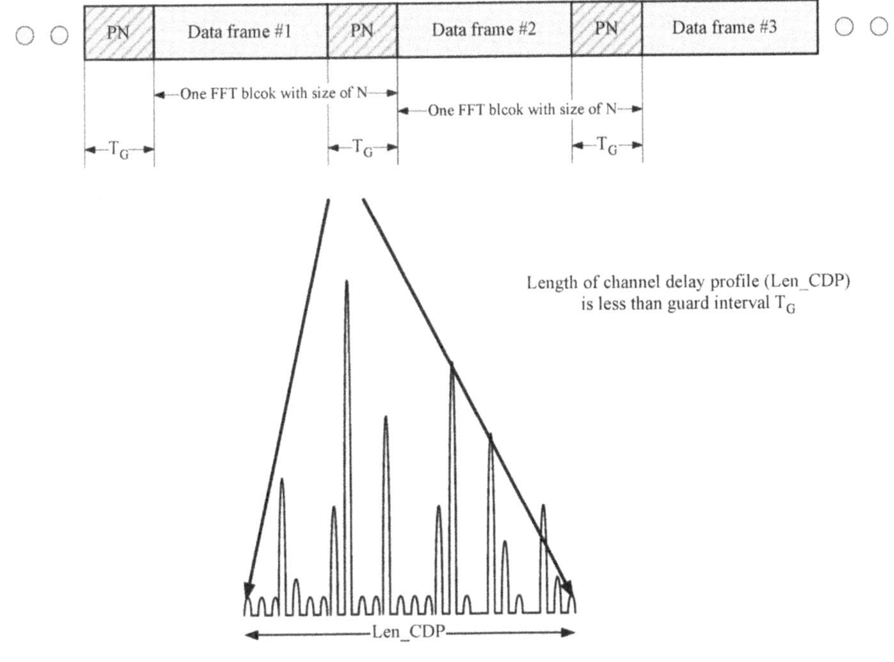

Figure 8.11. One example of channel delay profile

An example of channel delay profile is shown in Figure 8.11, where the PN sequence is used as the training sequence for estimating the channel delay profile. The power $|h(t)|^2$ of the output signal from the channel delay profile estimation block is defined as

$$|h(t)|^2 = I(t)^2 + Q(t)^2 \tag{8.1}$$

where $h(t) = I(t) + jQ(t)$ is the output signal from the channel delay profile estimation block.

Alternatively, $|h(t)|^2$ could be expressed as

$$|h(t)|^2 = \max(|I(t)|, |Q(t)|) + \frac{1}{2} \times \min(|I(t)|, |Q(t)|) \tag{8.2}$$

or

$$|h(t)|^2 = \max(|I(t)|, |Q(t)|) + \frac{1}{4} \times \min(|I(t)|, |Q(t)|) \tag{8.3}$$

As a result, the multiplication operations required to calculate the square of a complex-valued function is taken out and the hardware complexity is reduced.

The estimated channel delay profile of length Len_CDP is then padded with the N-Len_CDP zeros by the module of channel delay profile refinement, and the resultant N complex values after the FFT module then go to the channel equalizer for further processing. Note that the actual multi-path channel delay is determined by the dynamic wireless channel. If the estimated multi-path delay of Len_CDP is larger than the actual multi-path channel delay, more noise components will be included in the channel estimation and sent to the channel equalizer module, which will degrade the performance of the SC-FDE wireless system. On the other hand, if the estimated Len_CDP is smaller than the true multi-path channel delay, some significant paths may be lost, which also lead to performance degradation.

Besides that, even when the estimate of Len_CDP is quite accurate, some of the Len_CDP complex-valued components in the channel delay profile estimate may be very weak and mainly dominated by noise. These very small components can be eliminated by the refinement module before sending the channel delay profile estimate to the channel equalizer to further improve the performance. Thus, a predefined threshold Tn is introduced to reduce the effect of these noise components. If a component of the channel delay profile $|h(t)|^2$ is less than the threshold Tn, it is set to zero value. This process is illustrated in Figure 8.12. The processed Len_CDP channel delay profile components are then padded with N-Len_CDP zero components and sent to the FFT module of size N. Finally, the output of the FFT module is sent to the channel equalizer. As a consequence of removing those noise-dominated small components, the effect of the noise is reduced dramatically, and the performance of the SC-FDE wireless system is improved.

In addition, to guarantee a reliable channel estimation and equalization of SC-FDE wireless systems under dynamic wireless conditions, an adaptive tracking of the first significant path becomes very important. It is proposed to first obtain the correlation peak

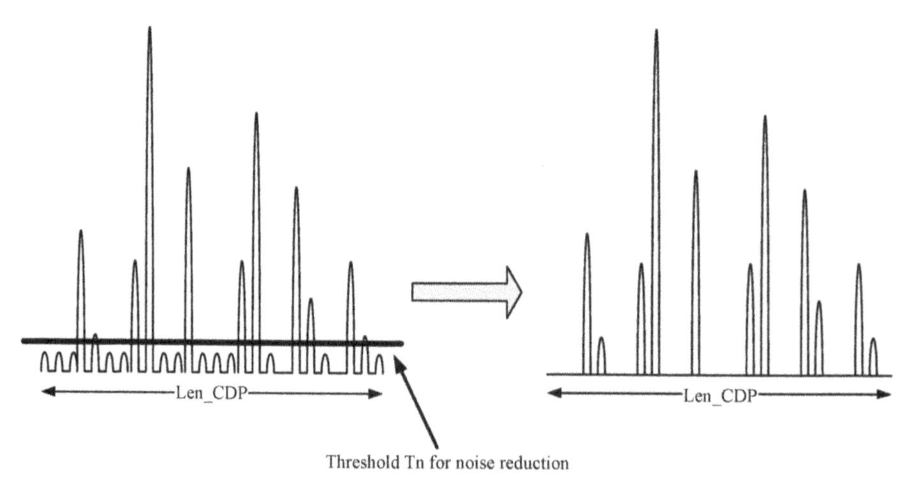

Figure 8.12. Noise reduction for SC-FDE

of the training sequence. The value of the correlation peak, Pmax, and its position are determined by the actual channel conditions. A threshold is defined as

$$\mathrm{Tp} = \mathrm{Pmax} - \mathrm{Hp} \tag{8.4}$$

whereby Hp is a fixed value which does not change with actual channel conditions and in most cases Tp $\geq$ Tn. The position of correlation peak is first determined by a forward searching, starting from the beginning of the sequence. Let the correlation peak occur at the position Len_p, as illustrated in Figure 8.13. The earliest path within this forward searching window of length Len_p that has a higher correlation value than the threshold Tp is chosen as the first path, and the position of this first significant path is identified as the beginning position of the channel estimation window. An illustration of this adaptive tracking of the first path is described in Figure 8.13.

Similarly, to further improve the performance of channel estimation and equalization under dynamic wireless conditions, the adaptive tracking of the late significant path has to be investigated. Another threshold is proposed as

$$\mathrm{Ta} = \mathrm{Pmax} - \mathrm{Ha} \tag{8.5}$$

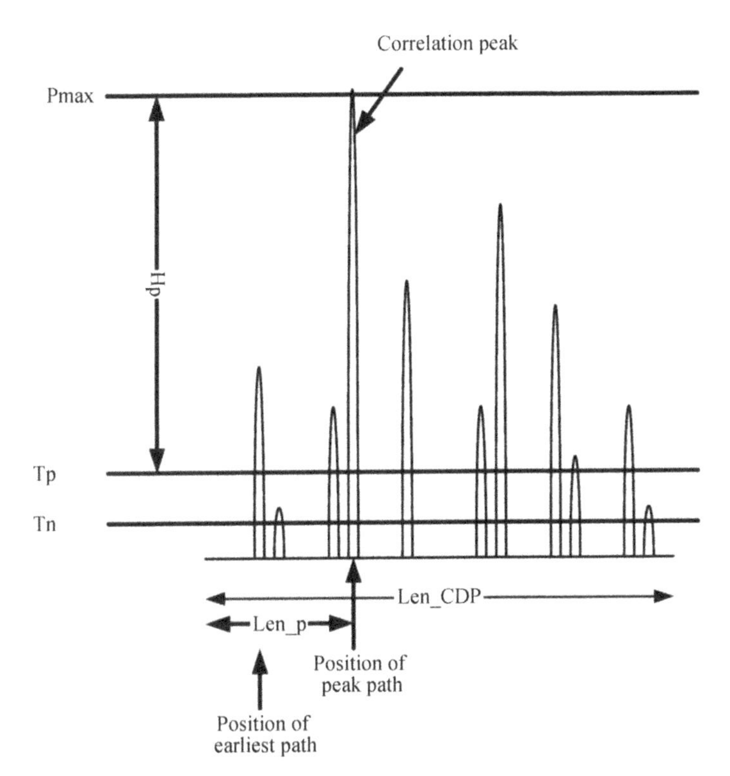

Figure 8.13. Earliest position search for actual channel estimation window

whereby Ha is a fixed value that does not change with actual channel conditions and in most cases Ta $\geq$ Tn. The position of the correlation peak can be determined by a backward searching, starting from the end position of the sequence. Let the correlation peak occurs at the position Len_a, as is shown in Figure 8.14. The latest path within this backward searching window of length Len_a that has a higher correlation value than the threshold Ta is chosen as the last path, and the position of this last significant path is identified as the end position of the channel estimation window. A description of this adaptive tracking of the last path is shown in Figure 8.14. It is apparent that

$$\text{Len\_p} + \text{Len\_a} = \text{Len\_CDP} \tag{8.6}$$

and the length of the actual channel estimation window, Len_est, is smaller than Len_CDP. Only the complex-valued channel taps within the actual channel estimation window having the length of Len_est, padded with N-Len_est zeros, are used as the channel estimate. The idea of adaptive channel estimation, as detailed in Figure 8.14, ensures that good performance can be achieved for a SC-FDE wireless system.

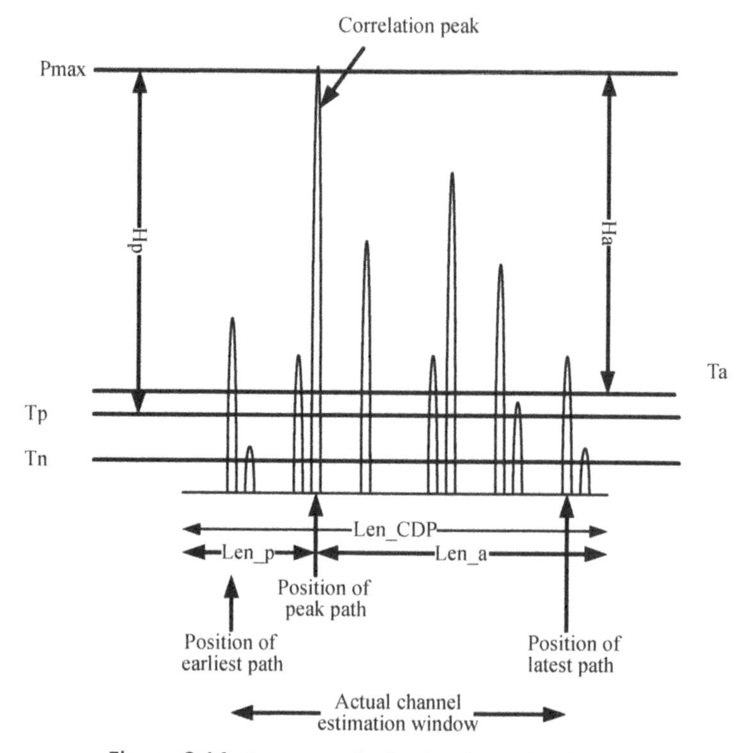

Figure 8.14. Summary of adaptive channel estimation

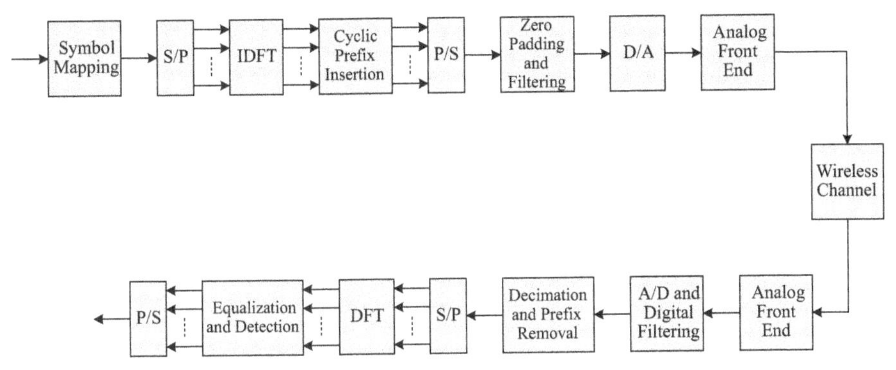

Figure 8.15. Block diagram of an OFDM system [25] (©2008 IEEE)

8.4 FREQUENCY DOMAIN EQUALIZATION

SC-FDE is the bridge between SC and OFDM. It is a form of SC with low PAPR. It is also a form of OFDM with Fourier spreading. Figure 8.15 and Figure 8.16 show the detailed comparison between multi-carrier OFDM systems and single-carrier systems with frequency domain equalization. It is evident that OFDM and SC-FDE systems have similar structures. The only difference is the placement of the inverse discrete Fourier transform (IDFT) block. In OFDM systems it is placed in the transmitter to map the data symbols onto subcarriers. However, in SC-FDE systems it is placed in the receiver to transform the equalized data into time domain for signal detection. The implementation complexities of these two systems are similar with the same FFT length.

Based on the relationship between OFDM and SC-FDE systems, it is convenient to achieve the conversion from one mode to the other. Since each terminal consists of a transmitter and a receiver, it can be reconfigured to handle either single-carrier or OFDM signals by switching the IFFT block between the transmitter and the receiver, as is illustrated in Figure 8.17.

It is advantageous to operate a dual-mode wireless system, wherein the base station uses an OFDM transmitter and a single-carrier receiver, and the handset uses a single-

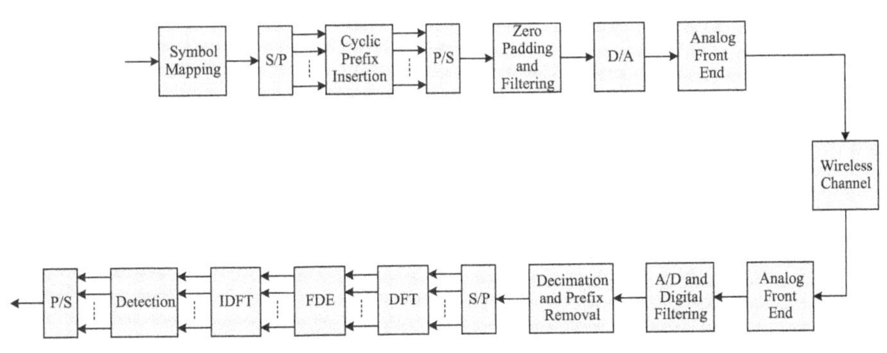

Figure 8.16. Block diagram of a SC-FDE system [25] (©2008 IEEE)

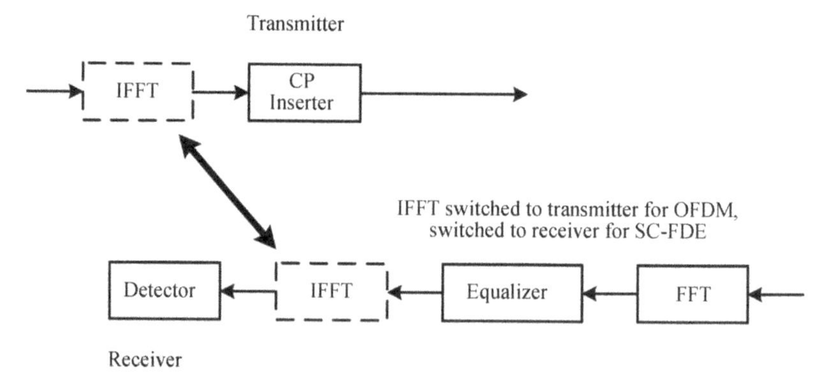

Figure 8.17. Conversion between OFDM and SC-FDE

carrier transmitter and an OFDM receiver, as shown in Figure 8.18. The use of the OFDM scheme in the downlink and single-carrier scheme in the uplink has two advantages:

- There are two IFFT blocks and one FFT block in the base station, while the handset has only one FFT block. Thus, most of the signal processing complexity is concentrated at the base station.
- Since the handset uses a single-carrier scheme, it is more efficient in terms of power consumption due to the inherent low PAPR ratio. This will reduce the cost of the PA, especially for millimeter wave rclated devices.

8.4.1 SC-FDE Based on MMSE Criteria

The frequency domain equalizer for single-carrier wireless systems is illustrated in Figure 8.19. The data $\{a_m\}_{m=0}^{M-1}$ is transmitted in blocks of M data symbols, the received signal is $\{r_m\}_{m=0}^{M-1}$, and the equalized signal is $\{z_m\}_{m=0}^{M-1}$. Assume that the cyclic prefix

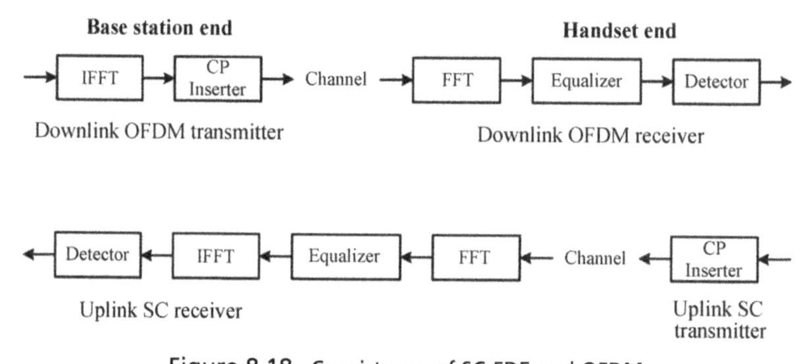

Figure 8.18. Coexistence of SC-FDE and OFDM

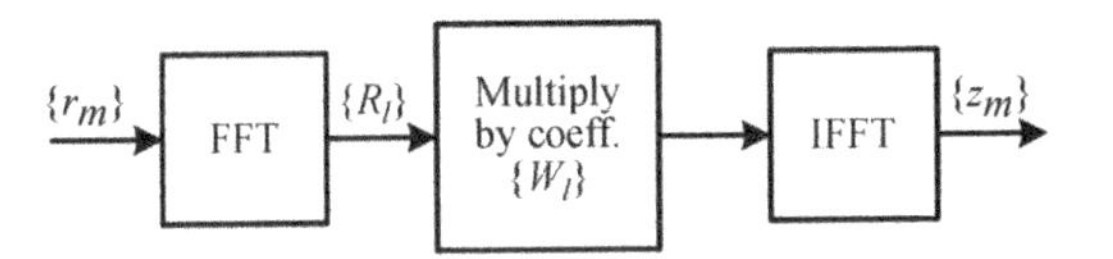

Figure 8.19. Single-carrier frequency domain equalization

(CP) is inserted and the multi-path time delay spread is shorter than the length of CP. The equalizer has M coefficients W_l, $l = 0, 1, \ldots, M-1$.

Because of the CP insertion, R_l, $0 \le l \le M-1$, can be expressed as

$$R_l = H_l A_l + V_l \tag{8.7}$$

where $\{H_l\}_{l=0}^{M-1}$ is the channel frequency response (CFR), $\{A_l\}_{l=0}^{M-1}$ is the discrete Fourier transformation (DFT) of the time domain data, and $\{V_l\}_{l=0}^{M-1}$ is the frequency domain noise. Specifically,

$$H_l = \sum_{m=0}^{M-1} h_m \exp\left(-j\frac{2\pi}{M} lm\right) \tag{8.8}$$

$$A_l = \sum_{m=0}^{M-1} a_m \exp\left(-j\frac{2\pi}{M} lm\right) \tag{8.9}$$

$$V_l = \sum_{m=0}^{M-1} n_m \exp\left(-j\frac{2\pi}{M} lm\right) \tag{8.10}$$

where h_m is the channel impulse response extended to M values and n_m is the noise components, $0 \le m \le M-1$. The time domain output of the equalizer $\{z_m\}_{m=0}^{M-1}$ can be expressed as

$$z_m = \frac{1}{M} \sum_{l=0}^{M-1} W_l R_l \exp\left(j\frac{2\pi}{M} lm\right) \tag{8.11}$$

The error signal at the detection point is

$$e_m = z_m - a_m \tag{8.12}$$

So the mean squared error (MSE) at the detection point is

$$\begin{aligned} MSE &= \mathrm{E}\left(e_m e_m^*\right) = \mathrm{E}\left[(z_m - a_m)(z_m - a_m)^*\right] \\ &= \frac{\sigma_a^2}{M}\sum_{l=0}^{M-1}|W_l|^2|H_l|^2 - 2\frac{\sigma_a^2}{M}\mathrm{Re}\sum_{l=0}^{M-1}W_l H_l + \sigma_a^2 + \frac{\sigma_n^2}{M}\sum_{l=0}^{M-1}|W_l|^2 \\ &= \frac{\sigma_a^2}{M}\sum_{l=0}^{M-1}|W_l H_l - 1|^2 + \frac{\sigma_n^2}{M}\sum_{l=0}^{M-1}|W_l|^2 \end{aligned} \tag{8.13}$$

where $\sigma_a^2 = E(|a_m|^2)\quad 0 \le m \le M-1$ is the power of the received signal and σ_n^2 is the power of the noise.

By setting the derivative of the MSE (8.13) with respect to W_l to zero, we have

$$\frac{\partial MSE}{\partial W_l} = \left[\frac{\sigma_a^2}{M}(W_l H_l - 1)H_l^* + \frac{\sigma_n^2}{M}W_l\right] = 0 \tag{8.14}$$

which leads to the MMSE solution for W_l given by

$$\hat{W}_l = \frac{\sigma_a^2 H_l^*}{\sigma_n^2 + \sigma_a^2|H_l|^2} \tag{8.15}$$

Assuming that the signal-to-noise ratio (SNR) is equal to $\frac{\sigma_a^2}{\sigma_n^2}$, we have

$$\hat{W}_l = \frac{H_l^*}{|H_l|^2 + \dfrac{1}{\mathrm{SNR}}} \tag{8.16}$$

To reduce the complexity of SNR estimation, a suboptimal MMSE equalizer is given by

$$\hat{W}_l = \frac{H_l^*}{|H_l|^2 + \frac{1}{\mathrm{SNR}_{\min}}} \tag{8.17}$$

whereby $\mathrm{SNR}_{\min}$ is the predefined and fixed value, which indicates the minimum required SNR value in order to achieve the necessary bit error rate (BER) performance for specific user scenarios. When the actual SNR is larger than $\mathrm{SNR}_{\min}$, some performance loss may occur. However, the actual performance of this suboptimal MMSE equalizer can still satisfy the user requirement for the range of SNR larger than $\mathrm{SNR}_{\min}$, which is illustrated in Figure 8.20. To further reduce the calculation complexity, the division operation could be implemented by a look-up table (LUT).

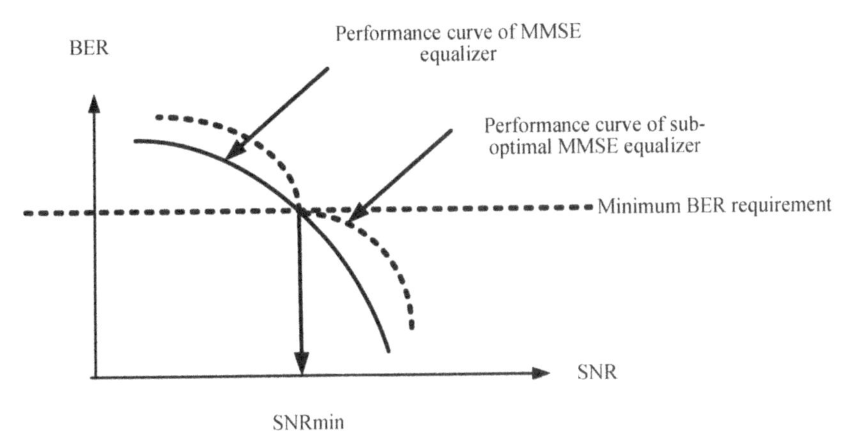

Figure 8.20. Performance comparison between MMSE and suboptimal MMSE equalizer

8.5 DECISION FEEDBACK EQUALIZATION

Although the frequency domain linear equalizer (FD-LE) of single-carrier systems discussed in Section 8.4 has obvious advantages, its performance will degrade in some cases due to noise enhancement. In millimeter wave wireless conditions, when strong reflection paths exist and the multi-path channel delay is over tens of symbol periods (e.g., for high rate applications beyond Gbps), there exist strong frequency selectivity and deep frequency nulls in the frequency response of the wireless channel. As a result, the noise components at these frequency nulls will be amplified greatly by the linear equalization process.

Compared to the FD-LE, the frequency domain decision feedback equalization (FD-DFE) gives better performance, in which already decided symbols are used for a further reduction of ISI without any additional noise enhancement by eliminating residual perturbing pulse response tails of the channel. In the conventional time domain DFE equalizers, the received signals are processed symbol by symbol. So the detected symbol is fed back immediately to remove its interference on subsequent symbols. However, when adopting decision feedback in the SC-FDE system, because of the processing delay of FFT and IFFT blocks, the DFE structure has to be changed and designed carefully.

Figure 8.21 shows a block diagram of a DFE equalizer implemented in the frequency domain, which feeds decisions back after a certain delay [30]. However, the impact of this delay on the performance of FD-DFE equalizer over fast fading wireless channels needs further investigation.

A more popular single-carrier FD-DFE structure is shown in Figure 8.22, where a hybrid time-frequency domain approach is used to avoid the feedback delay problem mentioned above. The frequency domain filtering is used only for the forward filter part of the DFE, and the conventional time domain transversal filter acts as the feedback part. The transversal feedback filter is relatively simple in any case, since it does not require complex multipliers, and it can be made to be short or long depending on the performance requirement. This approach could also limit the error propagation problem due to the

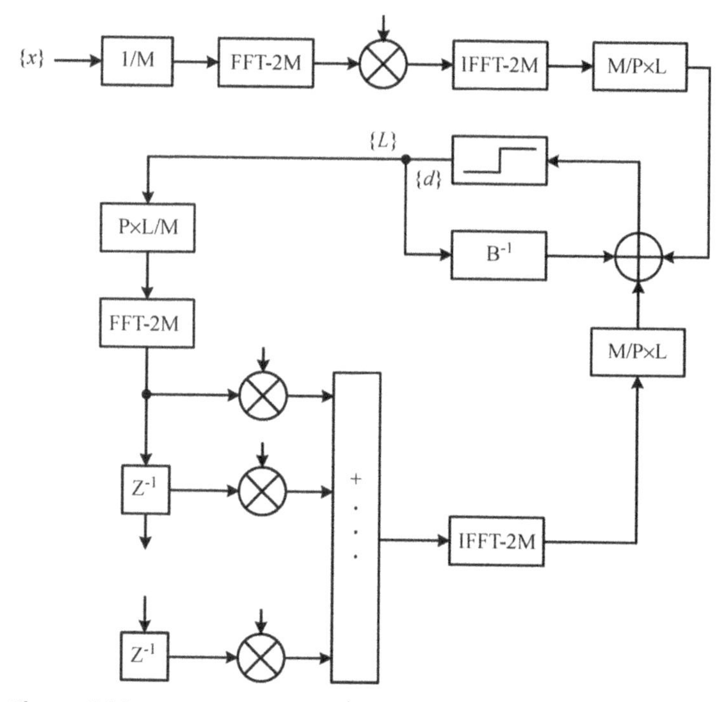

Figure 8.21. Block diagram of frequency domain DFE [30] (©1995 IEEE)

decision errors, because the feedback is performed block by block. Such a DFE system uses frequency domain processing to compute forward equalizer parameters and a small matrix inversion to compute the time domain feedback taps, where the training sequence or pilots are required [31].

In order to use frequency domain equalization for the single-carrier systems, data should be transmitted in blocks. There are two popular data structures to fulfil the requirement of FDE, that is, the cyclic prefix single carrier (CP-SC) and unique word single carrier (UW-SC), as shown in Figure 8.23, respectively. T_G is the length of CP or UW. T_{FFT} is the length of FFT block. CP-SC is similar to the conventional OFDM systems using cyclic prefix (CP-OFDM) and the FFT length is equal to the data block

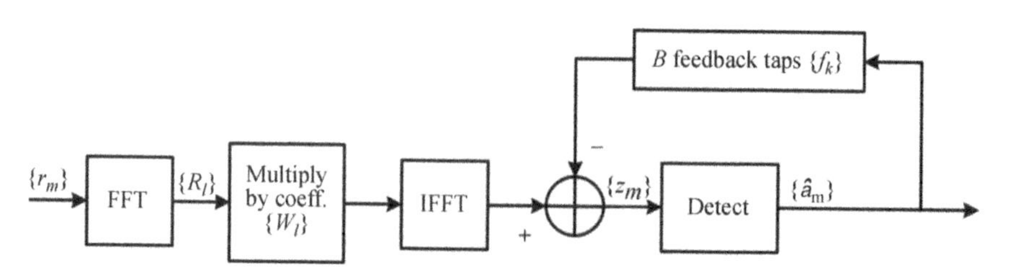

Figure 8.22. Structure of SC-FDE decision feedback equalizer

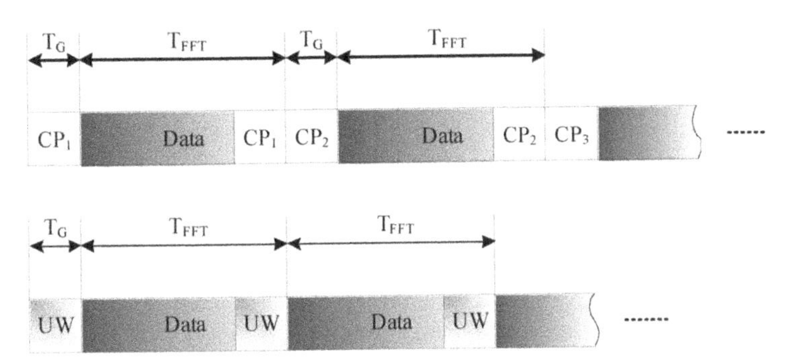

Figure 8.23. Data structures for single-carrier FDE

length. In UW-SC, the UW can be used for synchronization and channel estimation. Furthermore, it is treated as a guard interval, just like CP if the data block and the following UW are treated as a whole. So the FFT length in UW-SC is equal to the total length of data block plus UW [32].

In a UW-SC system, because UW is known in the receiver, the known UW symbols can be applied to the feedback filter directly instead of using decided symbols [32], as illustrated in Figure 8.24. In Figure 8.24, r_n is the received signal, y_n is the time domain output of the feed-forward equalizer, $\hat{s}_n$ is the input symbol of the feedback filter, $G_{FF,i}$ is the coefficient of the feed-forward filter in the frequency domain, and $g_{FB,j}$ is the coefficient of the feedback filter in the time domain. Using the known UW symbols in decision feedback removes the decision errors, and this can further improve the bit error rate (BER) performance. Therefore, the UW-SC is more suitable for the FD-DFE structure.

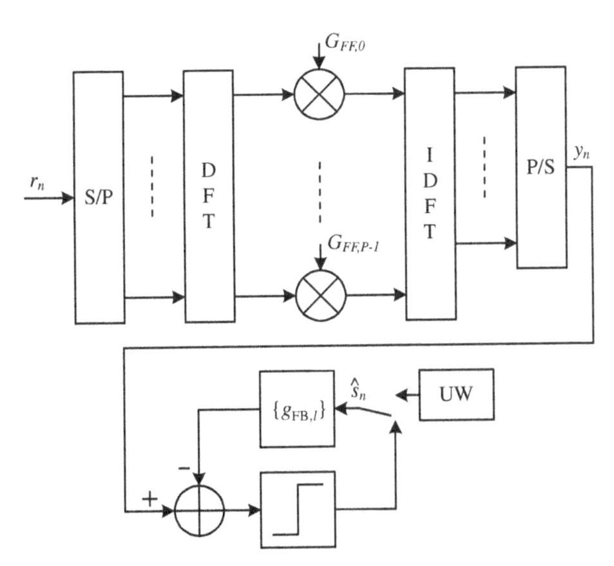

Figure 8.24. Structure of FD-DFE for UW-SC

8.5.1 Theoretical Analysis of the Hybrid Domain DFE

The data $\{a_k\}$ is transmitted in blocks of M data symbols. In the following analyses the CP-SC is assumed to be used. So the FFT point is same as the data block length. Assume that the maximum multi-path delay spread is shorter than CP and the ISI could be eliminated thoroughly. There is one sample in each received data symbol. Perfect synchronization and channel estimation are considered. The frequency domain feedforward filter in Figure 8.22 has M complex coefficients $\{W_l\}$. There are B complex feedback coefficients $\{f_k, k \in F_B\}$ in the system, where F_B is a set of nonzero indices that correspond to the delays of the B feedback taps. Normally, the indices F_B could correspond to the relative estimated delays of the channel impulse response. If $B = 0$, the FD-DFE is identical to FD-LE.

The time domain output of the forward filter $\{z_m\}$ can be expressed as

$$z_m = \frac{1}{M}\sum_{l=0}^{M-1} W_l R_l \exp\left(j\frac{2\pi}{M}lm\right) - \sum_{k \in F_B} f_k^* a_{m-k} \qquad m = 0, 1, 2, \cdots, M-1 \tag{8.18}$$

where $\{R_l\}$ is the DFT of the received symbol $\{r_m\}$ given by

$$R_l = \sum_{m=0}^{M-1} r_m \exp\left(-j\frac{2\pi}{M}lm\right), \qquad l = 0, 1, 2, \cdots, M-1 \tag{8.19}$$

Because of the cyclic prefix, the transmitted signal forces the linear convolution with the channel-impulse response to be circular [34]. So in the frequency domain, R_l can be express as

$$R_l = H_l A_l + V_l \tag{8.20}$$

where $\{H_l\}$ is the channel frequency response, $\{A_l\}$ is the DFT of the time domain data, and $\{V_l\}$ is the DFT of the time-domain noise samples $\{n_m\}$. Specifically,

$$H_l = \sum_{m=0}^{M-1} h_m \exp\left(-j\frac{2\pi}{M}lm\right) \tag{8.21}$$

$$A_l = \sum_{m=0}^{M-1} a_m \exp\left(-j\frac{2\pi}{M}lm\right) \tag{8.22}$$

$$V_l = \sum_{m=0}^{M-1} n_m \exp\left(-j\frac{2\pi}{M}lm\right) \tag{8.23}$$

From (8.21), (8.22), and (8.23), the autocorrelation of A_l, V_l, and R_l are [31]

$$E\left(A_{l_1}A_{l_2}^*\right) = M\sigma_a^2\delta(l_1-l_2) \tag{8.24}$$

$$E\left(V_{l_1}V_{l_2}^*\right) = M\sigma^2\delta(l_1-l_2) \tag{8.25}$$

$$E\left(R_{l_1}R_{l_2}^*\right) = M\sigma_a^2 H_{l_1}H_{l_2}^*\delta(l_1-l_2) + M\sigma^2\delta(l_1-l_2) \tag{8.26}$$

where $\sigma_a^2 = E(|a_m|^2)$ is the power of the received signal, σ^2 is the power of noise, $\delta(l)$ is the Kronecker delta function, and $0 \le l_1, l_2 \le M-1$.

The error signal at mth symbol is

$$e_m = z_m - a_m \tag{8.27}$$

Then, the MSE of error signal e_m is

$$\begin{aligned} E\left(|e_m|^2\right) &= \frac{1}{M}\sum_{l_1=0}^{M-1}\sum_{l_2=0}^{M-1} W_{l_1}W_{l_2}^*\left[\sigma_a^2 H_{l_1}H_{l_2}^*\delta(l_1-l_2) + \sigma^2\delta(l_1-l_2)\right] \\ &\quad - \frac{2\sigma_a^2}{M}\mathrm{Re}\left\{\sum_{l=0}^{M-1} W_l H_l\left[1 + \sum_{k\in F_B} f_k \exp\left(j\frac{2\pi}{M}lk\right)\right]\right\} + \sigma_a^2\left(1 + \sum_{k\in F_B}|f_k|^2\right) \end{aligned} \tag{8.28}$$

Equation (8.28) can be rewritten as

$$\mathrm{MSE} = E\left(|e_m|^2\right) = \frac{\sigma_a^2}{M}\sum_{l=0}^{M-1}|W_l H_l - F_l|^2 + \frac{\sigma^2}{M}\sum_{l=0}^{M-1}|W_l|^2 \tag{8.29}$$

where

$$F_l = 1 + \sum_{k\in F_B} f_k^* \exp\left(-j\frac{2\pi}{M}kl\right) \tag{8.30}$$

In order to solve the optimal W_l, we apply the MMSE criterion to (8.29) and obtain

$$\frac{\partial\left[E\left(|e_m|^2\right)\right]}{\partial W_l^*} = \left[\frac{\sigma_a^2}{M}(W_l H_l - F_l)H_l^* + \frac{\sigma^2}{M}W_l\right] = 0 \tag{8.31}$$

The optimum frequency domain forward filter coefficients can be expressed as [35]

$$\tilde{W}_l = \frac{\sigma_a^2 H_l^* F_l}{\sigma^2 + \sigma_a^2 |H_l|^2} = \frac{\sigma_a^2 H_l^* \left(1 + \sum\limits_{k \in F_B} f_k^* \exp(-j\frac{2\pi}{M}kl)\right)}{\sigma^2 + \sigma_a^2 |H_l|^2} = \frac{H_l^* F_l}{1/SNR + |H_l|^2} \tag{8.32}$$

where $\text{SNR} = \sigma_a^2/\sigma^2$. Inserting (8.32) into (8.29), we obtain the MSE given $\tilde{W}_l$

$$\text{MSE}|_{\tilde{W}_l} = E\left(|e_m|^2\right) = \frac{\sigma^2}{M} \sum_{l=0}^{M-1} \frac{|F_l|^2}{1/SNR + |H_l|^2} \tag{8.33}$$

which is the function of $\{f_l\}$. Applying the MMSE criterion to (8.33), we have

$$\frac{\partial\left[\text{MSE}|_{\tilde{W}_l}\right]}{\partial f_{k1}^*} = \frac{\sigma^2}{M} \sum_{l=0}^{M-1} \frac{\left(1 + \sum\limits_{k \in F_B} f_k \exp(j\frac{2\pi}{M}kl)\right) \exp(-j\frac{2\pi}{M}lk_1)}{1/SNR + |H_l|^2} \tag{8.34}$$

Rewrite (8.34) as follows:

$$\frac{\sigma^2}{M} \sum_{k \in F_B} f_k \sum_{l=0}^{M-1} \frac{\exp(-j\frac{2\pi}{M}l(k_1 - k))}{1/SNR + |H_l|^2} = -\frac{\sigma^2}{M} \sum_{l=0}^{M-1} \frac{\exp(-j\frac{2\pi}{M}lk_1)}{1/SNR + |H_l|^2} \qquad k_1 \in F_B \tag{8.35}$$

From (8.35) we get an equation group with B variables, which can be expressed as the following matrix form:

$$\mathbf{Vf} = -\mathbf{v} \tag{8.36}$$

where

$$\mathbf{f} = (f_{k1}, f_{k2}, \cdots, f_{kB})^T \tag{8.37}$$

$$\mathbf{v} = (v_{k1}, v_{k2}, \cdots, v_{kB})^T \tag{8.38}$$

$$\mathbf{V} = \begin{bmatrix} v_0 & v_{k1-k2} & \cdots & v_{k1-kB} \\ v_{k2-k1} & v_0 & v_{k2-k3} & \cdots \\ \vdots & \vdots & \ddots & \vdots \\ v_{kB-k1} & v_{kB-k2} & \cdots & v_0 \end{bmatrix} \tag{8.39}$$

$$v_k = \frac{\sigma^2}{M}\sum_{l=0}^{M-1}\frac{\exp(-j\frac{2\pi}{M}lk)}{1/SNR+|H_l|^2} \tag{8.40}$$

The optimal MMSE time-domain feedback filter coefficient vector is then given by

$$\hat{\mathbf{f}} = -\mathbf{V}^{-1}\mathbf{v}$$

Let

$$\hat{F}_l = 1 + \sum_{k\in F_B}\hat{f}_k^{*}\exp\left(-j\frac{2\pi}{M}kl\right)$$

Then, the optimal MMSE frequency-domain forward filter coefficients are given by

$$\hat{W}_l = \frac{H_l^*\hat{F}_l}{1/SNR+|H_l|^2}, \quad l = 0, 1, \cdots, M-1$$

The performance comparisons between the single-carrier FD-LE and FD-DFE as well as that of the OFDM systems were given in [36] with perfect channel knowledge. The multi-path fading channel model of "SUI-5" was chosen. "SUI-5" is one of the six channel models with three Rayleigh fading taps adopted by IEEE 802.16a for evaluating broadband wireless systems in 2–11 GHz bands [37]. The fading taps are at delays of 0, 5, and 10 µs, with relative powers of 0, −5, and −10 dB, respectively. This model is a high delay spread model associated with omnidirectional antennas in suburban hilly environments. The channel RMS delay spread is 3.05 µs.

Figure 8.25 shows the average BERs vs. SNR using $^1/_2$ convolutional code and with three modulation schemes, that is, QPSK, 16-QAM, and 64-QAM, respectively. The performance curves of single-carrier FD-LE, FD-DFE, and OFDM systems are given for each constellation mode. For a comparison the matched filter bound (MFB) is also given. It concludes that for QPSK modulation, OFDM, FD-LE and ideal FD-DFE (without decision errors) single-carrier systems all perform to within about 1.5–2 dB range of one another. The performance of ideal FD-DFE is around 1 dB worse than that of the ideal MFB case for QPSK. OFDM performs slightly better than FD-LE, and slightly worse than the ideal FD-DFE. For 16-QAM and 64-QAM cases, there is a larger spread among the results, but with the same relative rankings. The performance of FD-DFE is comparable to OFDM under various modulation schemes. For low rate applications using QPSK, FD-DFE is preferred to be adopted due to its low PAPR ratio.

Figure 8.26 shows the performance using QPSK and higher code rates over the SUI-5 channel. From the results we can see that for the higher code rate of 3/4 or 7/8, OFDM performs worse than FD-LE and FD-DFE. For OFDM systems, narrow band notches of channel transfer function would degrade the performance of the transmitted symbols assigned over the relevant subcarriers. Such kinds of degradation could be compensated for by the strong error control coding scheme together with suitable

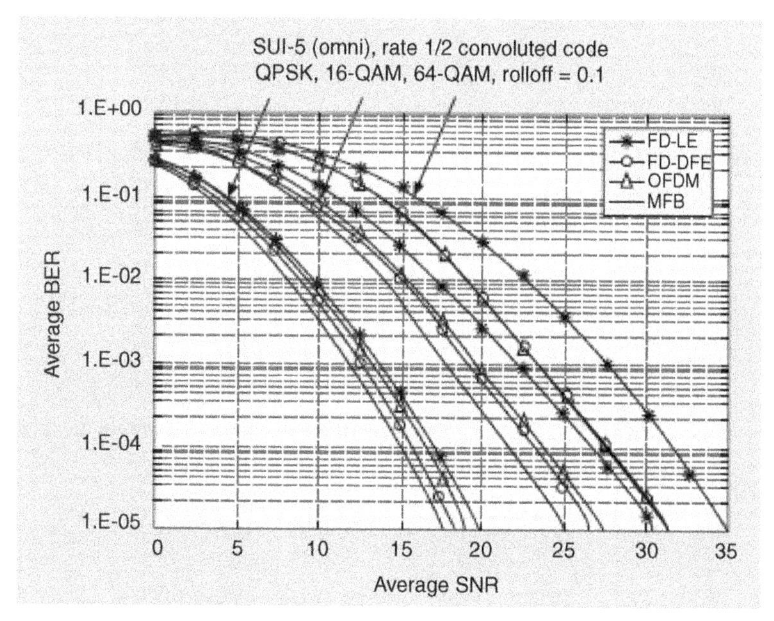

Figure 8.25. Performance comparison with perfect channel estimation over SUI-5 channel [36] (©2002 IEEE)

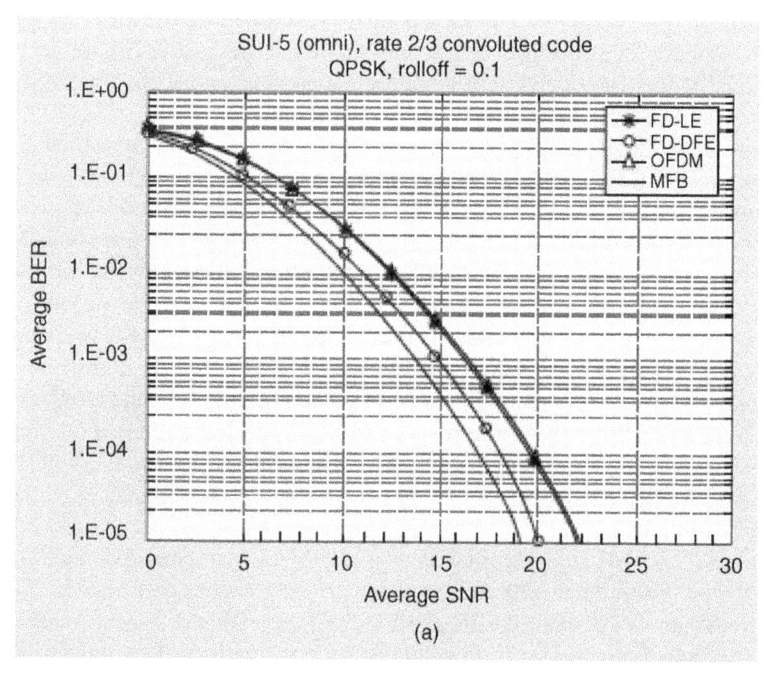

Figure 8.26. Performance comparison for QPSK and higher code rate: (a) 2/3 rate, (b) 3/4 rate, (c) 7/8 rate [36] (©2002 IEEE)

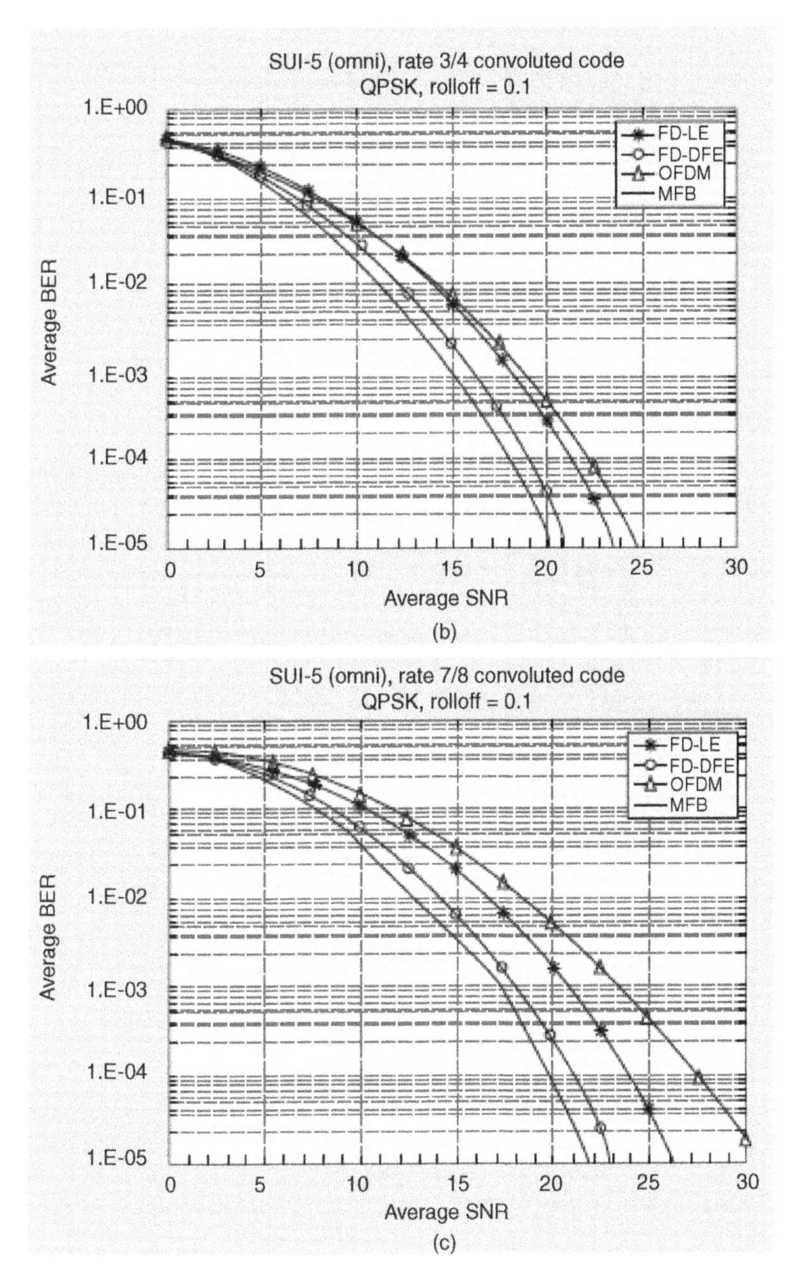

Figure 8.26. *(Continued)*

frequency domain interleaver. However, when a high code rate is applied, the error control decoding capabilities and the corresponding diversity gain from frequency domain become weak. Thus, the performance of OFDM systems is worse than that of FD-LE and FD-DFE.

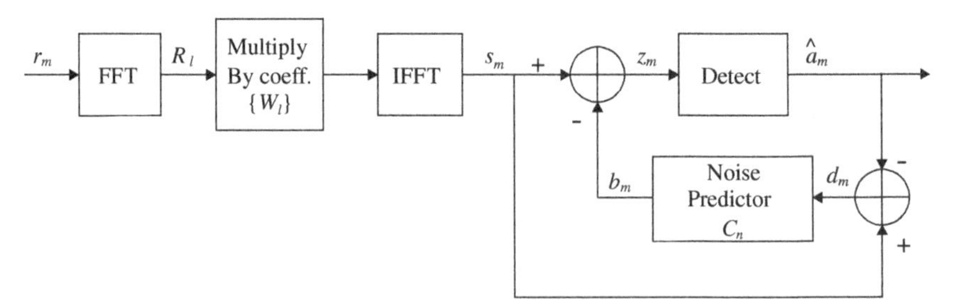

Figure 8.27. Block diagram of SC-FDE with noise prediction

8.5.2 SC-FDE with Noise Prediction

Another SC-FDE structure, which is similar to the FD-DFE discussed previously, consists of a frequency-domain linear equalizer with a time-domain noise prediction filter (FDE-NP) [38], as shown in Figure 8.27. Theoretical analysis and simulation results prove that the FD-DFE and FDE-NP have the same performance. However, in FDE-NP the parameters of a feed-forward filter and noise predictor are optimized separately, while in the FD-DFE these parameters are designed jointly. Furthermore, in the FDE-NP scheme, performance and complexity trade-off can be obtained by only changing the order of the noise predictor, while in the FD-DFE scheme both the FDE and the time domain feedback filter must be modified simultaneously. This makes the FDE-NP more suitable for adaptive designs.

We rewrite (8.20) in matrix form as

$$\mathbf{R} = \mathbf{HA} + \mathbf{V} \tag{8.41}$$

where $\mathbf{R} = [R_0, R_1, \ldots, R_{M-1}]^{\mathrm{T}}$, $\mathbf{A} = [A_0, A_1, \ldots, A_{M-1}]^{\mathrm{T}}$, and $\mathbf{V} = [V_0, V_1, \ldots, V_{M-1}]^{\mathrm{T}}$, while $\mathbf{H} = \mathrm{diag}\{H_0, H_1, \ldots, H_{M-1}\}$ denotes the $M \times M$ diagonal matrix with the diagonal elements $H_0, H_1, \ldots, H_{M-1}$.

The feed-forward equalizer output in time domain after the IFFT is

$$\mathbf{s} = \frac{1}{M}\mathbf{F}^{H}(\mathbf{WHA} + \mathbf{WV}) \tag{8.42}$$

where $\mathbf{s} = [s_0, s_1, \cdots, s_{M-1}]^{T}$, $\mathbf{W} = \mathrm{diag}\ \{W_0, W_1, \cdots, W_{M-1}\}$, and $\mathbf{F}$ is the FFT matrix with its entry $F_k = \exp\{-j2\pi kl/M\}, 0 \le k, l \le M-1$, while $\mathbf{F}^{\mathrm{H}}$/M denotes the IFFT transform matrix.

The error element going to the noise predictor is defined as $\mathbf{d} = \mathbf{s} - \hat{\mathbf{a}}$, where $\hat{\mathbf{a}}$ is the vector of the decision symbols. The noise predictor in Figure 8.27 has the B feedback taps c_i for $1 \le I \le B$ [38]. Then the output of the noise predictor is defined as

$$\mathbf{b} = (\mathbf{I} - \mathbf{c})\mathbf{d} = (\mathbf{I} - \mathbf{c})(\mathbf{s} - \hat{\mathbf{a}}) \tag{8.43}$$

where $\mathbf{I}$ is the M $\times$ M identity matrix, and $\mathbf{c}$ is a M $\times$ M circular matrix with the main diagonal value equal to 1 and the first row equal to $[1\,0\ldots 0\,c_B\,c_{B-1}\ldots c_1]$. Here the ideal feedback is assumed so that $\hat{\mathbf{a}} = \mathbf{a}$, where $\mathbf{a}$ denotes the transmitted true symbol vector. Then the signal input to the detector can be expressed as

$$\mathbf{z} = \mathbf{s} - \mathbf{b} = \mathbf{s} - (\mathbf{I} - \mathbf{c})(\mathbf{s} - \mathbf{a}) \tag{8.44}$$

The error signal between $\mathbf{z}$ and $\mathbf{a}$ is

$$\mathbf{e} = \mathbf{z} - \mathbf{a} = \mathbf{c}(\mathbf{s} - \mathbf{a}) \tag{8.45}$$

The autocorrelation matrix of the error signal is

$$\begin{aligned} \mathrm{E}(\mathbf{e}\mathbf{e}^H) &= \mathbf{c}\mathrm{E}\left[(\mathbf{s}-\mathbf{a})(\mathbf{s}-\mathbf{a})^H\right]\mathbf{c}^H \\ &= \frac{1}{M}\mathbf{c}\left(\mathbf{F}^H\mathbf{W}\mathbf{T}\mathbf{W}^H\mathbf{F} - \sigma_a^2\mathbf{F}^H\mathbf{W}\mathbf{H}\mathbf{F} - \sigma_a^2(\mathbf{W}\mathbf{H}\mathbf{F})^H\mathbf{F} + M\sigma^2\mathbf{I}\right)\mathbf{c}^H \end{aligned} \tag{8.46}$$

where

$$\mathbf{T} = \sigma_a^2\mathbf{H}\mathbf{H}^H + \sigma^2\mathbf{I} \tag{8.47}$$

By setting the derivative of the trace of (8.46) with respect to $\mathbf{W}$ to zero, we obtain the optimal $\mathbf{W}$ as

$$\hat{\mathbf{W}} = \sigma_a^2\mathbf{H}^H\mathbf{T}^{-1} \tag{8.48}$$

The diagonal elements of $\hat{\mathbf{W}}$ are

$$\hat{W}_l = \frac{\sigma_a^2 H_l^*}{\sigma_a^2|H_l|^2 + \sigma^2} \tag{8.49}$$

Then from (8.46) and (8.48), the MSE of the detected symbol is

$$MSE = \mathrm{tr}\left[\mathrm{E}(\mathbf{e}\mathbf{e}^H)\right] = \frac{\sigma_a^2}{M}\sum_{l=0}^{M-1}\sum_{k=0}^{M-1}\left|e^{j\frac{2\pi}{M}kl}\left(\sum_{n=0}^{B} c_n e^{-j\frac{2\pi}{M}nk}\right)\right|^2 Q_k \tag{8.50}$$

where $c_0 = 1$, $tr[\,]$ denotes the matrix trace operation and

$$Q_k = \frac{\sigma^2}{\sigma_a^2|H_k|^2 + \sigma^2} \tag{8.51}$$

By setting the derivative of (8.50) with respect to c_m to zero, we obtain the following equations for solving for the optimal c_m

$$\sum_{m=1}^{B}\sum_{k=0}^{M-1} c_m e^{j\frac{2\pi}{M}k(n-m)} Q_k = -\sum_{k=0}^{M-1} e^{j\frac{2\pi}{M}} Q_k \quad n = 1, \cdots, B \tag{8.52}$$

Comparing (8.52) with (8.35) we can see that the results between FD-DFE and FDE-NP are identical. This proves that the FDE-NP scheme has exactly the same MMSE performance as that of the FD-DFE. Furthermore, (8.48) shows that the coefficients of **W** are independent of the noise prediction. So the implementation of FDE-NP is more flexible than FD-DFE.

8.5.3 Iterative Block DFE (IBDFE) in the Frequency Domain

In both the FD-DFE and DFE-NP schemes, the error-propagation phenomenon exists and the computation complexity due to the matrix inverse is relatively high. Aiming at alleviating this, a new iterative block DFE in the frequency domain was proposed in [40]. In this structure, the hard decisions or soft decisions can be used as the input of the feedback filter. The decision errors are also considered when the parameters of the feedback filter are computed. As an extension of [32], transmitted data are arranged according to a format that allows the use of DFT both for the feedforward (FF) and feedback (FB) filters, that is, cancellation is performed in the frequency domain. Moreover, by using an iterative configuration similar to that proposed in [41], the filter design is performed directly in the frequency domain.

In the following analysis, the hard decision IBDFE (HD-IBDFE) is considered. The feedforward and feedback parts of the HD-IBDFE are shown in Figure 8.28 and Figure 8.29, respectively. The forward equalization and the feedback filter are all performed in the frequency domain based on FFT and IFFT blocks.

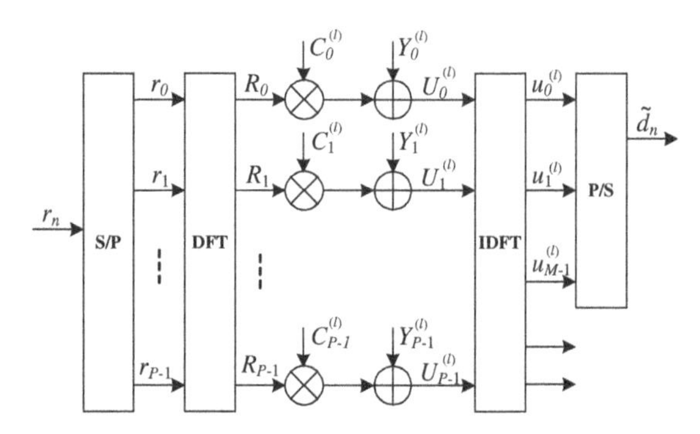

Figure 8.28. General architecture of the IBDFE [40] (©2005 IEEE)

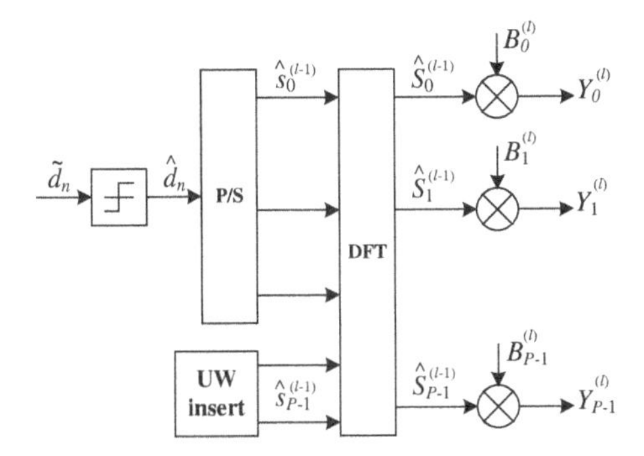

Figure 8.29. Feedback part of the IBDFE [40] (©2005 IEEE)

In the system, data block size is M and the extension length of CP or UW is L. So the DFT length is $P = M + L$. At the receiver, DFT is applied to successive blocks of P received samples. Denote

$$\mathbf{r} = [r_0, r_1, \cdots, r_{P-1}]^T \tag{8.53}$$

$$\mathbf{R} = [R_0, R_1, \cdots, R_{P-1}]^T = \mathbf{Fr} \tag{8.54}$$

and

$$R_p = H_p S_p + W_p \tag{8.55}$$

where $\mathbf{F}$ is the $P \times P$ DFT matrix, H_p is the CFR at the frequency point of p, and W_p is DFT of the p-th noise sample.

The equalizer consists of two parts: the feedforward filter with coefficients $\{C_p\}$, $p = 0, 1, \ldots, P-1$, in the frequency domain; and feedback filter with coefficients $\{B_p\}$, $p = 0, 1, \ldots, P-1$, in the frequency domain.

In the IBDFE, the design of the filter and data detection is iterated by N_I times. At iteration l ($l = 1, 2, \ldots, N_I$), the FF filter vector $\mathbf{C}^{(l)}$ is element-wise multiplied with $\mathbf{R}$ to yield the vector $\mathbf{Z}^{(l)}$, with elements

$$Z_p^{(l)} = C_p^{(l)} R_p, \quad p = 0, 1, \cdots, P-1 \tag{8.56}$$

In the FB filter, the detected data block, including UW sequence at iteration $(l-1)$, $\hat{\mathbf{s}}^{(l-1)}$ is transformed by DFT to yield $\hat{\mathbf{S}}^{(l-1)}$. The FB output signal $\mathbf{Y}^{(l)}$ has the components

$$Y_p^{(l)} = B_p^{(l)} \hat{S}_p^{(l-1)}, \quad p = 0, 1, \cdots, P-1 \tag{8.57}$$

The detection signal vector $\mathbf{U}^{(l)}$ is

$$\mathbf{U}^{(l)} = \mathbf{Z}^{(l)} + \mathbf{Y}^{(l)} \tag{8.58}$$

After the IDFT, the time domain vector signal at the detection point is

$$\mathbf{u}^{(l)} = \frac{1}{P}\mathbf{F}^H\mathbf{U}^{(l)} \tag{8.59}$$

The power of the involved signals in the frequency domain is introduced

$$M_{S_p} = \mathrm{E}\left(\left|S_p\right|^2\right) \tag{8.60}$$

$$M_{\hat{S}_p^{(l)}} = \mathrm{E}\left(\left|\hat{S}_p^{(l)}\right|^2\right) \tag{8.61}$$

And the correlation between the transmitted and detected data is

$$r_{S_p,\hat{S}_p^{(l-1)}} = \mathrm{E}\left(S_p\hat{S}_p^{(l-1)*}\right) \tag{8.62}$$

The MSE at the detection point is given by

$$J^{(l)} = \mathrm{E}\left(\left|\tilde{d}_n^{(l)} - d_n\right|^2\right) = \frac{1}{P}\sum_{i=0}^{P-1} E\left(\left|u_i^{(l)} - s_i\right|^2\right) \tag{8.63}$$

By applying Parseval's theorem, (8.63) can be rewritten as

$$\begin{aligned} J^{(l)} &= \frac{1}{P^2}\sum_{p=0}^{P-1}\left\{\left|C_p^{(l)}\right|^2 M_W + \left|C_p^{(l)}H_p - 1\right|^2 M_{S_p} + \left|B_p^{(l)}\right|^2 M_{\hat{S}_p^{(l-1)}}\right. \\ &\quad \left. + 2\mathrm{Re}\left[B_p^{(l)*}\left(C_p^{(l-1)}H_p - 1\right) r_{S_p,\hat{S}_p^{(l-1)}}\right]\right\} \end{aligned} \tag{8.64}$$

where $M_W = P\sigma_w^2$ is the noise power in the frequency domain.

To derive the filters that minimize (8.64), we also impose the constraint that the FB filter removes pre- and post-cursors, but does not remove the desired component, that is, it must be

$$\sum_{p=0}^{P-1} B_p^{(l)} = 0 \tag{8.65}$$

In order to minimize the MSE under the constraint of (8.65), we apply the Lag range multiplier method and minimize the function

$$\begin{aligned} f\left(\mathbf{C}^{(l)},\mathbf{B}^{(l)},\lambda^{(l)}\right) &= \frac{1}{P^2}\sum_{p=0}^{P-1}\left\{\left|C_p^{(l)}\right|^2 M_W + \left|C_p^{(l)}H_p-1\right|^2 M_{S_p} + \left|B_p^{(l)}\right|^2 M_{\hat{S}_p^{(l-1)}}\right. \\ &\quad \left. + 2\mathrm{Re}\left[B_p^{(l)*}\left(C_p^{(l-1)}H_p-1\right)r_{S_p,\hat{S}_p^{(l-1)}}\right]\right\} + \lambda^{(l)}\sum_{p=0}^{P-1}B_p^{(l)} \end{aligned} \tag{8.66}$$

where $\lambda^{(l)}$ is the Lagrange multiplier. Assume that the correlation $r_{S_p,\hat{S}_p^{(l-1)}}$ and powers M_{S_p} and $M_{\hat{S}_p^{(l)}}$ are independent of the frequency p. By setting the gradient of (8.66) with respect to $C_p^{(l)}$, $B_p^{(l)}$, and $\lambda^{(l)}$, respectively, to zero, we can get

$$\frac{\partial f\left(\mathbf{C}^{(l)},\mathbf{B}^{(l)},\lambda^{(l)}\right)}{\partial C_p^{(l)}} = \frac{1}{P}\left[C_p^{(l)}M_W + \left(C_p^{(l)}H_p-1\right)H_p^* M_{S_p} + B_p^{(l)}H_p^* r^*_{S_p,\hat{S}_p^{(l-1)}}\right] = 0 \tag{8.67}$$

$$\frac{\partial f\left(\mathbf{C}^{(l)},\mathbf{B}^{(l)},\lambda^{(l)}\right)}{\partial B_p^{(l)}} = \frac{1}{P}\left[B_p^{(l)}M_{\hat{S}_p^{(l-1)}} + \left(C_p^{(l)}H_p-1\right)r_{S_p,\hat{S}_p^{(l-1)}}\right] + \lambda^{(l)} = 0 \tag{8.68}$$

$$\frac{\partial f\left(\mathbf{C}^{(l)},\mathbf{B}^{(l)},\lambda^{(l)}\right)}{\partial \lambda} = \sum_{p=0}^{P-1}B_p^{(l)} = 0 \tag{8.69}$$

From (8.68) we can get

$$B_p^{(l)} = -\frac{1}{M_{\hat{S}_p^{(l-1)}}}\left[r_{S_p,\hat{S}_p^{(l-1)}}\left(C_p^{(l)}H_p-1\right) + P^2\lambda^{(l)}\right] \tag{8.70}$$

From (8.69) and (8.70)$B_p^{(l)}$ can be expressed as

$$B_p^{(l)} = -\frac{r_{S_p,\hat{S}_p^{(l-1)}}}{M_{\hat{S}_p^{(l-1)}}}\left(H_pC_p^{(l)}-\gamma^{(l)}\right) \tag{8.71}$$

where

$$\gamma^{(l)} = \frac{1}{P}\sum_{p=0}^{P-1}H_pC_p^{(l)} \tag{8.72}$$

From (8.67) and (8.71) we can get

$$C_p^{(l)} M_W + \left(C_p^{(l)} H_p - 1\right) H_p^* M_{S_p} - \frac{r_{S_p,\hat{S}_p^{(l-1)}}}{M_{\hat{S}_p^{(l-1)}}} \left(H_p C_p^{(l)} - \gamma^{(l)}\right) H_p^* r^*_{S_p,\hat{S}_p^{(l-1)}} = 0 \tag{8.73}$$

Therefore, $C_p^{(l)}$ can be expressed as

$$C_p^{(l)} = \frac{H_p^* M_{S_p} \left(1 - \frac{\left|r_{S_p,\hat{S}_p^{(l-1)}}\right|^2}{M_{\hat{S}_p^{(l-1)}} M_{S_p}} \gamma^{(l)}\right)}{M_W + M_{S_p} \left(1 - \frac{\left|r_{S_p,\hat{S}_p^{(l-1)}}\right|^2}{M_{\hat{S}_p^{(l-1)}} M_{S_p}}\right) \left|H_p\right|^2} \tag{8.74}$$

with $p = 0, 1, \ldots, P-1$.

Figure 8.30 shows the performance comparison of HD-IBDFE with iterations 1–4 and the other equalization schemes, whereby TD-DFE and TD-DFE-ID are the nonideal and ideal time domain decision feedback equalization, respectively, H-DFE is the hybrid DFE,which is the same as the FD-DFE with time domain feedback filter, and MF is the match filter bound. From Figure 8.30, we can see that the first two iterations of HD-IBDFE yield a significant gain in performance, while the gain of further iterations is diminished. Compared with H-DFE, HD-IBDFE is about 1 dB better at BER = 3E−4.

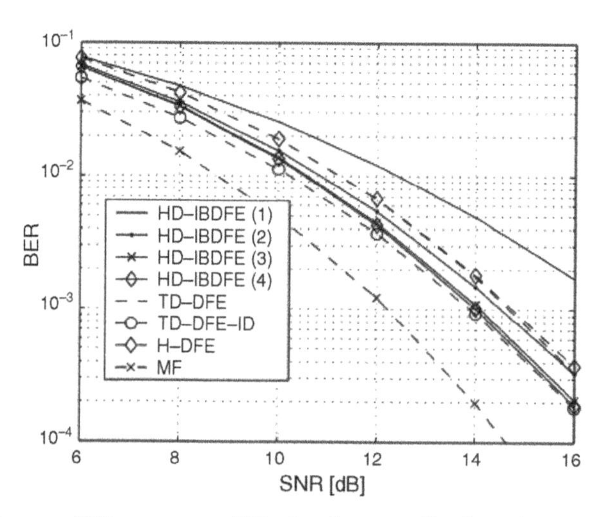

Figure 8.30. Average BER vs. average SNR of various equalization schemes [40] (©2005 IEEE)

8.5.4 Frequency Domain Decision Feedback Equalization for Uplink SC-FDMA

Single-carrier frequency division multiple access (SC-FDMA) has been extensively adopted as a typical uplink transmission method for multi-user access scenarios. Compared with the OFDMA scheme, the SC-FDMA system has its own advantages, such as low PAPR ratio. Low PAPR can allow the system to relax the specifications of linearity in the power amplifier of the mobile terminal, which will reduce cost and power consumption. Since SC-FDMA is an effective method used in wireless broadband communication systems, it has been chosen as the uplink transmission method in long-term evaluation (LTE) systems [39], which can support $1.25 \sim 20$ MHz bandwidth and high uplink transmission rate up to 50 Mbps. Note that SC-FDMA scheme is more suitable for the uplink channel due to its lower power consumption.

In the SC-FDMA system, the FD-LE could be adopted to perform the equalization after the DFT module, which transforms the received signal from time domain to frequency domain. However, in deep frequency selective fading channels spectral nulls of CFR appear at some specific frequencies. Since the noise at these spectral nulls is significantly amplified when the FD-LE is used, the performance of FD-LE deteriorates in highly frequency-selective and deep fading channels. This problem can be reduced by introducing DFE into the SC-FDMA system.

A frequency-domain DFE structure that has the unified form for different subcarrier allocations, including localized, distributed, and frequency hopping allocations, for multi-user SC-FDMA systems is introduced next.

The transmitter and receiver structures of the SC-FDMA system are shown in Figure 8.31 (a) and (b), respectively. It is assumed that the total number of users in SC-FDMA systems is K.

The number of subcarriers for each user is M. The total subcarriers used in the system is P. So we have $P = KM$. For the kth user, the input bit stream is converted to constellation symbols, and then grouped into a data block $\mathbf{s}^{(k)}$ of size M, that is,

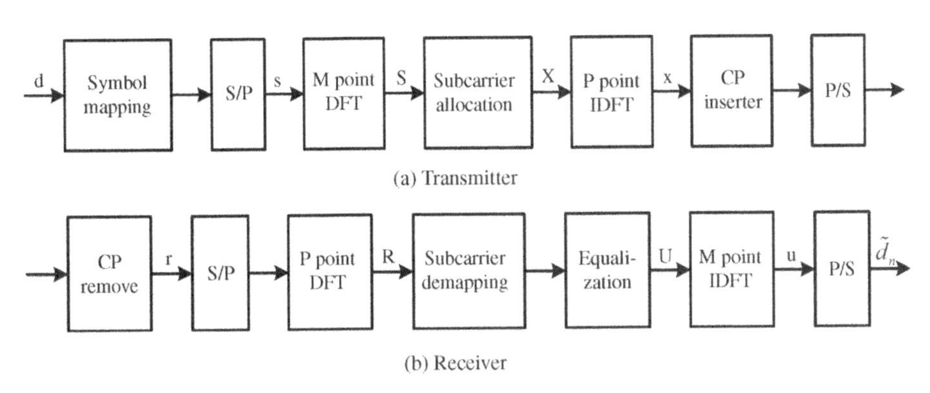

Figure 8.31. Block diagram of SC-FDMA systems

$\mathbf{s}^{(k)} = [s_0^{(k)}, s_1^{(k)}, \cdots, s_{M-1}^{(k)}]^T \cdot \mathbf{s}^{(k)}$ is then transformed to the frequency-domain symbol $\mathbf{S}^{(k)}$ by a M-point DFT. According to different resource allocation schemes, $\mathbf{S}^{(k)}$ is mapped to P subcarriers, which can be expressed as

$$\mathbf{X}^{(k)} = \mathbf{D}^{(k)}\mathbf{S}^{(k)} \qquad k = 1, 2, \cdots, K \tag{8.75}$$

where $\mathbf{D}^{(k)}$ is the resource allocation matrix with size of $P \times M$. For different schemes, $\mathbf{D}^{(k)}$ is expressed as follows:

$$\mathbf{D}_{Loc,(m,n)}^{(k)} = \begin{cases} 1 & m = (k-1)M + n, \ 0 \le n \le M-1 \\ 0 & \text{otherwise} \end{cases} \tag{8.76}$$

$$\mathbf{D}_{Dis,(m,n)}^{(k)} = \begin{cases} 1 & m = (k-1) + Kn, \ 0 \le n \le M-1 \\ 0 & \text{otherwise} \end{cases} \tag{8.77}$$

$$D_{FH,(m,n)}^{(k)} = \begin{cases} 1 & m = \mathrm{FH}^{(k)}(n), \ 0 \le n \le M-1 \\ 0 & \text{otherwise} \end{cases} \tag{8.78}$$

where $\mathbf{D}_{Loc}^{(k)}$, $\mathbf{D}_{Dis}^{(k)}$, and $\mathbf{D}_{FH}^{(k)}$ denote the resource allocation matrices of localized, distributed, and frequency hopping methods of the kth user, respectively, (m,n) is the element index with $0 \le m \le P-1$ and $0 \le n \le M-1$, and $\mathrm{FH}^{(k)}(n)$ represents the frequency hopping function of the kth user. From equations (8.76)–(8.78), a general expression of the mapping process is obtained as

$$\mathbf{D}_{(m,n)}^{(k)} = \begin{cases} 1 & m = \mathrm{MP}^{(k)}(n), \ 0 \le n \le M-1 \\ 0 & \text{otherwise} \end{cases} \tag{8.79}$$

where $\mathbf{D}_{(m,n)}^{(k)}$ is the (m,n) element of $\mathbf{D}^{(k)}$. $\mathrm{MP}^{(k)}(n)$ is the subcarrier allocation function of the kth user. For different users, their resource allocation matrices are orthogonal, and hence we have

$$\mathbf{D}^{(k_1)\mathrm{T}}\mathbf{D}^{(k_2)} = \begin{cases} \mathbf{I}_M & k_1 = k_2 \\ \mathbf{O}_M & k_1 \ne k_2 \end{cases} \tag{8.80}$$

where the superscript T denotes the matrix transpose, $\mathbf{I}_M$ is an identify matrix with size of $M \times M$, and $\mathbf{O}_M$ is the $M \times M$ zero matrix.

After the subcarrier mapping, the time domain data block $\mathbf{x}^{(k)}$ is obtained by a P-point IDFT transform. The SC-FDMA signal is generated after inserting CP and parallel to serial converting. The transmitted signal is described as follows:

$$\mathbf{x}^{(k)} = \mathbf{W}_P^H\mathbf{D}^{(k)}\mathbf{W}_M\mathbf{s}^{(k)} \tag{8.81}$$

where H denotes the Hermitian transpose, $\mathbf{W}_P$ is a $P\times P$ DFT matrix, that is, $W_{P,(m,n)} = e^{-j2\pi/P\cdot mn}$, $m, n=0, 1,\ldots, P-1$, and $\mathbf{W}_M$ is an $M\times M$ DFT matrix.

The received signal after removing CP can be treated as the circular convolution between the transmitted signal and channel impulse response. The received frequency domain signal after DFT is

$$\mathbf{R} = \sum_{m=1}^{K} \mathbf{H}^{(m)}\mathbf{D}^{(m)}\mathbf{W}_M\mathbf{s}^{(m)} + \mathbf{V} \tag{8.82}$$

where $\mathbf{H}^{(m)}$ is a $P\times P$ diagonal matrix with the CFR as its diagonal elements, that is, $\mathbf{H}^{(m)} = \text{diag}\left(\text{DFT}\left(\mathbf{h}^{(m)}\right)\right)$, and $\mathbf{V}$ is the noise vector with variance of $\sigma_V^2\mathbf{I}_P$ with $\mathbf{I}_P$ being the $P\times P$ identify matrix.

Owing to the orthogonality of (8.81), the received signal for kth user can be express as

$$\mathbf{R}^{(k)} = \mathbf{D}^{(k)\mathrm{T}}\mathbf{H}^{(k)}\mathbf{D}^{(k)}\mathbf{W}_M\mathbf{s}^{(k)} + \mathbf{V}^{(k)} \tag{8.83}$$

Let $\tilde{\mathbf{H}}^{(k)} = \mathbf{D}^{(k)\mathrm{T}}\mathbf{H}^{(k)}\mathbf{D}^{(k)}$, then $\tilde{\mathbf{H}}^{(k)}$ is a $M\times M$ CFR diagonal matrix that the signal from kth user has passed with the diagonal element $\tilde{H}_m^{(k)} = H^{(k)}_{\mathrm{MP}^{(k)}(m),m}$, $0\le m\le M-1$. So the received signal is

$$R_m^{(k)} = \tilde{H}_m^{(k)}\, S_m^{(k)} + V_m^{(k)} \tag{8.84}$$

The performance of FD-DFE suitable for SC-FDMA is analyzed, whereby its structure is shown in Figure 8.32 (a) and (b), respectively. Figure 8.32(a) describes the forward filter part and Figure 8.32(b) illustrates the feedback part. Here the nonideal feedback with decision errors is considered.

For the kth user, the output of the equalizer can be expressed as

$$u_m^{(k)} = \frac{1}{M}\sum_{l=0}^{M-1}\left(C_l^{(k)}R_l^{(k)} + B_l^{(k)}\hat{S}_l^{(k)}\right)e^{j\frac{2\pi}{M}lm} \tag{8.85}$$

where C_l denotes the coefficient of the feedforward filter and B_l denotes the coefficient of the feedback filter. In (8.85), the decision value of $\hat{S}_l^{(k)}$ given by

$$\hat{S}_l^{(k)} = \sum_{i=0}^{M-1}\hat{s}_i^{(k)}e^{-j\frac{2\pi}{M}il} \quad 0\le l\le M-1 \tag{8.86}$$

is used.

After the decision, the MSE is as follows:

$$MSE = E\left(\left|u_m^{(k)} - s_m^{(k)}\right|^2\right) \tag{8.87}$$

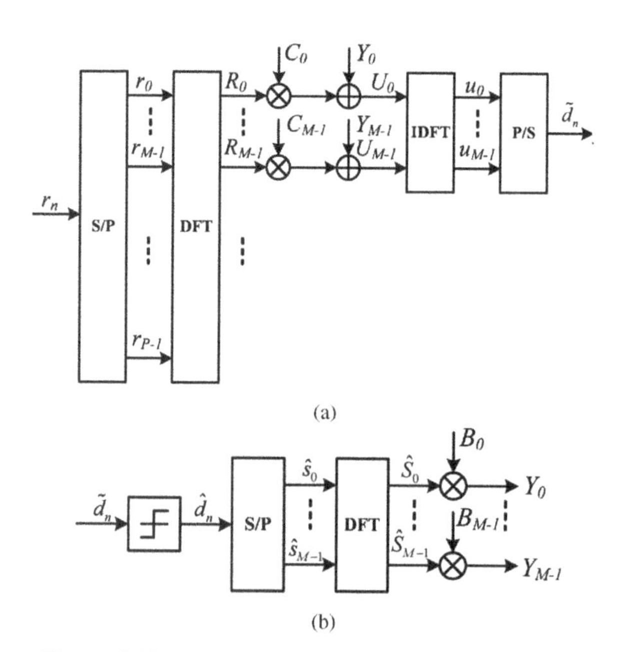

Figure 8.32. Structure of FD-DFE for SC-FDMA systems

So the MSE can be expressed as

$$MSE = \frac{P\sigma_V^2}{M^2}\sum_{l=0}^{M-1}\left|C_l^{(k)}\right|^2 + \frac{\sigma_s^2}{M}\sum_{l=0}^{M-1}\left|C_l^{(k)}\tilde{H}_l^{(k)}-1\right|^2 +$$

$$\frac{\sigma_{\hat{s}}^2}{M}\sum_{l=0}^{M-1}\left|B_l^{(k)}\right|^2 + 2\frac{1}{M}\mathrm{Re}\left[r_{s,\hat{s}}\sum_{l=0}^{M-1}B_l^{(k)*}\left(C_l^{(k)}\tilde{H}_l^{(k)}-1\right)\right] \tag{8.88}$$

where

$$\sigma_{\hat{s}}^2 = E\left(\hat{s}_l^{(k)}\hat{s}_l^{(k)*}\right),\ r_{s,\hat{s}} = E\left(s_l^{(k)}\hat{s}_l^{(k)*}\right) \tag{8.89}$$

With the constraint of $\sum_{l=0}^{M-1} B_l^{(k)} = 0$ [40], we use the Lagrange multiplier method to solve MMSE design. Construct the objective function $f\left(\mathbf{B}^{(k)}, \mathbf{C}^{(k)}, \lambda\right)$

$$f\left(\mathbf{B}^{(k)}, \mathbf{C}^{(k)}, \lambda\right) = \frac{P\sigma_V^2}{M^2}\sum_{l=0}^{M-1}\left|C_l^{(k)}\right|^2 + \frac{\sigma_s^2}{M}\sum_{l=0}^{M-1}\left|C_l^{(k)}\tilde{H}_l^{(k)}-1\right|^2 + \frac{\sigma_{\hat{s}}^2}{M}\sum_{l=0}^{M-1}\left|B_l^{(k)}\right|^2 +$$

$$2\frac{1}{M}\mathrm{Re}\left[r_{s,\hat{s}}\sum_{l=0}^{M-1}B_l^{(k)*}\left(C_l^{(k)}\tilde{H}_l^{(k)}-1\right)\right] + \lambda\cdot\sum_{l=0}^{M-1}B_l^{(k)} \tag{8.90}$$

By using the Lagrange multiplier method, we can obtain

$$\frac{\partial f\left(\mathbf{B}^{(k)},\mathbf{C}^{(k)},\lambda\right)}{\partial C_l^{(k)}} = \frac{P\sigma_V^2}{M^2}C_l^{(k)} + \frac{\sigma_s^2}{M}\left(C_l^{(k)}\tilde{H}_l^{(k)}-1\right)\tilde{H}_l^{(k)*} + \frac{1}{M}r_{s,\hat{s}}^{*}B_l^{(k)}\tilde{H}_l^{(k)*} = 0 \tag{8.91}$$

$$\frac{\partial f\left(\mathbf{B}^{(k)},\mathbf{C}^{(k)},\lambda\right)}{\partial B_l^{(k)}} = \frac{\sigma_{\hat{s}}^2}{M}B_l^{(k)} + \frac{r_{s,\hat{s}}}{M}\left(C_l^{(k)}\tilde{H}_l^{(k)}-1\right) + \lambda = 0 \tag{8.92}$$

$$\frac{\partial f\left(\mathbf{B}^{(k)},\mathbf{C}^{(k)},\lambda\right)}{\partial \lambda} = \sum_{i=0}^{M-1}B_i^{(k)} = 0 \tag{8.93}$$

From (8.92) we can obtain

$$B_l^{(k)} = -\frac{r_{s,\hat{s}}}{\sigma_{\hat{s}}^2}\left(C_l^{(k)}\tilde{H}_l^{(k)} - \beta^{(k)}\right) \tag{8.94}$$

where $\beta^{(k)}$ is given by

$$\beta^{(k)} = \frac{1}{M}\sum_{i=0}^{M-1}C_i^{(k)}\tilde{H}_i^{(k)} \tag{8.95}$$

Then put (8.94) into (8.91)

$$\frac{P\sigma_V^2}{M^2}C_l^{(k)} + \frac{\sigma_s^2}{M}\left|\tilde{H}_l^{(k)}\right|^2 C_l^{(k)} - \frac{\sigma_{\hat{s}}{}^2}{M}\tilde{H}_l^{(k)*} - \frac{\left|r_{s,\hat{s}}\right|^2}{M}\left(C_l^{(k)}\tilde{H}_l^{(k)} - \beta^{(k)}\right)\tilde{H}_l^{(k)*} = 0 \tag{8.96}$$

Then $C_l^{(k)}$ can be expressed as

$$C_l^{(k)} = \frac{M\sigma_s^2\left(1-\frac{\left|r_{s,\hat{s}}\right|^2}{\sigma_s^2\sigma_{\hat{s}}{}^2}\beta^{(k)}\right)\tilde{H}_l^{(k)}}{P\sigma_V^2 + M\sigma_s^2\left(1-\frac{\left|r_{s,\hat{s}}\right|^2}{\sigma_s^2\sigma_{\hat{s}}{}^2}\right)\left|\tilde{H}_l^{(k)}\right|^2} \tag{8.97}$$

By solving equations (8.95) and (8.97), $\beta^{(k)}$ can be rewritten as

$$\beta^{(k)} = \left(\sum_{i=0}^{M-1} \frac{\sigma_s^2 \left| \tilde{H}_i^{(k)} \right|^2}{P\sigma_V^2 + M\sigma_s^2 \left(1 - \frac{\left| r_{s,\hat{s}} \right|^2}{\sigma_s^2 \sigma_{\hat{s}}^2} \right) \left| \tilde{H}_i^{(k)} \right|^2} \right) \times \left(1 + \sum_{i=0}^{M-1} \frac{\frac{\left| r_{s,\hat{s}} \right|^2}{\sigma_{\hat{s}}^2} \left| \tilde{H}_i^{(k)} \right|^2}{P\sigma_V^2 + M\sigma_s^2 \left(1 - \frac{\left| r_{s,\hat{s}} \right|^2}{\sigma_s^2 \sigma_{\hat{s}}^2} \right) \left| \tilde{H}_i^{(k)} \right|^2} \right)^{-1} \tag{8.98}$$

We can use the methods in [40] to estimate the parameters $\sigma_{\hat{s}}^2$ and $r_{s,\hat{s}}$, respectively.

$$\sigma_{\hat{s}}^2 = \frac{1}{M^2} \sum_{l=0}^{M-1} \left| \hat{S}_l^{(k)} \right|^2 \tag{8.99}$$

$$r_{s,\hat{s}} = \frac{1}{M^2} \sum_{l=0}^{M-1} \frac{R_l^{(k)}}{\tilde{H}_l^{(k)}} \hat{S}_l^{(k)*} \tag{8.100}$$

Based on the above analyses, simulations are performed to evaluate the performance of the proposed frequency domain decision feedback equalization algorithms in multi-user SC-FDMA systems.

The number of subcarriers is $P = 1024$ and the length of CP is $P/8 = 128$. The number of users in this system is $K = 8$, and $M = 128$ subcarriers are allocated to each user. The QPSK constellation is selected and the comparison is conducted for both the uncoded and coded systems. Since the channel conditions for every user vary, the six-path channel model is chosen where the delay of the six paths distributes uniformly between $0 \sim 128$ taps and the path power spectrum has exponential distribution. During the simulations, channel estimation and synchronization are assumed ideal. Figure 8.33 and Figure 8.34 show the performance of the algorithm under different subcarrier allocations for the uncoded and coded systems, respectively.

The BER performance of uncoded systems is shown in Figure 8.33. From Figure 8.33 it is clear that for the three subcarrier allocation methods, the performance of the nonideal FD-DFE is around 3 dB better than the FD-LE at BER = 1.0E − 3. When a localized or distributed allocation scheme is adopted, the performance difference between the nonideal FD-DFE and the ideal FD-DFE is about 1.0 dB at BER = 1.0E - 3. However, in the FH allocation, this loss is much lower compared with the other allocation methods

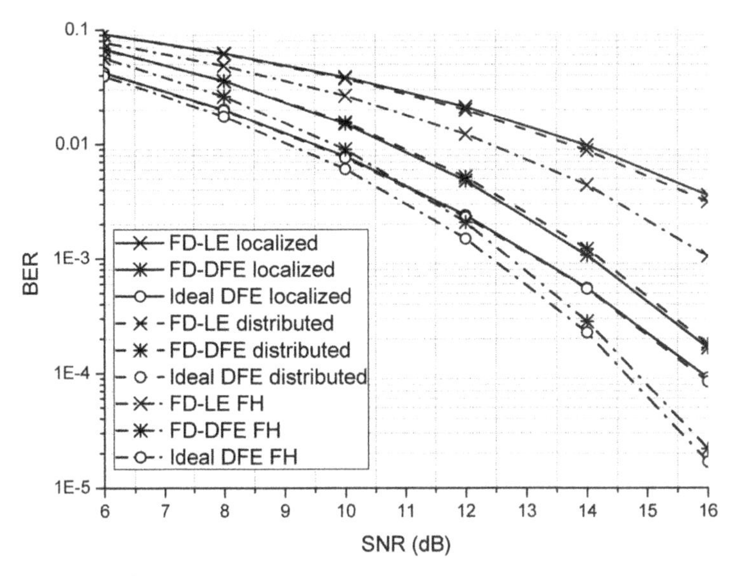

Figure 8.33. BER performance of uncoded systems

due to the random subcarrier allocation. For each equalizer, the performance of the FH allocation is better than those of the localized and distributed allocations.

Figure 8.34 shows the BER performances of the coded SC-FDMA system under various equalization schemes. A 1/2-rate convolutional encode (133,171) is used at the transmitter and a soft decision Viterbi decoder is used at the receiver. For the three

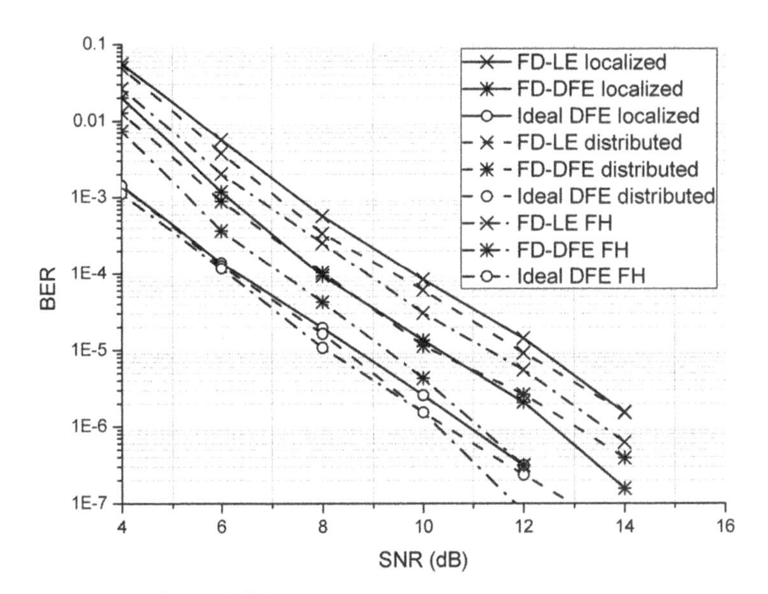

Figure 8.34. BER performance of coded systems

subcarrier allocation methods, the performance of the nonideal FD-DFE is around 2 dB better than the FD-LE at BER = 1.0E - 5. For the localized and distributed allocations, the performance difference between the nonideal FD-DFE and the ideal FD-DFE is about 2 dB at BER = 1.0E - 5. In the FH allocation, the loss due to the decision errors is about 1 dB. For nonideal feedback, the performance of FH allocation is about 1 dB better than those of the other allocations.

REFERENCES

[1] H. Sari, G. Karam, and I. Jeanclaude, "Transmission techniques for digital terrestrial TV broadcasting." *IEEE Commun. Mag.*, Vol. 33, No. 2, pp. 100–109, 1995.

[2] R. W. Chang, "Synthesis of band-limited orthogonal signals for multi-channel data transmission." *Bell Syst. Tech. J.*, Vol. 45, pp. 1775–1796, Apr. 1966.

[3] T. Walzman and M. Schwartz, "Automatic equalization using the discrete frequency domain." *IEEE Trans. Inform. Theory*, Vol. 19, No. 1, pp. 59–68, 1973.

[4] J. H. Tsai, Y. L. Lee, T. W. Huang, C. M. Yu, and J. G. J. Chern, "A 90-nm CMOS broadband and miniature Q-band balanced medium power amplifier." *IEEE/MTT-S International Microwave Symposium*, pp. 1129–1132, Jun. 2007.

[5] O. S. A. Tang, S. M. J. Liu, P. C. Chao, W. M. T. Kong, K. C. Hwang, K. Nichols, and J. Heaton, "Design and fabrication of a wideband 56- to 63-GHz monolithic power amplifier with very high power-added efficiency." *IEEE J. Solid-State Circuits*, Vol. 35, No. 9, pp. 1298–1306, 2000.

[6] D. Birru, R. Chen, C. T. Chou, et al., "SCBT based 60GHz PHY proposal." IEEE P802.15 Working Group for Wireless Personal Area Networks (WPANs), IEEE 802.15-0681-02-003c, pp. 87–90, 18 Jul. 2007.

[7] H. Yang, P. F. M. Smulders, and E.R. Fledderus, "Comparison of single- and multi-carrier block transmissions under the effect of nonlinear HPA." *14th IEEE Symposium on Communications and Vehicular Technology in the Benelux*, pp. 1–7, Nov. 2007.

[8] U. H. Rizvi, G. J. M. Janssen, and J. H. Weber, "Impact of RF circuit imperfections on multicarrier and single-carrier based transmissions at 60GHz." *2008 IEEE Radio and Wireless Symposium*, pp. 691–694, Jan. 2008.

[9] M. Lei et al., "Throughput comparison of multi-Gbps WPAN (IEEE 802.15.3c) PHY layer designs under non-linear 60GHz power amplifier:" *IEEE 18th International Symposium on Personal, Indoor and Mobile Radio Communications* (PIMRC'07), pp. 1–5, Sep. 2007.

[10] M. Noune and A. Nix, "Frequency-domain precoding for single carrier frequency- division multiple access." *IEEE Commun. Mag.* pp. 68–74, Jun. 2009.

[11] Y. Zhu and K. B. Letaief, "Single carrier frequency domain equalization with time domain noise prediction for wideband wireless communications." *IEEE Trans. Wireless Commun.*, Vol. 5, No. 12, pp. 3458–3467, 2006.

[12] F. Pancaldi, G. M. Vitetta, R. Kalbasi, N. Al-Dhahir, M. Uysal, and H. Mheidat, "Single-carrier frequency domain equalization." *IEEE Signal Process. Mag.*, pp. 37–56, Sep. 2008.

[13] A. Gusmao, P. Torres, R. Dinis, and N. Esteves, "On SC/FDE block transmission with reduced cyclic prefix assistance." *2006 IEEE International Conference on Communications* (ICC'06), Vol. 1, pp. 5058–5063, Jun. 2006.

[14] L. Deneire, "Training sequence vs. cyclic prefix a new look on single carrier communication." *2000 IEEE Global Telecommunications Conference* (Globecom'00), Vol. 2, pp. 1056–1060, Dec. 2000.

[15] Z. Wang and M. Uno,"Wireless system using a new type of preamble for a burst frame, Europe patent." European patent EP2099187A1.

[16] M. Rice and X. Y. Dang, "Aeronautical telemetry using offset QPSK in frequency selective multipath." *IEEE Trans. Aerosp. Electron. Syst.*, Vol. 41, No. 2, pp. 758–767, 2005.

[17] Y. S. Lee, H. C. Shin, and H. N. Kim, "Channel estimation based on a time-domain threshold for OFDM systems." *IEEE Trans. Broadcast.*, Vol. 55, No. 3, pp. 656–662, 2009.

[18] U. K. Kwon, D. Kim, and G. H. Im, "Frequency domain pilot multiplexing technique for channel estimation of SC-FDE." *IEE Electron. Lett.*, Vol. 44, No. 5, 2008.

[19] D. Falconer, S. L. Ariyavisitakul, A. B. Seeyar, and B. Eidson, "Frequency domain equalization for single-carrier broadband wireless systems." *IEEE Commun. Mag.*, pp. 58–66, 2002.

[20] H. Liu and P. Schniter, "Iterative frequency-domain channel estimation and equalization for single-carrier transmissions without cyclic-prefix." *IEEE Trans. Wireless Commun.*, Vol. 7, No. 10, pp. 3686–3691, 2008.

[21] P. H. Chiang, D. B. Lin, H. J. Li, and G. L. Stüber, "Joint estimation of carrier-frequency and sampling-frequency offsets for SC-FDE systems on multipath fading channels." *IEEE Trans. Commun.*, Vol. 56, No. 8, pp. 1231–1235, 2008.

[22] S. Chang, "LR-WPAN based on a novel SC-FDE for wideband wireless sensor networks." *IEEE Trans. Consum. Electron.*, Vol. 55, No. 3, pp. 1241–1245, 2009.

[23] Y. Zeng and T. S. Ng, "Pilot cyclic prefixed single carrier communication: channel estimation and equalization." *IEEE Signal Process. Lett.*, Vol. 12, No. 1, pp. 56–59, 2005.

[24] D. Kim, U. K. Kwon, and G. H. Im, "Pilot position selection and detection for channel estimation of SC-FDE." *IEEE Commun. Lett.*, Vol. 12, No. 5, pp. 350–352, 2008.

[25] F. Pancaldi, G. M. Vitetta, R. Kalbasi, N. Al-Dhahir, M. Uysal, and H. Mheidat, "Single-carrier frequency domain equalization." *IEEE Signal Process. Mag.*, Vol. 25, No. 5, pp. 37–56, 2008.

[26] B. Ng, C. T. Lam, and D. Falconer, "Turbo frequency domain equalization for single-carrier broadband wireless systems." *IEEE Trans. Wireless Commun.*, Vol. 6, No. 2, pp. 759–767, 2007.

[27] L. Deneire, B. Gyselinckx, and M. Engels, "Training sequence versus cyclic prefix - A new look on single carrier communication." *IEEE Commun. Lett.*, Vol. 5, No. 7, pp. 292–294, 2001.

[28] D. Wang, L. G. Jiang, and C. He, "Noise variance estimation and optimal sequences for channel estimation in SC-FDE UWB systems." *IEE Electron. Lett.*, Vol. 43, No. 11, 2007.

[29] H. G. Myung, J. Lim, and D. J. Goodman, "Single carrier FDMA for uplink wireless transmission." *IEEE Veh. Technol. Mag.*, Vol. 1, No. 3, pp. 30–38, 2006.

[30] K. Berberidis and J. Palicot, "A frequency domain decision feedback equalizer for multipath echo cancellation." *1995 IEEE Global Telecommunications Conference* (Globecom'95), pp. 98–102, Dec. 1995.

[31] D. Falconer, S. L. Ariyavisitakul, A. B. Seeyar, and B. Eidson,"White paper: frequency equalization for single-carrier broadband wireless systems." http://www.sce.carleton.ca/bbw/papers.

[32] H. Witschnig, M. Kemptner, R. Weigel, and A. Springer, "Decision feedback equalization for single carrier system with frequency domain equalization – An overall system approach." *1st International Symposium on Wireless Communication Systems*, pp. 26–30, Sep. 2004.

[33] N. Benvenuto and S. Tomasin, "On the comparison between OFDM and single carrier modulation with a DFE using a frequency-domain feedforward filter." *IEEE Trans. Commun.*, Vol. 5, No. 6, pp. 947–955, 2002.

[34] A. V. Oppenheim and R. W. Schafer, *Digital Signal Processing*, Englewood Cliffs, NJ: Prentice-Hall, 1975.

[35] M. V. Clark, "Adaptive frequency-domain equalization and diversity combining for broadband wireless communications." *IEEE J. Sel. Areas Commun.*, Vol. 16, No. 8, pp. 1385–1395, 1998.

[36] D. Falconer, S. L. Ariyavisitakul, A. B. Seeyar, and B. Eidson, "Frequency domain equalization for single-carrier broadband wireless systems." *IEEE Commun. Mag.* pp. 56–66, Apr. 2002.

[37] V. Erceg et al., "Channel models for fixed wireless applications." IEEE 802.16a cont. IEEE 802.16.3c-01/29r1, Feb. 2001.

[38] Y. Zhu and K. B. Letaief, "Single carrier frequency domain equalization with noise prediction for broadband wireless systems." *2004 IEEE Global Telecommunications Conference* (Globecom'04), Vol. 5, pp. 3098–3102, Dec. 2004.

[39] 3GPP, "Physical Channels and Modulation," TS 36.211 v8.5.0, 2008.

[40] N. Benvenuto and S. Tomasin, "Iterative design and detection of a DFE in frequency domain." *IEEE Trans. Commun.*, Vol. 53, No. 11, pp. 1867–1875, 2005.

[41] A. M. Chan and G. W. Wornell, "A class of block-iterative equalizers for inter-symbol interference channels: fixed channel results." *IEEE Trans. Commun.*, Vol. 49, No. 11, pp. 1966–1976, 2001.

APPENDIX
SIMULATION TOOLS

In general, millimeter wave communication systems are expensive to implement the hardware. To reduce unnecessary cost and design errors, it is important to simulate your design carefully before producing prototypes. Some simulation tools are advised in this appendix for readers' interests.

Electromagnetic simulation tools

- 3D full-wave electromagnetic field software, used to simulate millimeter wave bonding, package, vias, antenna etc.
  CST Microwave Studio, Computer Simulation Technology AG, Germany
  Ansoft HFSS, ANSYS, Inc., U.S.
  Sonnet, Sonnet Software, U.S.
- Planar simulation, used to simulate millimeter wave planar circuits, multilayer circuits, matching networks etc.
  IE3D, Zeland Software, Inc., U.S.
  Wipl-D, Wipl-D Ltd., Serbia

Communication system simulation

These tools are used to simulate the whole communication system with various modulations/demodulations.

Advanced Design System, Agilent Technologies Inc., U.S.
Matlab Simulink, MathWorks, Inc., U.S.
Microwave Office, AWR Corporation, U.S.

Millimeter Wave Communication Systems, by Kao-Cheng Huang and Zhaocheng Wang

INDEX

Millimeter Wave Communication Systems, by Kao-Cheng Huang and Zhaocheng Wang

IEEE PRESS SERIES ON DIGITAL AND MOBILE COMMUNICATION

SERIES EDITOR
John B. Anderson
University of Lund

Wireless Video Communications: Second to Third Generation and Beyond
Lajos Hanzo, Peter Cherriman, and Jurgen Streit

Wireless Communications in the 21st Century
Mansoor Sharif, Shigeaki Ogose, and Takeshi Hattori

Introduction to WLLs: Application and Deployment for Fixed and Broadband Services
Raj Pandya

Trellis and Turbo Coding
Christian Schlegel and Lance Perez

Theory of Code Division Multiple Access Communication
Kamil Sh. Zigangirov

Digital Transmission Engineering, Second Edition
John B. Anderson

Wireless Broadband: Conflict and Convergence
Vern Fotheringham and Sharma Chetan

Wireless LAN Radios: System Definition to Transistor Design
Arya Behzad

Millimeter Wave Communication Systems
Kao-Cheng Huang and Zhaocheng Wang

rthcoming Titles
annel Equalization: From Concepts to Detailed Mathematics
ndbook of Position Location: Theory, Practice and Advances
ndamentals of Convolutional Coding, Second Edition
n-Gaussian Statistical Communication Theory